CAMBRIDGE LIBRARY COLLECTION

Books of enduring scholarly value

Mathematical Sciences

From its pre-historic roots in simple counting to the algorithms powering
modern desktop computers, from the genius of Archimedes to the genius of
Einstein, advances in mathematical understanding and numerical techniques
have been directly responsible for creating the modern world as we know
it. This series will provide a library of the most influential publications and
writers on mathematics in its broadest sense. As such, it will show not only
the deep roots from which modern science and technology have grown, but
also the astonishing breadth of application of mathematical techniques in the
humanities and social sciences, and in everyday life.

Mathematical and Physical Papers

Sir George Stokes (1819-1903) established the science of hydrodynamics
with his law of viscosity describing the velocity of a small sphere through a
viscous fluid. He published no books, but was a prolific lecturer and writer
of papers for the Royal Society, the British Association for the Advancement
of Science, the Victoria Institute and other mathematical and scientific
institutions. These collected papers (issued between 1880 and 1905) are
therefore the only readily available record of the work of an outstanding and
influential mathematician, who was Lucasian Professor of Mathematics in
Cambridge for over fifty years, Master of Pembroke College, President of the
Royal Society (1885-90), Associate Secretary of the Royal Commission on
the University of Cambridge and a Member of Parliament for the University.

Cambridge University Press has long been a pioneer in the reissuing of out-of-print titles from its own backlist, producing digital reprints of books that are still sought after by scholars and students but could not be reprinted economically using traditional technology. The Cambridge Library Collection extends this activity to a wider range of books which are still of importance to researchers and professionals, either for the source material they contain, or as landmarks in the history of their academic discipline.

Drawing from the world-renowned collections in the Cambridge University Library, and guided by the advice of experts in each subject area, Cambridge University Press is using state-of-the-art scanning machines in its own Printing House to capture the content of each book selected for inclusion. The files are processed to give a consistently clear, crisp image, and the books finished to the high quality standard for which the Press is recognised around the world. The latest print-on-demand technology ensures that the books will remain available indefinitely, and that orders for single or multiple copies can quickly be supplied.

The Cambridge Library Collection will bring back to life books of enduring scholarly value (including out-of-copyright works originally issued by other publishers) across a wide range of disciplines in the humanities and social sciences and in science and technology.

Mathematical and Physical Papers

VOLUME 1

GEORGE GABRIEL STOKES

CAMBRIDGE
UNIVERSITY PRESS

CAMBRIDGE UNIVERSITY PRESS

Cambridge New York Melbourne Madrid Cape Town Singapore São Paolo Delhi

Published in the United States of America by Cambridge University Press, New York

www.cambridge.org
Information on this title: www.cambridge.org/9781108002622

© in this compilation Cambridge University Press 2009

This edition first published 1880
This digitally printed version 2009

ISBN 978-1-108-00262-2

This book reproduces the text of the original edition. The content and language reflect
the beliefs, practices and terminology of their time, and have not been updated.

MATHEMATICAL

AND

PHYSICAL PAPERS.

London:
CAMBRIDGE WAREHOUSE,
17, PATERNOSTER ROW.

Cambridge: DEIGHTON, BELL, AND CO.
Leipzig: F. A. BROCKHAUS.

MATHEMATICAL

AND

PHYSICAL PAPERS

BY

GEORGE GABRIEL STOKES, M.A., D.C.L., LL.D., F.R.S.,

FELLOW OF PEMBROKE COLLEGE AND LUCASIAN PROFESSOR OF MATHEMATICS
IN THE UNIVERSITY OF CAMBRIDGE.

Reprinted from the Original Journals and Transactions,
with Additional Notes by the Author.

VOL. I.

Cambridge :

AT THE UNIVERSITY PRESS.

1880

Cambridge:
PRINTED BY C. J. CLAY, M.A.
AT THE UNIVERSITY PRESS.

PREFACE.

It is now some years since I was requested by the Syndics of the University Press to allow my papers on mathematical and physical subjects, which are scattered over various Transactions and scientific Journals, to be reprinted in a collected form. Many of these were written a long time ago, and science has in the mean time progressed, and it seemed to me doubtful whether it was worth while now to reprint a series of papers the interest of which may in good measure be regarded as having passed away. However, several of my scientific friends, and among them those to whose opinions I naturally pay the greatest deference, strongly urged me to have the papers reprinted, and I have accordingly acceded to the request of the Syndics. I regret that in consequence of the pressure of other engagements the preparation of the first volume has been so long in hand.

The arrangement of the papers and the mode of treating them in other respects were left entirely to myself, but both the Syndics and my friends advised me to make the reprint full, leaning rather to the inclusion than exclusion of a paper in doubtful cases. I have acted on this advice, and in the first volume, now presented to the public, I have omitted nothing but a few papers which were merely controversial.

As to the arrangement of the papers, it seemed to me that the chronological order was the simplest and in many respects the best. Had an arrangement by subjects been attempted, not only would it have been difficult in some cases to say under what head a particular paper should come, but also a later paper on some one subject would in many cases have depended on a paper on some different subject which would come perhaps in some later volume, whereas in the chronological arrangement each paper reaches up to the level of the author's knowledge at the time, so that forward reference is not required.

Although notes are added here and there, I have not attempted to bring the various papers up to the level of the present time. I have not accordingly as a rule alluded to later researches on the same subject, unless for some special reason. The notes introduced in the reprint are enclosed in square brackets in order to distinguish them from notes belonging to the original papers. To the extent of these notes therefore, which were specially written for the reprint, the chronological arrangement is departed from. The same is the case as regards the last paper in the first volume, which suggested itself during the preparation for press of the paper to which it relates. In reprinting the papers, any errors of inadvertence which may have been discovered are of course corrected. Mere corrections of this kind are not specified, but any substantial change or omission is noticed in a foot-note or otherwise.

After full consideration, I determined to introduce an innovation in notation which was proposed a great many years ago, for at least partial use, by the late Professor De Morgan, in his article on the Calculus of Functions in the *Encyclopædia Metropolitana*, though the proposal seems never to have been taken up. Mathematicians have been too little in the habit of considering the mechanical difficulty of setting up in type the expressions which they so freely write with the pen ; and where the setting up can be facilitated with only a trifling departure from existing usage as regards the appearance of the expression, it seems advisable to make the change.

Now it seems to me preposterous that. a compositor should be called on to go through the troublesome process of what printers call *justification*, merely because an author has occasion to name some simple fraction or differential coefficient in the text, in which term I do not include the formal equations which are usually printed in the middle of the page. The difficulty may be avoided by using, in lieu of the bar between the numerator and denominator, some symbol which may be printed on a line with the type. The symbol ":" is frequently used in expressing ratios; but for employment in the text it has the fatal objection that it is appropriated to mean a colon. The symbol "÷" is certainly distinctive, but it is inconveniently long, and $dy \div dx$ for a differential coefficient would hardly be tolerated. Now simple fractions are frequently written with a slant line instead of the horizontal bar separating the numerator from the denominator, merely for the sake of rapidity of writing. If we simply consent to allow the same to appear *in print*, the difficulty will be got over, and a differential coefficient which we have occasion to name in the text may be printed as dy/dx. The type for the slant line already exists, being called a *solidus*.

On mentioning to some of my friends my intention to use the "solidus" notation, it met with a good deal of approval, and some of them expressed their readiness to join me in the use of it, amongst whom I may name Sir William Thomson and the late Professor Clerk Maxwell.

In the formal equations I have mostly preserved the ordinary notation. There is however one exception. It frequently happens that we have to deal with fractions of which the numerator and denominator involve exponentials the indices of which are fractions themselves. Such expressions are extremely troublesome to set up in type in the ordinary notation. But by merely using the solidus for the fractions which form the indices, the setting up of the expression is made comparatively easy, while yet there is not much departure from the appearance of the expressions according to the ordinary notation. Such exponential expressions are commonly associated with circular functions; and though it would not otherwise have been necessary, it seemed desirable to employ the solidus notation for the fraction under the symbol "sin" or "cos," in order to preserve the similarity of appearance between the exponential and circular functions.

In the use of the solidus it seems convenient to enact that it shall as far as possible take the place of the horizontal bar for which it stands, and accordingly that what stands immediately on the two sides of it shall be regarded as welded into one. Thus $\sin n\pi x/a$ means $\sin(n\pi x \div a)$, and not $(\sin n\pi x) \div a$. This welding action may be arrested when necessary by a stop: thus $\sin n\theta . /r^n$ means $(\sin n\theta) \div r^n$ and not $\sin(n\theta \div r^n)$.

The only objection that I have heard suggested against the solidus notation on the ground of its being already appropriated to something else, relates to a condensed notation sometimes employed for factorials, according to which $x(x + a) \ldots$ to n factors is expressed by $x^{n|a}$ or by $x^{n/a}$. I do not think the objection is a serious one. There is no risk of the solidus notation, as I have employed it, being mistaken for the expression of factorials; of the two factorial notations just given, that with the separating line vertical seems to be the more common, and might be adhered to when factorials are intended; and if a greater distinction were desired, a factorial might be printed in the condensed notation as $x^{n(a}$, where the "(" would serve to recall the parentheses in the expression written at length.

G. G. STOKES.

CAMBRIDGE,
　　August 16, 1880.

CONTENTS.

ERRATA.

P. 103, l. 14, *for* their *read* there.
P. 193, l. 3, for $p^{y=b}$ read $p_{y=b}$.

MATHEMATICAL AND PHYSICAL PAPERS.

[From the *Transactions of the Cambridge Philosophical Society*, Vol. VII. p. 439.]

ON THE STEADY MOTION OF INCOMPRESSIBLE FLUIDS.

[Read April 25, 1842.]

IN this paper I shall consider chiefly the steady motion of fluids in two dimensions. As however in the more general case of motion in three dimensions, as well as in this, the calculation is simplified when $udx + vdy + wdz$ is an exact differential, I shall first consider a class of cases where this is true. I need not explain the notation, except where it may be new, or liable to be mistaken.

To prove that $udx + vdy + wdz$ is an exact differential, in the case of steady motion, when the lines of motion are open curves, and when the fluid in motion has come from an expanse of fluid of indefinite extent, and where, at an indefinite distance, the velocity is indefinitely small, and the pressure indefinitely near to what it would be if there were no motion.

By integrating along a line of motion, it is well known that we get the equation

$$\frac{p}{\rho} = V - \tfrac{1}{2}(u^2 + v^2 + w^2) + C \ldots\ldots\ldots\ldots(1),$$

where $dV = Xdx + Ydy + Zdz$, which I suppose an exact differential. Now from the way in which this equation is obtained,

S.   1

it appears that C need only be constant for the same line of motion, and therefore in general will be a function of the parameter of a line of motion. I shall first shew that in the case considered C is absolutely constant, and then that whenever it is, $udx + vdy + wdz$ is an exact differential*.

To determine the value of C for any particular line of motion, it is sufficient to know the values of p, and of the whole velocity, at any point along that line. Now if there were no motion we should have

$$\frac{p_1}{\rho} = V + C_1 \dots\dots\dots\dots\dots\dots\dots(2),$$

p_1 being the pressure in that case. But considering a point in this line at an indefinite distance in the expanse, the value of p at that point will be indefinitely nearly equal to p_1, and the velocity will be indefinitely small. Consequently C is more nearly equal to C_1 than any assignable quantity: therefore C is equal to C_1; and this whatever be the line of motion considered; therefore C is constant.

In ordinary cases of steady motion, when the fluid flows in open curves, it does come from such an expanse of fluid. It is conceivable that there should be only a canal of fluid in this expanse in motion, the rest being at rest, in which case the velocity at an infinite distance might not be indefinitely small. But experiment shews that this is not the case, but that the fluid flows in from all sides. Consequently at an indefinite distance the velocity is indefinitely small, and it seems evident that in that case the pressure must be indefinitely near to what it would be if there were no motion.

Differentiating therefore (1) with respect to x, we get

$$\frac{1}{\rho}\frac{dp}{dx} = X - u\frac{du}{dx} - v\frac{dv}{dx} - w\frac{dw}{dx} \, ;$$

but

$$\frac{1}{\rho}\frac{dp}{dx} = X - u\frac{du}{dx} - v\frac{du}{dy} - w\frac{du}{dz} \, ;$$

whence

$$v\left(\frac{dv}{dx} - \frac{du}{dy}\right) + w\left(\frac{dw}{dx} - \frac{du}{dz}\right) = 0.$$

[* See note, page 3.]

Similarly,
$$w\left(\frac{dw}{dy} - \frac{dv}{dz}\right) + u\left(\frac{du}{dy} - \frac{dv}{dx}\right) = 0,$$

$$u\left(\frac{du}{dz} - \frac{dw}{dx}\right) + v\left(\frac{dv}{dz} - \frac{dw}{dy}\right) = 0 ;$$

whence*
$$\frac{dv}{dx} = \frac{du}{dy}, \quad \frac{dw}{dy} = \frac{dv}{dz}, \quad \frac{du}{dz} = \frac{dw}{dx},$$

and therefore $udx + vdy + wdz$ is an exact differential.

When $udx + vdy + wdz$ is an exact differential, equation (1) may be deduced in another way †, from which it appears that C is constant. Consequently, in any case, $udx + vdy + wdz$ is, or is not, an exact differential, according as C is, or is not, constant.

Steady Motion in Two Dimensions.

I shall first consider the more simple case, where $udx + vdy$ is an exact differential. In this case u and v are given by the equations

$$\frac{du}{dx} + \frac{dv}{dy} = 0 \quad\quad\quad\quad\quad\quad (3),$$

$$\frac{du}{dy} - \frac{dv}{dx} = 0 \quad\quad\quad\quad\quad\quad (4) ;$$

and p is given by the equation

$$\frac{p}{\rho} = V - \tfrac{1}{2}(u^2 + v^2) + C.$$

The differential equation to a line of motion is

$$\frac{dy}{dx} = \frac{v}{u}.$$

---

* [This conclusion involves an oversight (see *Transactions*, p. 465) since the three preceding equations are not independent, as may readily be seen. I have not thought it necessary to re-write this portion of the paper, since in the two classes of steady motion to which the paper relates, namely those of motion in two dimensions, and of motion symmetrical about an axis, the three analogous equations are reduced to one, and the proposition is true. None of the succeeding results are affected by this error, excepting that the second paragraph of p. 11 must be restricted to the two cases above mentioned.]

† See Poisson, *Traité de Mécanique*.

Now from equation (3) it follows that $udy - vdx$ is always the exact differential of a function of x and y. Putting then

$$dU = udy - vdx,$$

$U = C$ will be the equation to the system of lines of motion, C being the parameter. U may have any value which allows dU/dy and $-dU/dx$ to satisfy the equations which u and v satisfy. The first equation has been already introduced; the second leads to the equation which U is to satisfy; viz.

$$\frac{d^2U}{dx^2} + \frac{d^2U}{dy^2} = 0 \quad\dots\dots\dots\dots\dots\dots\dots\dots(5).$$

The integral of this equation may be put under different forms. By integrating according to the general method, we get

$$U = F(x + \sqrt{-1}\,y) + f(x - \sqrt{-1}\,y).$$

Now it will be easily seen that U must be wholly real for all values of x and y, at least within certain limits. But $F(\alpha)$ may be put under the form $F_1(\alpha) + \sqrt{-1}\,F_2(\alpha)$, where $F_1(\alpha)$ and $F_2(\alpha)$ are wholly real. Making this substitution in the value of U, we get a result, which, without losing generality, may be put under the form

$$U = F(x + \sqrt{-1}\,y) + F(x - \sqrt{-1}\,y)$$
$$+ \sqrt{-1}\,\{f(x + \sqrt{-1}\,y) - f(x - \sqrt{-1}\,y)\},$$

changing the functions.

If we develope these functions in series ascending according to integral powers of y, by Taylor's Theorem, which can always be done as long as the origin is arbitrary, we get a series which I shall write for shortness,

$$U = 2\cos\left(\frac{d}{dx}\,y\right)F(x) - 2\sin\left(\frac{d}{dx}\,y\right)f(x),$$

the same result as if we had integrated at once by series by Maclaurin's Theorem.

It has been proved that the general integral of (5) may be put under the form

$$U = \Sigma A^{\alpha x + \beta y},$$

where $\alpha^2 + \beta^2 = 0$. Consequently α and β must be, one real, the other imaginary, or both partly real and partly imaginary. Putting then $\alpha = \alpha_1 + \sqrt{-1}\,\alpha_2$, $\beta = \beta_1 + \sqrt{-1}\,\beta_2$, introducing the condition that $\alpha^2 + \beta^2 = 0$, and replacing imaginary exponentials by sines and cosines, we find that the most general value of U is of the form

$$U = \Sigma A e^{n(\cos\gamma \cdot x - \sin\gamma \cdot y + a)} \cdot \cos n\,(\sin\gamma \cdot x + \cos\gamma \cdot y + b),$$

where A, n, γ, a and b have any real values, the value of U being supposed to be real.

If we take the value of U,

$$U = 2\cos\left(\frac{d}{dx}\,y\,\right) F(x) - 2\sin\left(\frac{d}{dx}\,y\,\right) f(x),$$

and develope each term, such as ax^n, in $F(x)$ or $f(x)$, in a series, and then sum the series by the formula

$$\cos n\theta + \sqrt{-1}\,\sin n\theta = \cos^n\theta\left(1 + \frac{n}{1}\sqrt{-1}\,\tan\theta - \dots\right),$$

we find that the general value of U takes the form

$$U = \Sigma A r^n \cos(n\theta + B).$$

As long as the origin of x is arbitrary, only integral powers of x will enter into the development $F(x)$ and $f(x)$, and therefore the above series will contain only integral values of n. For particular positions of the origin however, fractional powers may enter. The equation

$$\frac{d^2U}{dr^2} + \frac{1}{r}\frac{dU}{dr} + \frac{1}{r^2}\frac{d^2U}{d\theta^2} = 0,$$

which (5) becomes when transferred to polar co-ordinates, is satisfied by the above value of U, whatever n be, even if it be imaginary, in which case the value of U takes the form

$$U = \Sigma A r^m e^{n\theta}\cos(m\theta - \log_e r^n + B).$$

We may employ equation (5), to determine whether a proposed system of lines can be a system in which fluid can move, the motion being of the kind for which $udx + vdy$ is an exact differential.

Let $f(x, y) = U_1 = C$ be the equation to the system, C being the parameter. Then, if the motion be possible, some value of

U which satisfies (5) must be constant for all values of x and y for which U_1 is constant. Consequently this value must be a function of U_1. Let it $= \phi(U_1)$. Then, substituting this value in (5), and performing the differentiations, we get

$$\phi''(U_1)\left\{\left(\frac{dU_1}{dx}\right)^2 + \left(\frac{dU_1}{dy}\right)^2\right\} + \phi'(U_1)\left\{\frac{d^2U_1}{dx^2} + \frac{d^2U_1}{dy^2}\right\} = 0,$$

or

$$\frac{\phi''(U_1)}{\phi'(U_1)} + \frac{\dfrac{d^2U_1}{dx^2} + \dfrac{d^2U_1}{dy^2}}{\left(\dfrac{dU_1}{dx}\right)^2 + \left(\dfrac{dU_1}{dy}\right)^2} = 0 \dots\dots\dots\dots(6).$$

Now, if the motion be possible, the second term of this equation must be a function of U_1; x, y and U_1 being connected by the equation $f(x, y) = U_1$. Consequently, if by means of this latter equation we eliminate x or y from the second term of (6), the other must disappear. If it does not, the motion is impossible; if it does, the integration of equation (6), in which the variables are separated, will give $\phi(U_1)$ under the form

$$\phi(U_1) = A F(U_1) + B,$$

A and B being the arbitrary constants. The values of u and v will immediately be got by differentiation, and then p will be known. Nothing will be left arbitrary but a constant multiplying the values of u and v, and another added to the value of p.

I shall mention a few examples. Let $U = ar^{\frac{1}{2}} \cos \frac{1}{2}\theta$. In this case the lines of motion are similar parabolas about the same focus. The velocity at any point varies inversely as the square root of the distance from the focus.

Again, let $U = axy$. In this case the lines of motion are rectangular hyperbolas about the same asymptotes. Also,

$$u = \frac{dU}{dy} = ax, \text{ and } v = -\frac{dU}{dx} = -ay.$$

In this case therefore the velocity varies as the distance from the centre, and the particles in a section parallel to either of the axes remain in a section parallel to that axis.

I shall now consider the general case, where $udx + vdy$ need not be an exact differential.

In this case p, u and v, are given by the equations

$$\frac{1}{\rho}\frac{dp}{dx} = X - u\frac{du}{dx} - v\frac{du}{dy} \quad \dots\dots\dots\dots(7),$$

$$\frac{1}{\rho}\frac{dp}{dy} = Y - u\frac{dv}{dx} - v\frac{dv}{dy} \quad \dots\dots\dots\dots(8),$$

$$\frac{du}{dx} + \frac{dv}{dy} = 0 \dots\dots\dots\dots\dots\dots(9).$$

We still have $\dfrac{dy}{dx} = \dfrac{v}{u}$, for the differential equation to a line of motion, where $u\,dy - v\,dx$ is still an exact differential, on account of equation (9). Eliminating p by differentiation from (7) and (8), and expressing the result in terms of U, we get the equation which U is to satisfy, viz.

$$\frac{dU}{dy}\frac{d}{dx}\left(\frac{d^2U}{dx^2} + \frac{d^2U}{dy^2}\right) - \frac{dU}{dx}\frac{d}{dy}\left(\frac{d^2U}{dx^2} + \frac{d^2U}{dy^2}\right) = 0,$$

or, for shortness,

$$\left(\frac{dU}{dy}\frac{d}{dx} - \frac{dU}{dx}\frac{d}{dy}\right)\left(\frac{d^2U}{dx^2} + \frac{d^2U}{dy^2}\right) = 0\dots\dots\dots(10)^*.$$

* [This equation may be applied to prove an elegant theorem due to Mr F. D. Thomson {see the *Oxford, Cambridge, and Dublin Messenger of Mathematics*, Vol. III. (1866), p. 238, and Vol. IV. p. 37}, that if a vessel bounded by a cylindrical surface of any kind and by two planes perpendicular to its generating lines be filled with homogeneous liquid, and the whole be revolving uniformly about a fixed axis parallel to its generating lines, then if the vessel be suddenly arrested the motion of the liquid will be steady.

If ω be the angular velocity, we shall have for the motion before impact

$$U = -\int(\omega y\,dy + \omega x\,dx) = -\tfrac{1}{2}\omega(x^2 + y^2) = -\tfrac{1}{2}\omega r^2,$$

omitting the constant as unnecessary. If u, v be the components of the change of velocity produced by impact, it follows from the equations of impulsive motion that $u\,dx + v\,dy$ will be a perfect differential $d\phi$, where ϕ satisfies the partial differential equation $\nabla\phi = 0$, ∇ standing for $\dfrac{d^2}{dx^2} + \dfrac{d^2}{dy^2}$. If U' be the U-function corresponding to this motion—and such a function exists by virtue of the equation of continuity whether the motion be steady or not—we have

$$U' = \int\left(\frac{d\phi}{dx}\,dy - \frac{d\phi}{dy}\,dx\right) = \int\left(\frac{d\phi}{dr}\,r\,d\theta - \frac{1}{r}\frac{d\phi}{d\theta}\,dr\right),$$

where the quantity under the sign $\int$ is a perfect differential by virtue of the equa-

In this case, since $p = \int \left(\dfrac{dp}{dx}\, dx + \dfrac{dp}{dy}\, dy \right)$, equations (7) and (8) give

tion $\nabla \phi = 0$; and we see at once that $\nabla U' = 0$. Hence for the whole motion just after impact

$$\nabla (U + U') = \nabla U = -2\omega,$$

which satisfies the equation of steady motion (10); and as the condition at the boundary, namely that the fluid shall slide along it, is satisfied, being satisfied initially, it follows that the initial motion after impact will be continued as steady motion.

To actually determine the function ϕ or U', and thereby the motion in any given case, we must satisfy not only the general equation $\nabla \phi = 0$ but also the equation of condition at the boundary, namely that there shall be no velocity in a direction normal to the surface, which gives

$$\left(\frac{d\phi}{dx} - \omega y \right) dy - \left(\frac{d\phi}{dy} - \omega x \right) dx = 0 \quad \ldots\ldots\ldots\ldots\ldots\ldots\ldots\ldots (a),$$

at any point of the boundary. If $f(x, y) = 0$ be the equation of the boundary, we must substitute $- df / dx \div df / dy$ for dy / dx in (a), and the resulting equation will have to be satisfied when $f = 0$ is satisfied.

There are but few forms of boundary for which the solution of the problem can be actually effected analytically, among which may be mentioned in particular the case of a rectangle. But by taking particular solutions of the equation $\nabla \phi = 0$, substituting in (a) and integrating, which gives

$$- \tfrac{1}{2} \omega r^2 + U' = C \quad \ldots\ldots\ldots\ldots\ldots\ldots\ldots\ldots\ldots\ldots\ldots\ldots (\beta),$$

or what comes to the same thing taking particular solutions of the equation $\nabla U' = 0$ and substituting in (β), which gives the general equation of the lines of motion, we may synthetically obtain an infinity of examples in which the conditions of the problem are satisfied, any one of the lines of motion being taken as the boundary of the fluid.

Thus for $U' = kr^3 \cos 3\theta$ we have for the lines of motion

$$- \tfrac{1}{2} \omega r^2 + kr^3 \cos 3\theta = C \quad \ldots\ldots\ldots\ldots\ldots\ldots\ldots\ldots\ldots\ldots (\gamma),$$

or

$$- \tfrac{1}{2} \omega r^2 + k \{ 4 (r \cos \theta)^3 - 3r^2 . r \cos \theta \} = C \quad \ldots\ldots\ldots\ldots (\delta),$$

which therefore are cubic curves, recurring when θ is increased by 120°. (δ) is satisfied by

$$r \cos \theta = a,$$

giving a straight line, provided

$$k = - \frac{\omega}{6a}, \qquad C = 4ka^3 = - \tfrac{2}{3} \omega a^2.$$

Hence when k has the above value the cubic curve (γ) breaks up, for the particular value of the parameter C above written, into three straight lines forming the sides of an equilateral triangle, and the vessel may therefore be supposed to be an equilateral triangular prism. The various lines of motion correspond to values of the parameter C from 0 to $- \tfrac{2}{3} \omega a^2$. This case is given by Mr Thomson.

$U' = kr^2 \cos 2\theta$ leads to the case of steady motion in similar and concentric ellipses considered in the text a little further on, which therefore may be conceived to have been produced from motion about a fixed axis as pointed out by Mr Thomson. In fact, any case of steady motion in two dimensions in which $\nabla U = \text{const.}$ may be conceived to have been so produced.]

$$\frac{p}{\rho} = V - \int \left\{ \left(\frac{dU}{dy} \frac{d^2U}{dx\,dy} - \frac{dU}{dx} \frac{d^2U}{dy^2} \right) dx \right.$$

$$\left. + \left(\frac{dU}{dx} \frac{d^2U}{dx\,dy} - \frac{dU}{dy} \frac{d^2U}{dx^2} \right) dy \right\}.$$

Now $\frac{1}{2} d \left\{ \left(\frac{dU}{dx} \right)^2 + \left(\frac{dU}{dy} \right)^2 \right\} = \left(\frac{dU}{dx} \frac{d^2U}{dx^2} + \frac{dU}{dy} \frac{d^2U}{dx\,dy} \right) dx$

$$+ \left(\frac{dU}{dx} \frac{d^2U}{dx\,dy} + \frac{dU}{dy} \frac{d^2U}{dy^2} \right) dy \, ;$$

whence,

$$\frac{dU}{dy} \frac{d^2U}{dx\,dy} dx + \frac{dU}{dx} \frac{d^2U}{dx\,dy} dy - \frac{dU}{dx} \frac{d^2U}{dy^2} dx - \frac{dU}{dy} \frac{d^2U}{dx^2} dy$$

$$= \frac{1}{2} d \left\{ \left(\frac{dU}{dx} \right)^2 + \left(\frac{dU}{dy} \right)^2 \right\} - \left(\frac{d^2U}{dx^2} + \frac{d^2U}{dy^2} \right) \left(\frac{dU}{dx} dx + \frac{dU}{dy} dy \right) ;$$

and therefore

$$\frac{p}{\rho} = V - \frac{1}{2} \left\{ \left(\frac{dU}{dx} \right)^2 + \left(\frac{dU}{dy} \right)^2 \right\} + \int \left(\frac{d^2U}{dx^2} + \frac{d^2U}{dy^2} \right) \left(\frac{dU}{dx} dx + \frac{dU}{dy} dy \right),$$

$$= V - \frac{1}{2} (v^2 + u^2) + \int \left(\frac{d^2U}{dx^2} + \frac{d^2U}{dy^2} \right) dU.$$

It will be observed that $\dfrac{d^2U}{dx^2} + \dfrac{d^2U}{dy^2} = \chi(U)$, is a first integral of (10). Consequently this latter term, which is the value of C in (1), comes out a function of the parameter of a line of motion as it should.

We may employ equation (10), precisely as before, to enquire whether a proposed system of lines can, under any circumstances, be a system of lines of motion. Let $f(x, y) = U_1 = C$, be the equation to the system; then, putting as before, $U = \phi(U_1)$, we get

$$\phi''(U_1) \left(\frac{dU_1}{dy} \frac{d}{dx} - \frac{dU_1}{dx} \frac{d}{dy} \right) \left\{ \left(\frac{dU_1}{dx} \right)^2 + \left(\frac{dU_1}{dy} \right)^2 \right\}$$

$$+ \phi'(U_1) \left(\frac{dU_1}{dy} \frac{d}{dx} - \frac{dU_1}{dx} \frac{d}{dy} \right) \left(\frac{d^2U_1}{dx^2} + \frac{d^2U_1}{dy^2} \right) = 0 \, ;$$

or, $\qquad\qquad P\phi''(U_1) + Q\phi'(U_1) = 0,$ suppose.

Hence, as before, if we express y in terms of x and U_1, from the equation $f(x, y) = U_1$, and substitute that value in $\dfrac{Q}{P}$, the result must not contain x. If it does, the proposed system of lines cannot be a system of lines of motion; if not, the integration of the above equation will give $\phi(U_1)$, under the form

$$\phi(U_1) = AF(U_1) + B,$$

and we can immediately get the values of u, v and p, with the same arbitrary constants as in the previous case.

One case in which the motion is possible is where the lines of motion are a system of similar ellipses or hyperbolas about the same centre, or a system of equal parabolas having the same axis. In the case of the ellipse, the particles in a radius vector at any time remain in a radius vector, and the value of p has the form

$$\rho V + A + B\,(x^2 + y^2).$$

When however the ellipse becomes a circle, P and Q vanish in the equation $P\phi''(U_1) + Q\phi'(U_1) = 0$. Consequently the form of ϕ may be any whatever. The value of U_1 being $x^2 + y^2$, we have

$$u = 2\phi'(U_1)\, y, \quad v = -2\phi'(U_1)\, x;$$

whence, $u^2 + v^2 = 4\,\{\phi'(U_1)\}^2\,(x^2 + y^2) = 4U_1\,\{\phi'(U_1)\}^2.$

Hence, the velocity may be any function of the distance from the centre. It is evident that we may conceive cylindrical shells of fluid, having a common axis, to be revolving about that axis with any velocities whatever, if we do not consider friction, or whether such a mode of motion would be stable. The result is the same if we enquire in what way fluid can move in a system of parallel lines.

In any case where the motion in a certain system of lines is possible, if we suppose two of these lines to be the bases of bounding cylindrical surfaces, and if we suppose the velocity and direction of motion, at each point of a section of the entering, and also of the issuing fluid, to be what that case requires, I have not proved that the fluid *must* move in that system of lines. When the above conditions are given there may still perhaps be different modes of steady motion; and of these some may be stable, and others unstable. There may even be no stable steady mode of motion possible, in which case the fluid would continue perpetually eddying.

In the case of rectangular hyperbolas, the fluid appeared, on making the experiment, to move in hyperbolas when the end at which the fluid entered was broad and the other end narrow, but not when the end by which the fluid entered was narrow. This may, I think, in some measure be accounted for. Suppose fluid to flow out of a vessel where the pressure is p_1 into one where it is p_2, through a small orifice. Then, the motion being steady, we have, along the same line of motion, $p/\rho = C - \frac{1}{2}v^2$, where v is the whole velocity. At a distance from the orifice, in the first vessel, the pressure will be approximately p_1, and the velocity nothing. At a distance in the second vessel, the pressure will be approximately p_2, and therefore the velocity $= \sqrt{\dfrac{2(p_1 - p_2)}{\rho}}$, nearly. The result is the same if forces act on the fluid. Hence the velocity must be approximately constant; and therefore, the fluid which came from the first vessel, instead of spreading out, must keep to a canal of its own of uniform breadth. This is found to agree with experiment. Hence we might expect that in the case of the hyperbolas, if the end at which the fluid entered were narrow, the entering fluid would have a tendency to keep to a canal of its own, instead of spreading out.

In ordinary cases of steady motion, when the lines of motion are open curves, the fluid is supplied from an expanse of fluid, and consequently $u\,dx + v\,dy + w\,dz$ is an exact differential. Consequently, cases of open curves for which it is not an exact differential do not ordinarily occur. We may, however, conceive such cases to occur; for we may suppose the velocity and direction of motion, at each point of a section of the entering, and also of the issuing stream, to be such as any case requires, by supposing the fluid sent in and drawn out with the requisite velocity and in the requisite direction through an infinite number of infinitely small tubes.

In the case of closed curves however, in whatever manner the fluid may have been put in motion, it seems probable that, if we neglect the friction against the sides of the vessel, the fluid will have a tendency to settle down into some steady mode of motion. Consequently, taking account of the friction against the sides of the vessel, it seems probable that the motion may in some cases become approximately steady, before the friction has caused it to cease altogether.

Motion symmetrical about an axis, the lines of motion being in planes passing through the axis.

Before considering this case, it may be well to prove a principle which will a little simplify our equations.

The general equations of motion are,

$$\frac{1}{\rho}\frac{dp}{dx} = X - u\frac{du}{dx} - v\frac{du}{dy} - w\frac{du}{dz} \dots\dots\dots\dots(11),$$

$$\frac{1}{\rho}\frac{dp}{dy} = Y - u\frac{dv}{dx} - v\frac{dv}{dy} - w\frac{dv}{dz} \dots\dots\dots\dots(12),$$

$$\frac{1}{\rho}\frac{dp}{dz} = Z - u\frac{dw}{dx} - v\frac{dw}{dy} - w\frac{dw}{dz} \dots\dots\dots\dots(13).$$

And the equation of continuity is

$$\frac{du}{dx} + \frac{dv}{dy} + \frac{dw}{dz} = 0 \dots\dots\dots\dots\dots(14).$$

Putting $\varpi_1, \varpi_2, \varpi_3,$ for the last three terms in (11), (12), (13), respectively, we have

$$\frac{p}{\rho} = V - \int (\varpi_1 dx + \varpi_2 dy + \varpi_3 dz).$$

Hence the pressure consists of two parts, the first, ρV, the same as if there were no motion, the second, the part due to the velocity. Now the velocities are given by equation (14), and by the three equations which result on eliminating p from (11), (12), and (13). These latter equations, as well as (14), will be the same as if there were no forces since

$$\frac{dX}{dy} = \frac{dY}{dx}, \quad \frac{dX}{dz} = \frac{dZ}{dx}, \quad \text{and} \quad \frac{dY}{dz} = \frac{dZ}{dy};$$

and therefore we shall not lose generality by omitting the forces in (11), (12) and (13), since we shall only have to add ρV to the value of p so determined.

When the motion is symmetrical about an axis, and in planes passing through that axis, let z be measured along the axis, and

r be the perpendicular distance from the axis, and s be the velocity perpendicular to the axis. Then, transforming the co-ordinates to z and r, and omitting the forces, it will be found that equations (11), (12) and (13) are equivalent to only two separate equations, which are

$$\frac{1}{\rho}\frac{dp}{dr} = -s\frac{ds}{dr} - w\frac{ds}{dz} \quad\ldots\ldots\ldots\ldots(15),$$

$$\frac{1}{\rho}\frac{dp}{dz} = -s\frac{dw}{dr} - w\frac{dw}{dz} \quad\ldots\ldots\ldots\ldots(16),$$

and the equation of continuity becomes

$$\frac{ds}{dr} + \frac{s}{r} + \frac{dw}{dz} = 0 \quad\ldots\ldots\ldots\ldots(17).$$

In the case where $udx + vdy + wdz$ is an exact differential, it will be found that the three equations

$$\frac{du}{dy} = \frac{dv}{dx}, \quad \frac{du}{dz} = \frac{dw}{dx}, \quad \frac{dv}{dz} = \frac{dw}{dy},$$

are equivalent to only one equation, which is

$$\frac{ds}{dz} = \frac{dw}{dr} \quad\ldots\ldots\ldots\ldots(18).$$

In the general case we get, by eliminating p from (15) and (16),

$$\frac{d}{dz}\left(s\frac{ds}{dr} + w\frac{ds}{dz}\right) = \frac{d}{dr}\left(s\frac{dw}{dr} + w\frac{dw}{dz}\right),$$

or $\quad\dfrac{ds}{dr}\dfrac{ds}{dz} + \dfrac{ds}{dz}\dfrac{dw}{dz} + s\dfrac{d^2s}{drdz} + w\dfrac{d^2s}{dz^2}$

$$= \frac{dw}{dr}\frac{dw}{dz} + \frac{dw}{dr}\frac{ds}{dr} + w\frac{d^2w}{drdz} + s\frac{d^2w}{dr^2}\ldots\ldots\ldots(19).$$

The differential equation, between z and r, to a line of motion is

$$\frac{dz}{dr} = \frac{w}{s}.$$

Let μ be a factor which renders $sdz - wdr$ an exact differential, then

$$\frac{d\mu s}{dr} + \frac{d\mu w}{dz} = 0,$$

or

$$\mu\left(\frac{ds}{dr} + \frac{dw}{dz}\right) + s\frac{d\mu}{dr} + w\frac{d\mu}{dz} = 0,$$

or, using (17), $$s\frac{d\mu}{dr} + w\frac{d\mu}{dz} = \mu\frac{s}{r};$$

whence we easily see that $\mu = r$ is one such factor.

Let then $$dU = rsdz - rwdr,$$

so that $$s = \frac{1}{r}\frac{dU}{dz}, \quad w = -\frac{1}{r}\frac{dU}{dr}.$$

The equation which U is to satisfy will be got by expressing s and w in terms of U, and substituting in (19) in the general case, or by substituting in (18), in the case where $udx + vdy + wdz$ is an exact differential.

In the latter case the equation which U is to satisfy is

$$\frac{d^2U}{dz^2} + \frac{d^2U}{dr^2} - \frac{1}{r}\frac{dU}{dr} = 0 \ldots\ldots\ldots\ldots\ldots(20).$$

In the general case, the equation is what I shall write

$$\left(\frac{dU}{dz}\frac{d}{dr} - \frac{dU}{dr}\frac{d}{dz}\right)\left\{\frac{1}{r^2}\left(\frac{d^2U}{dz^2} + \frac{d^2U}{dr^2} - \frac{1}{r}\frac{dU}{dr}\right)\right\} = 0 \ldots(21).$$

The value of p is given by the equation

$$\frac{p}{\rho} = -\int\left\{\left(s\frac{ds}{dr} + w\frac{ds}{dz}\right)dr + \left(s\frac{dw}{dr} + w\frac{dw}{dz}\right)dz\right\}.$$

Now

$$\tfrac{1}{2}d(s^2 + w^2) = s\frac{ds}{dr}dr + w\frac{dw}{dz}dz + s\frac{ds}{dz}dz + w\frac{dw}{dr}dr;$$

and therefore

$$\left(s\frac{ds}{dr} + w\frac{ds}{dz}\right)dr + \left(s\frac{dw}{dr} + w\frac{dw}{dz}\right)dz$$

$$= \tfrac{1}{2}d(s^2 + w^2) + \frac{ds}{dz}(wdr - sdz) + \frac{dw}{dr}(sdz - wdr)$$

$$= \tfrac{1}{2}d(s^2 + w^2) + \left(\frac{dw}{dr} - \frac{ds}{dz}\right)\frac{1}{r}dU;$$

whence $$\frac{p}{\rho} = -\tfrac{1}{2}(s^2 + w^2) + \int\left(\frac{ds}{dz} - \frac{dw}{dr}\right)\frac{1}{r}dU$$

$$= -\frac{1}{2r^2}\left\{\left(\frac{dU}{dz}\right)^2 + \left(\frac{dU}{dr}\right)^2\right\} + \int\frac{1}{r^2}\left(\frac{d^2U}{dz^2} + \frac{d^2U}{dr^2} - \frac{1}{r}\frac{dU}{dr}\right)dU\ldots(22).$$

Hence the quantity under the integral sign must be a function of U. And in fact, we can easily shew by trial that

$$\frac{1}{r^2}\left(\frac{d^2 U}{dz^2} + \frac{d^2 U}{dr^2} - \frac{1}{r}\frac{dU}{dr}\right) = \psi\,(U)$$

is a first integral of (21). The last term of (22) is the value of the constant in (1).

By expanding U in a series ascending according to integral powers of z, which may be done as long as the origin is arbitrary, it will be found that the integral of (20) may be written under the form

$$U = \cos\,(\nabla z)\, F\,(r) + \sin\,(\nabla z)\, \nabla^{-1} f\,(r),$$

where $\nabla^2 F(r)$ denotes $\left(\dfrac{d^2}{dr^2} - \dfrac{1}{r}\dfrac{d}{dr}\right) F\,(r)$, and $\nabla^{2n} F(r)$ denotes that the operation $\dfrac{d^2}{dr^2} - \dfrac{1}{r}\dfrac{d}{dr}$ is repeated n times on $F\,(r)$.

We may employ equations (21) or (20) just as before, to determine whether the motion in a proposed system of lines is possible. If $F(r, z) = U_1 = C$ be the equation to the system, we must have, as before, $U = \phi\,(U_1)$; whence we get, in the general case,

$$\phi''\,(U_1)\left\{\left(\frac{dU_1}{dz}\frac{d}{dr} - \frac{dU_1}{dr}\frac{d}{dz}\right)\left[\frac{1}{r^2}\left(\frac{dU_1}{dz}\right)^2 + \left(\frac{dU_1}{dr}\right)^2\right]\right\}$$

$$+ \phi'\,(U_1)\left\{\left(\frac{dU_1}{dz}\frac{d}{dr} - \frac{dU_1}{dr}\frac{d}{dz}\right)\left[\frac{1}{r^2}\left(\frac{d^2 U_1}{dz^2} + \frac{d^2 U_1}{dr^2} - \frac{1}{r}\frac{dU_1}{dr}\right)\right]\right\} = 0,$$

and in the more restricted case where $u\,dx + v\,dy + w\,dz$ is an exact differential, we get

$$\phi''\,(U_1)\left\{\left(\frac{dU_1}{dz}\right)^2 + \left(\frac{dU_1}{dr}\right)^2\right\} + \phi'\,(U_1)\left(\frac{d^2 U_1}{dz^2} + \frac{d^2 U_1}{dr^2} - \frac{1}{r}\frac{dU_1}{dr}\right) = 0.$$

As before, the ratio of the coefficients of $\phi''\,(U_1)$ and $\phi'\,(U_1)$ must be a function of U_1 alone, when z, r and U_1 are connected by the equation $F(r, z) = U_1$. If the motion be possible, it will in general be determinate, U being of the form $Af(r, z) + B$. If $U_1 = r$ however, the form of ϕ remains arbitrary. In this case the fluid may be conceived to move in cylindrical shells parallel to the axis, the velocity being any function of the distance from the axis.

Particular cases are, where the lines of motion are right lines directed to a point in the axis, and where they are equal parabolas having the axis of z for a common axis.   In these cases

$$udx + vdy + wdz$$

is an exact differential.

We may employ equations (20) and (21) to determine whether the hypothesis of parallel sections can be strictly true in any case. In this case, the sections being perpendicular to the axis of z, we must have

$$w = -\frac{1}{r}\frac{dU}{dr} = F(z) \; ;$$

$$\frac{dU}{dr} = -rF(z);$$

$$U = -\tfrac{1}{2} r^2 F(z) + f(z).$$

Substituting this value in (21), we find, by equating to zero coefficients of different powers of r, that the most general case corresponds to

$$U = (a + bz + cz^2) r^2 + ez + f.$$

If $udx + vdy + wdz$ be an exact differential, the most general case corresponds to

$$U = (a + bz) r^2 + c + ez.$$

[From the *Transactions of the Cambridge Philosophical Society*,
Vol. VIII. p. 105.]

ON SOME CASES OF FLUID MOTION.

[Read May 29, 1843.]

THE equations of Hydrostatics are founded on the principles
that the mutual action of two adjacent elements of a fluid is normal
to the surface which separates them, and that the pressure is equal
in all directions. The latter of these is a necessary consequence
of the former, as has been shewn by Mr Airy*. An exactly similar proof may be employed in Hydrodynamics, by which it may
be shewn that, if the mutual action of two adjacent elements of a
fluid in motion is normal to their common surface, the pressure
must be equal in all directions, in order that the accelerating force
which acts on the centre of gravity of an element may not become
infinite, when we suppose the dimensions of the element indefinitely diminished. In Hydrostatics, the accurate agreement of the
results of our calculations with experiments, (those phenomena
which depend on capillary attraction being excepted), fully justifies
our fundamental assumption. The same assumption is made in
Hydrodynamics, and from it are deduced the fundamental equations of fluid motion. But the verification of our fundamental law
in the case of a fluid at rest, does not at all prove it to be true
in the case of a fluid in motion, except in the very limited case of
a fluid moving as if it were solid. Thus, oil is sufficiently fluid to
obey the laws of fluid equilibrium, (at least to a great extent),
yet no one would suppose that oil in motion ought to be considered
a perfect fluid. It would appear from the following consideration,
that the fluidity of water and other such fluids is not quite perfect.

---

* See also Professor Miller's *Hydrostatics*, page 2.

S.

2

When a mass of water contained in a vessel of the form of a solid of revolution is stirred round, and then left to itself, it presently comes to rest. This, no doubt, is owing to the friction against the sides of the vessel. But if the fluidity of water were perfect, it does not appear how the retardation due to this friction could be transmitted through the mass. It would appear that in that case a thin film of fluid close to the sides of the vessel would remain at rest, the remaining part of the fluid being unaffected by it. And in this respect, that part of Poisson's solution of the problem of an oscillating sphere, which relates to friction, appears to me in some degree unsatisfactory. A term enters into the equation of motion of the sphere depending on the friction of the fluid on the sphere, while no such term enters into the equations of motion of the fluid, to express the equal and opposite friction of the sphere on the fluid. In fact, as long as we regard the fluidity of the fluid as perfect, no such term can enter. The only way by which to estimate the extent to which the imperfect fluidity of fluids may modify the laws of their motion, without making any hypothesis as to the molecular constitution of fluids, appears to be, to calculate according to the hypothesis of perfect fluidity some cases of fluid motion, which are of such a nature as to be capable of being accurately compared with experiment. The cases of that nature which have hitherto been calculated, are by no means numerous. My object in the present paper which I have the honour to lay before the Society, has been partly to calculate some such cases which may be useful in determining how far we are justified in regarding fluids as perfectly fluid, and partly to give examples of the methods by which the solution of problems depending on partial differential equations may be effected.

In the first seven articles, I have mentioned and explained some general principles, which are afterwards applied. Some of these are not new, but it was convenient to state them for the sake of reference. Others are I believe new, at least in their development. In the remaining articles, I have given different problems, of which I have succeeded in obtaining the solutions. As the problem to be solved is usually stated at the head of each article, I shall here only mention some of the results. As a particular case of the problem given in Art. 8, I find that, when a cylinder oscillates in an infinitely extended fluid, the effect of the inertia of the fluid is to increase the mass of the cylinder by that of the fluid displaced. In part of Art. 9, I find that when a ball pendulum oscillates in a concentric spherical envelope, the effect of the inertia of the fluid is to increase the mass of the ball by $\dfrac{b^3 + 2a^3}{2(b^3 - a^3)}$ times that of the fluid displaced, a being the radius of the ball, b that of the envelope. Poisson, in his solution of the problem of the sphere, arrives at the strange result that the envelope does not at all retard the oscillating sphere. I have pointed out the erroneous step by which he was led to this conclusion, which I am clearly called upon to do, in venturing to differ from so high an authority. Of the different cases of fluid motion which I have given, that which appears to be capable of the most accurate and varied comparison with experiment, is the motion of fluid in a rectangular box which is closed on all sides, given in Art. 13. The experiment consists in comparing the calculated and observed times of oscillation. I find that when the motion is small, the effect of the fluid on the motion of the box is the same as that of a solid having the same mass, centre of gravity, and principal axes, but having different moments of inertia, these moments being given by infinite series, which converge with great rapidity. I have also in Art. 11, given some cases of progressive motion, deduced on the supposition that the same particles of fluid remain in contact with the solid, which do not at all agree with experiment.

In almost all the cases given in this paper, the problem of finding the permanent state of temperature in the several solids considered, supposing the surfaces of those solids kept up to constant temperatures varying from point to point, may be solved by a similar analysis. I find that some of these cases have been already solved by M. Duhamel in a paper inserted in the 22nd *Cahier* of the *Journal de l'Ecole Polytechnique*. The cases alluded to are those of the temperature in a solid sphere, and in a rectangular parallelepiped. Since, however, the application of the formulæ in the two cases of fluid motion and of the permanent state of temperature is different, as well as the formulæ themselves to a certain extent, I thought it might be worth while to give them.

---

1. The investigations in this paper apply directly to *incompressible* fluids, as the fluids spoken of will be supposed to be,

unless the contrary is stated. The motions of elastic fluids may in most cases be divided into two classes, one consisting of those condensations on which sound depends, the other, of those motions which the fluid takes in consequence of the motion of solid bodies in it. Those motions of the fluid, which take place in consequence of very rapid motions of solids, (such as those of bullets), form a connecting link between these two classes. The motions of the second class are, it is true, accompanied by condensations, and propagated with the velocity of sound, but if the motions of the solids are not great we may, without sensible error, suppose the motions of the fluid propagated instantaneously to distances where they cease to be sensible, and may neglect the condensation. The investigations in this paper will apply without sensible error to this kind of motion of elastic fluids.

In all cases also the motion will be supposed to begin from rest, which allows us to suppose that $udx + vdy + wdz$ is an exact differential $d\phi$, where u, v and w are the components, parallel to the axes of x, y, and z, of the whole velocity of any particle. In applying our investigations however to fluids such as they exist in nature, this principle must not be strained too far. When a body is made to revolve continually in a fluid, the parts of the fluid near the body will soon acquire a rotatory motion, in consequence, in all probability, of the mutual friction of the parts of the fluid; so that after a time $udx + vdy + wdz$ could no longer be taken an exact differential. It is true that in motion in two dimensions there is one sort of rotatory motion for which that quantity is an exact differential; but if a close vessel, filled with fluid at first at rest, be made to revolve uniformly round a fixed axis, the fluid will soon do so too, and therefore that quantity will cease to be an exact differential. For the same reason, in the progressive motion of a solid in a fluid, the effect of friction continually accumulating, the motion might at last be sensibly different from what it would be if there were no friction, and that, even if the friction were very small. In the case of small oscillatory motions however it would appear that the effect of friction in the forward oscillation, supposing that friction small, would be counteracted by its effect in the backward oscillation, at least if the two were symmetrical. In this case then we might expect our results to agree very nearly with experiment, so far at least as the *time* of oscillation is concerned.

The forces which act on the fluid are supposed in the following investigations to be such that $Xdx + Ydy + Zdz$ is the exact differential of a function of x, y and z, where X, Y, Z, are the components, parallel to the axes, of the acccelerating force acting on the particle whose co-ordinates are x, y, z. The only effect of such forces, in the case of a homogeneous, incompressible fluid, being to add the quantity $\rho \int (Xdx + Ydy + Zdz)$ to the pressure, the forces, as well as the pressure due to them, will for the future be omitted for the sake of simplicity.

2. It is a recognized principle, and one of great importance in these investigations, that when a problem is determinate any solution which satisfies all the requisite conditions, no matter how obtained, is the solution of the problem. In the case of fluid motion, when the initial circumstances and the conditions with respect to the boundaries of the fluid are given, the problem is determinate. If it were required to find what sort of steady motion could take place between given surfaces, the problem would not be determinate, since different kinds of steady motion might result from different initial circumstances.

It may be well here to enumerate the conditions which must be satisfied in the case of a homogeneous incompressible fluid without a free surface, the case which is considered in this paper. We have first the equations,

$$\frac{1}{\rho}\frac{dp}{dx} = -\varpi_1, \quad \frac{1}{\rho}\frac{dp}{dy} = -\varpi_2, \quad \frac{1}{\rho}\frac{dp}{dz} = -\varpi_3, \ldots\ldots\ldots(A);$$

putting ϖ_1 for $\dfrac{du}{dt} + u\dfrac{du}{dx} + v\dfrac{du}{dy} + w\dfrac{du}{dz}$, and ϖ_2, ϖ_3, for the corresponding quantities for y and z, and omitting the forces.

We have also the equation of continuity,

$$\frac{du}{dx} + \frac{dv}{dy} + \frac{dw}{dz} = 0 \ldots\ldots\ldots\ldots\ldots(B);$$

(A) and (B) hold at all times for all points of the fluid mass.

If σ be the velocity of the point (x, y, z) of the surface of a solid in contact with the fluid resolved along the normal, and ν the velocity, resolved along the same normal, of the fluid particle, which at the time t is in contact with the above point of the solid, we must have

$$v = \sigma \dots\dots\dots\dots\dots\dots\dots\dots\dots\dots(a)^*,$$

at all times and for all points of the fluid which are in contact with a solid.

If the fluid extend to infinity, and the motion at first be zero at an infinite distance, we must have

$$u = 0, \quad v = 0, \quad w = 0, \text{ at an infinite distance.}\dots\dots\dots(b).$$

An analagous condition is, that the motion shall not become infinitely great about a particular point, as the origin.

Lastly, if u_0, v_0, w_0, be the initial velocities, subject of course to satisfy equations (B) and (a), we must have

$$u = u_0, \quad v = v_0, \quad w = w_0, \text{ when } t = 0\dots\dots\dots\dots\dots(c).$$

In the most general cases the equations which u, v and w are to satisfy at every point of the mass and at every time are (B) and the three equations

$$\frac{d\varpi_1}{dy} = \frac{d\varpi_2}{dx}, \quad \frac{d\varpi_2}{dz} = \frac{d\varpi_3}{dy}, \quad \frac{d\varpi_3}{dx} = \frac{d\varpi_1}{dz}\dots\dots\dots\dots(C).$$

These equations being satisfied, the quantity $\varpi_1 dx + \varpi_2 dy + \varpi_3 dz$ will be an exact differential, whence p may be determined by integrating the value of dp given by equations (A). Thus the condition that these latter equations shall be satisfied is equivalent to the condition that the equations (C) shall be satisfied.

In nearly all the cases considered in this paper, and in all those of which the complete solution is given, the motion is such that $u dx + v dy + w dz$ is an exact differential $d\phi$. This being the case, the equations (C) are, as it is well known, always satisfied, the value of p being given by the equation

$$\frac{p}{\rho} = \psi(t) - \frac{d\phi}{dt} - \frac{1}{2}\left\{\left(\frac{d\phi}{dx}\right)^2 + \left(\frac{d\phi}{dy}\right)^2 + \left(\frac{d\phi}{dz}\right)^2\right\}\dots\dots\dots\dots(D),$$

---

* For greater clearness, those equations which must hold for all values of the variables within limits depending on the problem are denoted by capitals, while those which hold only for certain values of the variables, or of some of them, are denoted by small letters. The latter class serve to determine the forms of the arbitrary functions contained in the integrals of the former.

$\psi(t)$ being an arbitrary function of t, which may if we please be included in ϕ. In this case, therefore, the *single* condition which has to be satisfied at all times, and at every point of the mass is (B), which becomes in this case

$$\frac{d^2\phi}{dx^2} + \frac{d^2\phi}{dy^2} + \frac{d^2\phi}{dz^2} = 0 \dots\dots\dots\dots\dots (E).$$

In the case of impulsive motion, if u_0, v_0, w_0, be the velocities just before impact, u, v, w, the velocities just after, and q the impulsive pressure, the equations (A) are replaced by the equations

$$\frac{1}{\rho}\frac{dq}{dx} = -u + u_0, \quad \frac{1}{\rho}\frac{dq}{dy} = -v + v_0, \quad \frac{1}{\rho}\frac{dq}{dz} = -w + w_0 \dots(F);$$

and in order that these equations may be satisfied it is necessary and sufficient that $(u - u_0)\,dx + (v - v_0)\,dy + (w - w_0)\,dz$ be an exact differential $d\phi$, which gives

$$q = C - \rho\phi.$$

The only equation which must be satisfied at every point of the mass is (B), which is equivalent to (E), since by hypothesis u_0, v_0, and w_0 satisfy (B). The conditions (a) and (b) remain the same as before.

One observation however is necessary here. The values of u, v and w are always supposed to alter continuously from one point in the interior of a fluid mass to another. At the extreme boundaries of the fluid they may however alter abruptly. Suppose now values of u, v and w to have been assigned, which do not alter abruptly, which satisfy equations (B) and (C) as well as the conditions (a), (b) and (c), or, to take a particular case, values which do not alter abruptly, which satisfy the equation (B) and the same conditions, and which render $u\,dx + v\,dy + w\,dz$ an exact differential. Then the values of dp/dx, dp/dy and dp/dz will alter continuously from one point to another, but it does not follow that the value of p itself cannot alter abruptly. Similarly in impulsive motion the value of q may alter abruptly, although those of dq/dx, dq/dy and dq/dz alter continuously. Such abrupt alterations are, however, inadmissible; whence it follows as an additional condition to be satisfied,

> that the value of p or q, obtained by integrating equations (A) or (F), shall not alter abruptly from one point of the fluid to another. $\Big\}$ ....(d).

24   ON SOME CASES OF FLUID MOTION.

An example will make this clearer. Suppose a mass of fluid
to be at rest in a finite cylinder, whose axis coincides with that of
z, the cylinder being entirely filled, and closed at both ends. Suppose the cylinder to be moved by impact with an initial velocity C
in the direction of x; then shall

$$u = C, \ v = 0, \ w = 0.$$

For these values render $u\,dx + v\,dy + w\,dz$ an exact differential $d\phi$,
where ϕ satisfies (E); they also satisfy (a); and, lastly, the value
of q obtained by integrating equations (F), namely, $C' - C\rho x$, does
not alter abruptly. But if we had supposed that ϕ was equal
to $Cx + C'\theta$, where $\theta = \tan^{-1} y/x$, the equation (E) and the condition (a) would still be satisfied, but the value of q would be
$C'' - \rho(Cx + C'\theta)$, in which the term $\rho C'\theta$ alters abruptly from
$2\pi\rho C'$ to 0, as θ passes through the value 2π. The condition (d)
then alone shews that the former and not the latter is the true
solution of the problem.

The fact that the analytical conditions of a problem in fluid
motion, as far as those conditions depend on the velocities, may be
satisfied by values of those velocities, which notwithstanding correspond to a pressure which alters abruptly, may be thus explained.
Conceive two masses of the same fluid contained in two similar
and equal close vessels A and B. For more simplicity, suppose
these vessels and the fluid in them to be at first at rest. Conceive
the fluid in B to be divided by an infinitely thin lamina which is
capable of assuming any form, and, at the same time, of sustaining
pressure. Suppose the vessels A and B to be moved in exactly
the same manner, the lamina in B being also moved in any arbitrary manner. It is clear that, except for one particular motion
of the lamina, the motion of the fluid in B will be different from
that of the fluid in A. The velocities $u, v, w,$ will in general be
different on opposite sides of the lamina in B. For particular
motions of the lamina however the velocities $u, v, w,$ may be the
same on opposite sides of it, while the pressures are different.
The motion which takes place in B in this case might, only for
the condition (d), be supposed to take place in A.

It is true that equations (A) or (F), could not strictly speaking
be said to hold good at those surfaces where such a discontinuity
should exist. Still, to avoid the liability to error, it is well to
state the condition (d) distinctly.

When the motion begins from rest, not only must $udx+vdy+wdz$ be an exact differential $d\phi$, and u, v, w, not alter abruptly, but also ϕ must not alter abruptly, provided the particles in contact with the several surfaces remain in contact with those surfaces; for if this condition be not fulfilled, the surface for which it is not fulfilled will as it were cut the fluid into two. For it follows from the equation (D) that $d\phi/dt$ must not alter abruptly, since otherwise p would alter abruptly from one point of the fluid to another; and $d\phi/dt$ neither altering abruptly nor becoming infinite, it follows that ϕ will not alter abruptly. Should an impact occur at any period of the motion, it follows from equations (F) that that cannot cause the value of ϕ to alter abruptly, since such an abrupt alteration would give a corresponding abrupt alteration in the value of q.

3. A result which follows at once from the principle laid down in the beginning of the last article is this, that when the motion of a fluid in a close vessel which is at rest, and is completely filled, is of such a kind that $udx + vdy + wdz$ is an exact differential, it will be steady. For let u, v, w, be the initial velocities, and let us see if the velocities at the same point can remain u, v, w. First, $udx + vdy + wdz$ being an exact differential, equations (A) will be satisfied by a suitable value of p, which value is given by equation (D). Also equation (B) is satisfied since it is so at first. The condition (a) becomes $v = 0$, which is also satisfied since it is satisfied at first. Also the value of p given by equation (D) will not alter abruptly, for $d\phi/dt = 0$, or a function of t, and the velocities $d\phi/dx$ &c., are supposed not to alter abruptly. Hence, all the requisite conditions are satisfied; and hence, (Art. 2) the hypothesis of steady motion is correct*.

4. In the case of an incompressible fluid, either of infinite extent, or confined, or interrupted in any manner by any solid bodies, if the motion begin from rest, and if there be none of the cutting motion mentioned in Art. 2, the motion at the time t will be the

---

* [N.B. It is only within a space which is at least doubly connected that such a motion is possible. Thus in the example given in the preceding article, the axis of the cylinder, where the velocity becomes infinite, may be regarded as an infinitely slender core which we are forbidden to cross, and which renders the space within the cylinder virtually ring-shaped.]

same as if it were produced instantaneously by the impulsive motion of the several surfaces which bound the fluid, including among these surfaces those of any solids which may be immersed in it. For let u, v, w, be the velocities at the time t. Then by a known theorem $u dx + v dy + w dz$ will be an exact differential $d\phi$, and ϕ will not alter abruptly (Art. 2). ϕ must also satisfy the equation (E), and the conditions (a) and (b). Now if u', v', w', be the velocities on the supposition of an impact, these quantities must be determined by precisely the same conditions as u, v and w. But the problem of finding u', v' and w', being evidently determinate, it follows that the identical problem of finding u, v and w is also determinate, and therefore the two problems have the same solution; so that

$$u = u', \quad v = v', \quad w = w'.$$

This principle has been mentioned by M. Cauchy, in a memoir entitled *Mémoire sur la Théorie des Ondes*, in the first volume of the *Mémoires des Savans Étrangers* (1827), page 14. It will be employed in this paper to simplify the requisite calculations *by enabling us to dispense with all consideration of the previous motion*, in finding the motion of the fluid at any time in terms of that of the bounding surfaces. One simple deduction from it is that, when all the bounding surfaces come to rest, each element of the fluid will come to rest. Another is, that if the velocities of the bounding surfaces are altered in any ratio the value of ϕ will be altered in the same ratio.

5. *Superposition of different motions.*

In calculating the initial motion of a fluid, corresponding to given initial motions of the bounding surfaces, we may resolve the latter into any number of systems of motions, which when compounded give to each point of each bounding surface a velocity, which when resolved along the normal is equal to the given velocity resolved along the same normal, provided that, if the fluid be enclosed on all sides, each system be such as not to alter its volume. For let u', v', w', ν', σ', be the values of u, v, &c., corresponding to the first system of motions; u'', v'', &c., the values of those quantities corresponding to the second system, and so on; so that

$$u = u' + u'' + \ldots, \quad v = v' + v'' + \ldots, \quad w = w' + w'' + \ldots,$$
$$\nu = \nu' + \nu'' + \ldots, \quad \sigma = \sigma' + \sigma'' + \ldots.$$

Then since we have by hypothesis $u'dx + v'dy + w'dz$ an exact differential $d\phi'$, $u''dx + v''dy + w''dz$ an exact differential $d\phi''$, and so on, it follows that $udx + vdy + wdz$ is an exact differential. Again by hypothesis $v' = \sigma'$, $v'' = \sigma''$, &c., whence $v = \sigma$. Also, if the fluid extend to an infinite distance, u, v, and w must there vanish, since that is the case with each of the systems u', v', w', &c. Lastly, the quantities ϕ', ϕ'', &c., not altering abruptly, it follows that ϕ, which is equal to $\phi' + \phi'' + \dots$, will not alter abruptly. Hence the compounded motion will satisfy all the requisite conditions, and therefore (Art. 2) it is the actual motion.

It will be observed that the pressure p will not be obtained by adding together the pressures due to each of the above systems of velocities. To find p we must substitute the complete value of ϕ in equation (D). If, however, the motion be very small, so that the square of the velocity is neglected, it will be sufficient to add together the several pressures just mentioned.

In general the most convenient systems into which to decompose the motion of the bounding surfaces are those formed by considering the motion of each surface, or of a certain portion of each surface, separately. Such a portion may be either finite or infinitesimal. In fact, in some of the cases of motion that will be presently given, where ϕ is expressed by a double integral with a function under the integral sign expressing the motion of the bounding surfaces, it will be found that each element of the integral gives a value of ϕ such that, except about the corresponding element of the bounding surface, the motion of all particles in contact with those surfaces is tangential.

A result which follows at once from this principle, and which appears to admit of comparison with experiment, is the following. Conceive an ellipsoid, or any body which is symmetrical with respect to three planes at right angles to each other, to be made to oscillate in a fluid in the direction of each of its three axes in succession, the oscillations being very small. Then, in each case, as may be shewn by the same sort of reasoning as that employed in Art. 8, in the case of a cylinder, the effect of the inertia of the fluid will be to increase the mass of the solid by a mass having a certain unknown ratio to that of the fluid displaced. Let the axes of co-ordinates be parallel to the axes of the solid; let x, y, z, be the co-ordinates of the centre of the solid, and let M, M', M'', be the imaginary masses which we must suppose added to that of the solid when it oscillates in the direction of the axes of x, y, z, respectively. Let it now be made to oscillate in the direction of a line making angles α, β, γ, with the axes, and let s be measured along this line. Then the motions of the fluid due to the motions of the solid in the direction of the three axes will be superimposed. The motion being supposed to be small, the resultant of the pressures of the fluid on the solid will be three forces, equal to

$$M \cos \alpha \frac{d^2s}{dt^2}, \quad M' \cos \beta \frac{d^2s}{dt^2}, \quad M'' \cos \gamma \frac{d^2s}{dt^2},$$

respectively, in the directions of the three axes. The resultant of these in the direction of the motion will be $M_{,} d^2s/dt^2$ where

$$M_{,} = M \cos^2 \alpha + M' \cos^2 \beta + M'' \cos^2 \gamma.$$

Each of the quantities M, M', M'' and $M_{,}$, may be determined by observation, and we may find whether the above relation holds between them. Other relations of the same nature may be deduced from the principle explained in this article.

6. *Reflection.*

Conceive two solids, A and B, immersed in a fluid of infinite extent, the whole being at rest. Suppose A to be moved in any manner by impulsive forces, while B is held at rest. Suppose the solids A and B of such forms that, if either were removed, and the several points of the surface of the other moved instantaneously in any given manner, the motion of the fluid could be determined: then the actual motion can be approximated to in the following manner. Conceive the place of B to be occupied by fluid, and A to receive its given motion; then by hypothesis the initial motion of the fluid can be determined. Let the velocity with which the fluid in contact with that which is supposed to occupy B's place penetrates into the latter be found, and then suppose that the several points of the surface of B are moved with normal velocities equal and opposite to those just found, A's place being supposed to be occupied by fluid. The motion of the fluid corresponding to the velocities of the several points of the surface of B can then be found, and A must now be treated as B has been, and so on. The system of velocities of the particles of the fluid corresponding to the first system of velocities of the particles of the surface of B, form what may be called *the motion of A reflected from B*; the motion of the fluid arising from the second system of velocities of the particles of the surface of A may be called *the motion of A reflected from B and again from A*, and so on. It must be remembered that all these motions take place simultaneously. It is evident that these reflected motions will rapidly decrease, at least if the distance between A and B is considerable compared with their diameters, or rather with the diameter of either. In this case the calculation of one or two reflections will give the motion of the fluid due to that of A with great accuracy. It is evident that the principle of reflection will extend to any number of solid bodies immersed in a fluid; or again, the body B may be supposed to be hollow, and to contain the fluid and A, or else A to contain B. In some cases the series arising from the successive reflections can be summed, in which case the motion will be determined exactly. The principle explained in this article has been employed in other subjects, and appears likely to be of great use in this. It is the same for instance as that of *successive influences* in Electricity.

7. If a mass of fluid be at rest or in motion in a close vessel which it entirely fills, the vessel being either at rest or moving in any manner, any additional motion of translation communicated to the vessel will not affect the relative motion of the fluid. For it is evident that on the supposition that the relative motion is not affected the equation (B) and the condition (a) will still be satisfied. Also, if $\varpi_1, \varpi_2, \varpi_3$, be the components of the effective force of any particle in the first case, and U, V, W, be the components of the velocity of translation, then

$$\varpi_1 + \frac{dU}{dt}, \quad \varpi_2 + \frac{dV}{dt}, \quad \varpi_3 + \frac{dW}{dt},$$

will be the components of the effective force of the same particle in the second case. Now since by hypothesis $\varpi_1 dx + \varpi_2 dy + \varpi_3 dz$ is an exact differential, as follows from equations (C), and U, V, W, are functions of t only, it follows at once that

$$\left(\varpi_1 + \frac{dU}{dt}\right) dx + \left(\varpi_2 + \frac{dV}{dt}\right) dy + \left(\varpi_3 + \frac{dW}{dt}\right) dz$$

is an exact differential, where x, y, z, are the co-ordinates of any particle referred to the old axes, which are themselves moving in space with velocities U, V, W. But if x_1, y_1, z_1, be the co-ordinates of the same particle referred to parallel axes fixed in space, we have

$$x_1 = x + \int U dt, \quad y_1 = y + \int V dt, \quad z_1 = z + \int W dt,$$

whence, supposing the time constant, $dx = dx_1$, $dy = dy_1$, $dz = dz_1$, and therefore

$$\left(\varpi_1 + \frac{dU}{dt}\right) dx_1 + \left(\varpi_2 + \frac{dV}{dt}\right) dy_1 + \left(\varpi_3 + \frac{dW}{dt}\right) dz_1$$

is an exact differential. Hence, equations (A) can be satisfied by a suitable value of p. Denoting by p the pressure about the particle whose co-ordinates are x, y, z, in the first case, the pressure about the same particle in the second case will be

$$p + \psi(t) - \rho \left\{ \frac{dU}{dt} x + \frac{dV}{dt} y + \frac{dW}{dt} z \right\},$$

none of the terms of which will alter abruptly, since by hypothesis p does not.

Since then the present hypothesis satisfies all the requisite conditions, it follows from Art. 2 that that hypothesis is correct. If F be the additional effective force of any particle of the vessel in consequence of the motion of translation, and we take new axes of x', y', z', of which the first is in the direction of F, the additional term introduced into the value of the pressure will be $-\rho F x'$, omitting the arbitrary function of the time. The resultant of the additional pressures on the sides of the vessel will be equal to F multiplied by the mass of the fluid, and will pass through the centre of gravity of the fluid, and act in the directon of $-x'$.

8. *Motion between two cylindrical surfaces having a common axis.*

Let us conceive a mass of fluid at rest, bounded by two cylindrical surfaces having a common axis, these surfaces being either infinite or bounded by two planes perpendicular to their axis. Let us suppose the several generating lines of these cylindrical surfaces to be moved parallel to themselves in any given manner consistent with the condition that the volume of the fluid be not altered:

it is required to determine the initial motion at any point of the mass.

Since the motion will take place in two dimensions, let the fluid be referred to polar co-ordinates r, θ, in a plane perpendicular to the axis, r being measured from the axis.   Let a be the radius of the inner surface, b that of the outer, $f(\theta)$ the normal velocity of any point of the inner surface, $F(\theta)$ the corresponding quantity for the outer.

Since for any particular radius vector between a and b the value of ϕ is a periodic function of θ which does not become infinite, (for the motion at each point of each bounding surface is supposed to be finite), and which does not alter abruptly, it may be expanded in a converging series of sines and cosines of θ and its multiples.   Let then

$$\phi = P_0 + \Sigma_1^\infty \left(P_n \cos n\theta + Q_n \sin n\theta \right) \dots\dots\dots(1).$$

Substituting the above value in the equation

$$r \frac{d}{dr} \left(r \frac{d\phi}{dr} \right) + \frac{d^2\phi}{d\theta^2} = 0 \dots\dots\dots\dots(2),$$

which ϕ is to satisfy, and equating to zero the coefficients of corresponding sines and cosines, which is allowable, since a given function can be expanded in only one series of the form (1), we find that P_0 must satisfy the equation

$$r \frac{d}{dr} \left(r \frac{dP_0}{dr} \right) = 0,$$

of which the general integral is

$$P_0 = A \log r + B,$$

the base being e, and P_n and Q_n must both satisfy the same equation, viz.

$$r \frac{d}{dr} \left(r \frac{dP_n}{dr} \right) - n^2 P_n = 0,$$

of which the general integral is

$$P_n = C r^{-n} + C' r^n.$$

We have then, omitting the arbitrary constant in ϕ, as will

32 QN SOME CASES OF FLUID MOTION.

be done for the future, since we have occasion to use only the differential coefficients of ϕ,

$$\phi = A_0 \log r + \Sigma_1^\infty \{(A_n r^{-n} + A'_n r^n) \cos n\theta$$
$$+ (B_n r^{-n} + B'_n r^n) \sin n\theta\} \ldots\ldots(3),$$

with the conditions

$$\frac{d\phi}{dr} = f(\theta) \quad \text{when } r = a \ldots\ldots(4),$$

$$\frac{d\phi}{dr} = F(\theta) \quad \text{when } r = b \ldots\ldots(5).$$

Let $\quad f(\theta) = C_0 + \Sigma_1^\infty (C_n \cos n\theta + D_n \sin n\theta),$
$\quad F(\theta) = C'_0 + \Sigma_1^\infty (C'_n \cos n\theta + D'_n \sin n\theta);$

so that

$$C_0 = \frac{1}{2\pi}\int_0^{2\pi} f(\theta')\, d\theta', \quad C_n = \frac{1}{\pi}\int_0^{2\pi} f(\theta') \cos n\theta'\, d\theta,$$

$$D_n = \frac{1}{\pi}\int_0^{2\pi} f(\theta') \sin n\theta'\, d\theta',$$

with similar expressions for C'_0, &c. Then the condition (4) gives

$$\frac{A_0}{a} + \Sigma_1^\infty n\{(-A_n a^{-(n+1)} + A'_n a^{n-1}) \cos n\theta$$
$$+ (-B_n a^{-(n+1)} + B'_n a^{n-1}) \sin n\theta\} = C_0 + \Sigma_1^\infty (C_n \cos n\theta + D_n \sin n\theta);$$

whence,

$$A_0 = a C_0,$$

$$A_n a^{-(n+1)} - A'_n a^{n-1} = -\frac{1}{n} C_n,$$

$$B_n a^{-(n+1)} - B'_n a^{n-1} = -\frac{1}{n} D_n.$$

Similarly, from the condition (5), we get

$$A_0 = b C'_0,$$

$$A_n b^{-(n+1)} - A'_n b^{n-1} = -\frac{1}{n} C'_n,$$

$$B_n b^{-(n+1)} - B'_n b^{n-1} = -\frac{1}{n} D'_n.$$

It will be observed that $aC_0 = bC'_0$, by the condition that the volume of fluid remains unchanged, which gives

$$a \int_0^{2\pi} f(\theta') \, d\theta' = b \int_0^{2\pi} F(\theta') \, d\theta'.$$

From the above equations we easily get

$$A_n = \frac{a^{2n} b^{2n}}{n (b^{2n} - a^{2n})} \{b^{-n+1} C'_n - a^{-n+1} C_n\},$$

and, changing the sign of n,

$$A'_n = \frac{1}{n (b^{2n} - a^{2n})} \{b^{n+1} C'_n - a^{n+1} C_n\},$$

with similar expressions for B_n and B'_n, involving D in place of C.

We have then

$$\phi = aC_0 \log r + \Sigma_1^{\infty} \frac{1}{n} (b^{2n} - a^{2n})^{-1} \{[(b^{-n+1} C'_n - a^{-n+1} C_n) \cos n\theta$$
$$+ (b^{-n+1} D'_n - a^{-n+1} D_n) \sin n\theta] a^{2n} b^{2n} r^{-n}$$
$$+ [(b^{n+1} C'_n - a^{n+1} C_n) \cos n\theta$$
$$+ (b^{n+1} D'_n - a^{n+1} D_n) \sin n\theta] r^n\} \dots\dots\dots\dots(6),$$

which completely determines the motion.

It will be necessary however, (Art. 2), to shew that this value of ϕ does not alter abruptly for points within the fluid, as may be easily done. For the quantities C_n, D_n cannot be greater than $\frac{1}{\pi} \int_0^{2\pi} \pm f(\theta) \, d\theta$, where each element of the integral is taken positively; and since by hypothesis $f(\theta)$ is finite for all values of θ from 0 to 2π, it follows that neither C_n nor D_n can be numerically greater than a constant quantity which is independent of n. The same will be true of C'_n and D'_n. Remembering then that $r > a$ and $< b$, it can be easily shewn that the series which occur in (6) have their terms numerically less than those of eight geometric series respectively whose ratios are less than unity; and since moreover the terms of the former set of series do not alter abruptly, it follows that ϕ cannot alter abruptly. The same may be proved in a similar manner of the differential coefficients of ϕ. The other infinite series expressing the value of ϕ which occur in this paper may be treated in the same way: and in Art. 10, where ϕ is expressed by a definite integral, the value of ϕ and its differential coefficients will alter continuously, since that is the case with each element of the integral. It will be unnecessary therefore to refer again to the condition (d).

If the fluid be infinitely extended, we must suppose C'_n and D'_n to vanish in (6), since the velocity vanishes at an infinite distance; we must then make b infinite, which reduces the above equation to

$$\phi = aC_0 \log r - \Sigma_1^\infty \frac{a^{n+1}}{nr^n} \{C_n \cos n\theta + D_n \sin n\theta\} \quad .....(7).$$

This value of ϕ may be put under the form of a definite integral: for, replacing C_0, C_n and D_n by their values, it becomes

$$\frac{a}{2\pi} \log r \int_0^{2\pi} f(\theta') d\theta' - \frac{a}{\pi} \Sigma_1^\infty \frac{1}{n}\left(\frac{a}{r}\right)^n \int_0^{2\pi} f(\theta') \cos n\,(\theta - \theta')\, d\theta',$$

which becomes on summing the series

$$\frac{a}{2\pi} \log r \int_0^{2\pi} f(\theta') d\theta' + \frac{a}{\pi} \int_0^{2\pi} \log \left\{1 - 2\frac{a}{r}\cos(\theta - \theta') + \frac{a^2}{r^2}\right\}^{\frac{1}{2}} f(\theta')\, d\theta';$$

whence

$$\frac{d\phi}{dr} = \frac{a}{\pi r} \int_0^{2\pi} \left\{\frac{1}{2} + \frac{ar \cos(\theta - \theta') - a^2}{r^2 - 2ar\cos(\theta - \theta') + a^2}\right\} f(\theta')\, d\theta'.$$

If we suppose r to become equal to a the quantity under the integral sign vanishes, except for values of θ', which are indefinitely near to θ. The value of the integral itself becomes $\pi f(\theta)$*. Hence it appears, that to the disturbance of each element of the surface, there corresponds a normal velocity of the particles in contact with the surface, which is zero, except just about the disturbed element. The whole disturbance of the fluid will be the aggregate of the disturbances due to those of the several elements of the surface. The case of the initial motion of fluid within a cylinder, and the analogous cases of motion within and without a sphere, which will be given in the next article, may be treated in the same manner.

The velocity in the direction of r given by the equation (7), $(= d\phi/dr)$,

$$= \frac{aC_0}{r} + \Sigma_1^\infty \left(\frac{a}{r}\right)^{n+1} \{C_n \cos n\theta + D_n \sin n\theta\},$$

---

* Poisson, *Théorie de la Chaleur*, Chap. VII.

and that perpendicular to r, and reckoned positive in the same direction as θ, $(= d\phi/rd\theta)$,

$$= \Sigma_1^\infty \left(\frac{a}{r}\right)^{n+1} \{C_n \sin n\theta - D_n \cos n\theta\}.$$

Conceive a mass of fluid comprised between two infinite parallel planes, and suppose that a certain portion of this fluid contains solid bodies bounded by cylindrical surfaces perpendicular to these planes. The whole being at first at rest, suppose that the surfaces of these solids are moved in any manner, the motion being in two dimensions. Conceive a circular cylindrical surface described perpendicular to the parallel planes, and with a radius so large that all the solids are comprised with it. Then, (Art. 4), we may suppose the motion of the fluid at any time to have been produced directly by impact. On this supposition the initial motion of the part of the fluid without the above cylindrical surface will be determined in terms of the normal motion of the fluid forming that surface, as has just been done. If C_0 be different from zero, then, at a great distance in the fluid, the velocity will be ultimately aC_0/r, and directed to or from the axis of the cylinder, and alike in all directions. Since the rate of increase of volume of a length l of the cylinder is equal to

$$la \int_0^{2\pi} f(\theta') \, d\theta' = 2\pi la C_0,$$

it appears that the velocity at a great distance is proportional to the expansion or contraction of a unit of length of the solids. If however there should be no expansion or contraction, or if the expansion of some of the solids should make up for the contraction of the rest, then in general the most important part of the motion at a great distance will consist of a velocity $C' \cos \theta_1 . /r^2$ directed to or from the centre, and another $C' \sin \theta_1 . /r^2$ perpendicular to the radius vector, the value of C' and the direction from which θ_1 is measured varying from one instant to another. The resultant of these velocities will vary inversely as the square of the distance.

Resuming the value of ϕ given by equation (6), let us suppose that the interior cylindrical surface is rigid, and moved with a velocity C in the direction from which θ is measured, the outer surface being at rest: then $f(\theta) = C \cos \theta$, $F(\theta) = 0$; whence $C_1 = C$, and the other coefficients are each zero. We have then

$$\phi = - \frac{Ca^2}{b^2 - a^2} \left(\frac{b^2}{r} + r \right) \cos \theta \dots \dots \dots \dots (8).$$

Suppose now that the inner cylinder has a small oscillatory motion about an axis parallel to the axis of the cylinders, the cylinders having their axes coincident in the position of equilibrium. Let ψ be the angle which a plane drawn through the axis of rotation, and that of the solid cylinder at any time makes with a vertical plane drawn through the former. The motion of translation of the axis of the cylinder will differ from a rectilinear motion by quantities depending on ψ^2: the motion of rotation about its axis will be of the order ψ, but will have no effect on the fluid. Therefore in considering the motion of the fluid we may, if we neglect squares of ψ, consider the motion of the cylinder rectilinear. The expression given for ϕ by equation (8) will be accurately true only for the instant when the axes of the cylinders coincide; but since the whole resultant pressure on the solid cylinder in consequence of the motion is of the order ψ, we may, if we neglect higher powers of ψ than the first, employ the approximate value of ϕ given by equation (8). Neglecting the square of the velocity, we have

$$p = - \rho \frac{d\phi}{dt}.$$

In finding the complete value of $d\phi/dt$ it would be necessary to express ϕ by co-ordinates referred to axes fixed in space, which after differentiation we might suppose to coincide with others fixed in the body. But the additional terms so introduced depending on the square of the velocity, which by hypothesis is neglected, we may differentiate the value of ϕ given by equation (8) as if the axes were fixed in space. We have then, to the first order of approximation,

$$\frac{d\phi}{dt} = - \frac{a^2 \dfrac{dC}{dt}}{b^2 - a^2} \left\{ \frac{b^2}{r} + r \right\} \cos \theta.$$

If l be the length of the cylinder, the pressure on the element $la d\theta$, resolved parallel to x and reckoned positive when it acts in the direction of x,

$$= -\frac{\rho l a^3 \frac{dC}{dt}}{b^2 - a^2} \left\{ \frac{b^2}{a} + a \right\} \cos^2 \theta d\theta;$$

and integrating from $\theta = 0$ to $\theta = 2\pi$, we have the whole resultant pressure parallel to x

$$= -\frac{b^2 + a^2}{b^2 - a^2} \pi \rho l a^2 \frac{dC}{dt}.$$

Since dC/dt is the effective force of the axis, parallel to x, and that parallel to y is of the order ψ^2, we see that the effect of the inertia of the fluid is to increase the mass of the cylinder by $\frac{b^2 + a^2}{b^2 - a^2} \mu$, where μ is the mass of the fluid displaced. This imaginary additional mass must be supposed to be collected at the axis of the cylinder.

If the cylinder oscillate in an infinitely extended fluid $b = \infty$, and the additional mass becomes equal to that of the fluid displaced. This appears to be a result capable of being compared with experiment, though not with very great accuracy. Two cylinders of the same material, and of the same radius, but whose lengths differ by several radii, might be made to oscillate in succession in a fluid, at a depth sufficiently great to allow us to neglect the motion of the surface of the fluid. The time of oscillation of each might then be calculated as if the cylinder oscillated in vacuum, acted on by a moving force equal to its weight *minus* that of the fluid displaced, acting downwards through its centre of gravity, and having its mass increased by an unknown mass collected in the axis. Equating the time of oscillation so calculated to that given by observation, we should determine the unknown mass. The difference of these masses would be very nearly equal to the mass which must be added to that of a cylinder whose length is equal to the difference of the lengths of the first two, when the motion is in two dimensions. This evidently comes to supposing that, at a distance from the middle of the longer cylinder not greater than half the difference of the lengths of the two, the motion may be taken as in two dimensions. The ends of the cylinders may be of any form, provided that they are all of the same. They may be suspended by fine equal wires, in which case we should have a compound pendulum, or attached to a rigid body oscillating above the fluid by means of thin flat bars of metal, whose plane is in the plane of motion. Another way of getting rid of the motion in three dimensions about the ends would be, to make those ends plane, and to fix two rigid planes parallel to the plane of motion, which should be almost in contact with the ends of the cylinder.

9. *Motion between two concentric spherical surfaces.—Motion of a ball pendulum enclosed in a spherical case.*

Let a mass of fluid be at rest, comprised between two concentric spherical surfaces. Let the several points of these surfaces be moved in any manner consistent with the condition that the volume of the fluid be not changed : it is required to determine the initial motion at any point of the mass.

Let a, b, be the radii of the inner and outer spherical surfaces respectively ; then employing the co-ordinates r, θ, ω, where r is the distance from the centre, θ the angle which r makes with a fixed line passing through the centre, ω the angle which a plane passing through these two lines makes with a fixed plane through the latter, the value of ϕ corresponding to any radius vector comprised between a and b can be expanded in a converging series of Laplace's coefficients. Let then

$$\phi = V_0 + V_1 \ldots\ldots + V_n + \ldots\ldots,$$

V_n being a Laplace's coefficient of the nth order.

Substituting in the equation,

$$r\frac{d^2 r\phi}{dr^2} + \frac{1}{\sin\theta}\frac{d}{d\theta}\left(\sin\theta\,\frac{d\phi}{d\theta}\right) + \frac{1}{\sin^2\theta}\frac{d^2\phi}{d\omega^2} = 0,$$

which ϕ is to satisfy, employing the equation

$$n\,(n+1)\,V_n + \frac{1}{\sin\theta}\frac{d}{d\theta}\left(\sin\theta\,\frac{dV_n}{d\theta}\right) + \frac{1}{\sin^2\theta}\frac{d^2 V_n}{d\omega^2} = 0\ldots(9),$$

and then equating to zero the Laplace's coefficients of the several orders, we find

$$r\frac{d^2 r V_n}{dr^2} - n\,(n+1)\,V_n = 0.$$

The general integral of this equation is

$$V_n = Cr^n + \frac{C'}{r^{n+1}},$$

where C and C' are functions of θ and ω. Substituting in the equation (9), and equating coefficients of the two powers of r which enter into it separately to zero, we find that both C and C' satisfy it, and therefore are both Laplace's coefficients of the nth order. We have then

$$\phi = \Sigma_0^\infty \left(Y_n r^n + Z_n r^{-(n+1)} \right) \dots\dots\dots(10),$$

where Y_n and Z_n are each Laplace's coefficients of the nth order, and do not contain r. Let $f(\theta, \omega)$ be the normal velocity of the point of the inner surface corresponding to θ and ω, $F(\theta, \omega)$ the corresponding quantity for the outer; then the conditions which ϕ is to satisfy are that

$$\frac{d\phi}{dr} = f(\theta, \omega) \text{ when } r = a,$$

$$\frac{d\phi}{dr} = F(\theta, \omega) \text{ when } r = b.$$

Let $f(\theta, \omega)$, expanded in a series of Laplace's coefficients, be

$$P_0 + P_1 \dots + P_n + \dots$$

which expansion may be performed by the usual formula, if not by inspection: then the first condition gives

$$\Sigma_0^\infty \left\{ n Y_n a^{n-1} - (n+1) Z_n a^{-(n+2)} \right\} = \Sigma_0^\infty P_n \, ;$$

and equating Laplace's coefficients of the same order, we get

$$n Y_n a^{n-1} - (n+1) Z_n a^{-(n+2)} = P_n \dots\dots\dots(11).$$

Let $F(\theta, \omega)$, expanded in a series of Laplace's coefficients, be

$$P'_0 + P'_1 \dots P'_n + \dots \, ;$$

then from the second condition, we get

$$n Y_n b^{n-1} - (n+1) Z_n b^{-(n+2)} = P'_n \dots\dots\dots(12).$$

From (11) and (12) we easily get

$$Y_n = \frac{P'_n b^{n+2} - P_n a^{n+2}}{n \left(b^{2n+} - a^{2n+1} \right)} \, ,$$

$$Z_n = \frac{a^{2n+1} b^{2n+1} \left\{ P'_n b^{-(n-1)} - P_n a^{-(n-1)} \right\}}{(n+1) \left(b^{2n+1} - a^{2n+1} \right)} \, ,$$

provided n be greater than 0. If $n = 0$, we have

$$-a^{-2} Z_0 = P_0, \quad -b^{-2} Z_0 = P'_0.$$

But the condition that the volume of the fluid be not altered, gives

$$a^2 \int_0^\pi \int_0^{2\pi} f(\theta,\ \omega) \sin\theta\, d\theta\, d\omega = b^2 \int_0^\pi \int_0^{2\pi} F(\theta,\ \omega) \sin\theta\, d\theta\, d\omega,$$

or
$$4\pi a^2 P_0 = 4\pi b^2 P'_0,$$

which reduces the two equations just given to one.

We have then, omitting the constant Y_0,

$$\phi = -\frac{P_0 a^2}{r} + \Sigma_1^\infty \{b^{2n+1} - a^{2n+1}\}^{-1} \left\{ \frac{1}{n} (P'_n b^{n+2} - P_n a^{n+2}) r^n \right.$$
$$\left. + \frac{a^{2n+1} b^{2n+1}}{n+1} (P'_n b^{-(n-1)} - P_n a^{-(n-1)}) r^{-(n+1)} \right\} \dots (13),$$

which determines the motion.

When the fluid is infinitely extended, we have $P'_n = 0$ since the velocity vanishes at an infinite distance, and $b = \infty$, whence

$$\phi = -\frac{P_0 a^2}{r} - \Sigma_1^\infty \frac{a^{n+2} P_n}{(n+1) r^{n+1}}.$$

It may be proved, precisely as was done, (Art. 8), for motion in two dimensions, that if any portion of an infinitely extended fluid be disturbed by the motion of solid bodies, or otherwise, if all the fluid beyond a certain distance from the part disturbed were at first at rest, the velocity at a great distance will ultimately be directed to or from the disturbed part, and will be the same in all directions, and will vary as r^{-2}. The coefficient of r^{-2} will be proportional to the rate of gain or loss of volume of the part disturbed. If however this rate should be zero, then the most important part of the velocity at a great distance will in general be that depending on the term $-\frac{1}{2} a^3 P_1 \cdot r^{-2}$ in ϕ. Since the general form of P_1 is

$$A \cos\theta + B \sin\theta \cos\omega + C \sin\theta \sin\omega,$$

we easily find, by making use of rectangular co-ordinates, changing the direction of the axes, and then again adopting polar co-ordinates, that the above term in ϕ takes the form $D \cos\theta_1 \cdot r^{-2}$, θ_1 being measured from some line passing through the origin. The motion will therefore be the same as that round a ball pendulum in an incompressible fluid, the centre of the ball being in the origin; a case of motion which will be considered immediately. In order to represent the motion at different times, we must suppose the velocity and direction of motion of the ball to change with the time.

The value of ϕ given by equation (13) is applicable to the determination of the motion of a ball pendulum enclosed in a spherical case which is concentric with the ball in its position of equilibrium. If C be the velocity of the centre of the ball at the instant when the centres of the ball and case coincide, and if θ be measured from the direction in which it is moving, we shall have

$$f(\theta) = C \cos \theta, \quad F(\theta) = 0;$$

$$\therefore \ P_0 = 0, \quad P_1 = C \cos \theta, \quad P_2 = 0, \ \&c., \quad P'_0 = 0, \ \&c.,$$

and the value of ϕ for this instant is accurately

$$-\frac{Ca^3}{b^3 - a^3}\left(r + \frac{b^3}{2r^2}\right)\cos\theta,$$

which, when $b = \infty$, becomes

$$-\frac{Ca^3 \cos \theta}{2r^2},$$

which is the known expression for the value of ϕ for a sphere oscillating in an infinitely extended, incompressible fluid.

It may be shewn, by precisely the same reasoning as was employed in the case of the cylinder, that in calculating the small oscillations of the sphere the value of $d\phi/dt$ to be employed is

$$-\frac{a^3 \dfrac{dC}{dt}}{b^3 - a^3}\left(a + \frac{b^3}{2a^2}\right)\cos\theta;$$

and from the equation $p = -\rho\, d\phi/dt$, we easily find that the whole resultant pressure on the sphere in the direction of its centre, and tending to retard it is

$$\frac{4}{3}\frac{\pi\rho a^5}{b^3 - a^3}\left(a + \frac{b^3}{2a^2}\right)\frac{dC}{dt},$$

and that perpendicular to this direction is zero. Since dC/dt is the effective force of the centre in the direction of the motion, and that perpendicular to this direction is of the second order, the effect of the inertia of the fluid will be to increase the mass of the sphere by a mass

$$= \frac{4}{3}\frac{\pi\rho a^5}{b^3 - a^3}\left(a + \frac{b^3}{2a^2}\right) = \frac{b^3 + 2a^3}{b^3 - a^3}\frac{\mu}{2},$$

μ being the mass of the fluid displaced; so that the effect of the case is, to increase the mass which we must suppose added to that of the ball in the ratio of $b^3 + 2a^3$ to $b^3 - a^3$.

Poisson, in his solution of the problem of the oscillating sphere given in the *Mémoires de l'Académie, Tome* XI. arrives at a different conclusion, viz. that the case does not at all affect the motion of the sphere. When the elimination which he proposes at p. 563 is made, the last term of equation (f), p. 550, becomes

$$\frac{\delta \gamma}{2a^2 c \lambda \, (l - \delta \gamma)} \left(\frac{d^3 \zeta}{dt^3} + \frac{d^3 \zeta'}{dt^3} \right),$$

where a is the velocity of propagation of sound, and δ the ratio of the density of air to that of the ball, ζ and ζ' being functions derived from others which enter into the value of ϕ by putting $r = c$, where c is the radius of the ball. He then argues that this term may be neglected as insensible, since it involves δ in the numerator and a^2 in the denominator, tacitly assuming that $\frac{d^3 \zeta}{dt^3} + \frac{d^3 \zeta'}{dt^3}$ is not large since ϕ is not large. Now for the disturbances of the air which have the same period as those of the pendulum $d\phi/dt$ is not large compared with ϕ, as it is for those on which sound depends. Let then Poisson's solution of equation (a), p. 547 of the volume already mentioned, be put under the form

$$\phi = \frac{1}{r^2} \left\{ f\left(t - \frac{r}{a}\right) + F\left(t + \frac{r}{a}\right) \right\} + \frac{1}{ar} \left\{ f'\left(t - \frac{r}{a}\right) - F'\left(t + \frac{r}{a}\right) \right\},$$

f' and F' denoting the derived functions, and all the Laplace's coefficients except those of the first order being omitted, the value of ϕ just given being supposed to be a Laplace's coefficient of that order. Then if we expand the above functions in series ascending according to powers of r/a, we find

$$\phi = \frac{1}{r^2} \left\{ f(t) + F(t) \right\} - \frac{1}{2a^2} \left\{ f''(t) + F''(t) \right\}$$

$$+ \frac{r}{3a^3} \left\{ f'''(t) - F'''(t) \right\} + \dots;$$

and in order that when $a = \infty$ this equation may coincide with (10), when all the Laplace's coefficients except those of the first order are omitted in that equation, it will be seen that it is

ON SOME CASES OF FLUID MOTION.

43

necessary to suppose $f'''(t) - F'''(t)$, and therefore $f(t) - F(t)$, to be of the order a^3, while $f(t) + F(t)$ is not large. Putting then

$$f(t) = \chi(t) + a^3 \varpi(t),$$

$$F(t) = \chi(t) - a^3 \varpi(t),$$

we shall have

$$\zeta + \zeta' = \chi\left(t - \frac{c}{a}\right) + \chi\left(t + \frac{c}{a}\right) + a^3 \left\{ \varpi\left(t - \frac{c}{a}\right) - \varpi\left(t + \frac{c}{a}\right) \right\};$$

so that $\dfrac{d^3(\zeta + \zeta')}{dt^3}$ will contain a term of the order a^2, and the term which Poisson proposes to leave out will be of the same order of magnitude as those retained.

In making the experiment of determining the resistance of the air to an oscillating sphere, it would appear to be desirable to enclose the sphere in a concentric spherical case, which would at the same time exclude currents of air, and facilitate in some measure the experiment by increasing the small quantity which is the subject of observation. The radius of the case however ought not to be nearly as small as that of the ball, for if it were, in the first place a small error in the position of the centre of the ball when at rest might not be insensible, and in the second place the oscillations would have to be inconveniently small, in order that the value of ϕ which has been given might be sufficiently approximate. The effect of a small slit in the upper part of the case, sufficient to allow the wire by which the ball is supported to oscillate, would evidently be insensible, for the condensation being insensible in a vertical plane passing through the axis of rotation, since the alteration of pressure in that plane is insensible, the air would not have a tendency alternately to rush in and out at the slit.

10. *Effect of a distant rigid plane on the motion of a ball pendulum.*

Although this problem may be more easily solved by an artifice, it may be well to give the direct solution of it by the method mentioned in Article 6. In order to calculate the motion reflected from the plane, it will be necessary to solve the following problem :

*To find the initial motion at any point of a mass of fluid in-
finitely extended, except where it is bounded by an infinite solid but
not rigid plane, the initial motion of each point of the solid plane
being given.*

It is evident that motion directed to or from a centre situated
in the plane, the velocity being the same in all directions, and
varying inversely as the square of the distance from that centre,
would satisfy the condition that $udx + vdy + wdz$ is an exact
differential, and would give to the particles in contact with the
plane a velocity directed *along* the plane, except just about the
centre. Let us see if the required motion can be made up of an
infinite number of such motions directed to or from an infinite
number of such centres.

Let $x, y, z,$ be the co-ordinates of any particle of fluid, the
plane xy coinciding with the solid plane, and the axis of z being
directed into the fluid. Let $x', y',$ be the co-ordinates of any point
in the solid plane : then the part of ϕ corresponding to the motion
of the element $dx'dy'$ of the plane will be

$$\frac{\psi(x', y')\, dx'dy'}{\sqrt{(x-x')^2 + (y-y')^2 + z^2}},$$

and therefore the complete value of ϕ will be given by the equation

$$\phi = \int_{-\infty}^{\infty}\int_{-\infty}^{\infty} \frac{\psi(x', y')dx'dy'}{\sqrt{\{(x-x')^2 + (y-y')^2 + z^2\}}} \ldots\ldots\ldots\ldots(14).$$

The velocity parallel to z at any point $= d\phi/dz$

$$= -\int_{-\infty}^{\infty}\int_{-\infty}^{\infty} \frac{\psi(x', y')z\,dx'dy'}{\{(x-x')^2 + (y-y')^2 + z^2\}^{\frac{3}{2}}}.$$

Now when z vanishes the quantity under the integral signs
vanishes, except for values of x' and y' indefinitely near to x and y
respectively, the function $\psi(x', y')$ being supposed to vanish when
x' or y' is infinite. Let then $x' = x + \xi,$ $y' = y + \eta,$ then, $\xi,$ and $\eta,$
being as small as we please, the value of the above expression
when $z = 0$ becomes

$$-\text{ the limit of}\int_{-\xi}^{\xi}\int_{-\eta}^{\eta} \frac{z\psi(x+\xi, y+\eta)\,d\xi d\eta}{(\xi^2 + \eta^2 + z^2)^{\frac{3}{2}}}\text{ when } z = 0.$$

Now if $\psi(x', y')$ does not alter abruptly between the limits $x-\xi,$

and $x+\xi$, of x', and $y-\eta$, and $y+\eta$, of y', the above expression may be replaced by

$$-\psi(x,y) \times \text{the limit of} \int_{-\xi_{}}^{\xi_{}} \int_{-\eta_{}}^{\eta_{}} \frac{z\,d\xi\,d\eta}{(\xi^2+\eta^2+z^2)^{\frac{3}{2}}},$$

which is $= -2\pi\psi(x,y)$.

If now $f(x',y')$ be the given normal velocity of any point (x',y') of the solid plane, the expression for ϕ given by equation (14) may be made to give the required normal velocity of the fluid particles in contact with the solid plane by assuming

$$\psi(x',y') = -\frac{1}{2\pi}f(x',y'),$$

whence

$$\phi = -\frac{1}{2\pi}\int_{-\infty}^{\infty}\int_{-\infty}^{\infty} \frac{f(x',y')\,dx'dy'}{\{(x-x')^2+(y-y')^2+z^2\}^{\frac{1}{2}}}.$$

This expression will be true for any point at a finite distance from the plane xy even when $f(x',y')$ does alter abruptly; for we may first suppose it to alter continuously, but rapidly, and may then suppose the rapidity of alteration indefinitely increased: this will not cause the value of ϕ just given to become illusory for points situated without the plane xy.

If it be convenient to use polar co-ordinates in the plane xy, putting $x=q\cos\omega$, $y=q\sin\omega$, $x'=q'\cos\omega'$, $y'=q'\sin\omega'$, and replacing $f(x',y')$ by $f(q',\omega')$, the equation just given becomes

$$\phi = -\frac{1}{2\pi}\int_0^\infty \int_0^{2\pi} \frac{f(q',\omega')q'\,dq'd\omega'}{\{q'^2+q^2-2qq'\cos(\omega-\omega')+z^2\}^{\frac{1}{2}}}.$$

To apply this to the case of a sphere oscillating in a fluid perpendicularly to a fixed rigid plane, let a be the radius of the sphere, and let its centre be moving towards the plane with a velocity C at the time t. Then, (Art. 4), we may calculate the motion as if it were produced directly by impact. Let h be the distance of the centre of the sphere from the fixed plane at the time t, and let the line h be taken for the axis of z, and let r, θ, be the polar co-ordinates of any point of the fluid, r being the distance from the centre of the sphere, and θ the angle between the lines r and h. Then if the fluid were infinitely extended around the sphere we should have

$$\phi = -\frac{Ca^3\cos\theta}{2r^2} \quad\ldots\ldots\ldots\ldots\ldots(15).$$

The velocity of any particle, resolved in a direction towards the plane, $= d\phi/dr \cdot \cos\theta - d\phi/rd\theta \cdot \sin\theta$

$$= \frac{Ca^3}{r^3} \{\cos^2\theta - \tfrac{1}{2}\sin^2\theta\}.$$

For a particle in the plane xy we have

$$r\cos\theta = h, \quad r\sin\theta = q',$$

and the above velocity becomes

$$\frac{Ca^3(2h^2 - q'^2)}{2(h^2 + q'^2)^{\frac{5}{2}}}.$$

We must now, according to the method explained in (Art. 6), suppose the several points of the plane xy moved with the above velocity parallel to z. We have then

$$f(q', \omega') = \frac{Ca^3(2h^2 - q'^2)}{2(h^2 + q'^2)^{\frac{5}{2}}};$$

whence, for the motion of the sphere reflected from the plane,

$$\phi = -\frac{Ca^3}{4\pi}\int_0^\infty \int_0^{2\pi} \frac{(2h^2 - q'^2)q'dq'd\omega'}{(h^2 + q'^2)^{\frac{5}{2}}\{q^2 + q'^2 - 2qq'\cos(\omega - \omega') + z^2\}^{\frac{1}{2}}}\ldots(16).$$

We must next find the velocity, corresponding to this value of ϕ, with which the fluid penetrates the surface of the sphere. We have in general

$$z = h - r\cos\theta, \quad q = r\sin\theta,$$

whence

$$\{q^2 + q'^2 - 2qq'\cos(\omega - \omega') + z^2\}^{-\frac{1}{2}}$$
$$= \{h^2 + r^2 + q'^2 - 2hr\cos\theta - 2q'r\sin\theta\cos(\omega - \omega')\}^{-\frac{1}{2}}.$$

Now supposing the ratio of a to h to be very small, and retaining the most important term, the value of $d\phi/dr$ when $r = a$ will be equal to the coefficient of r when ϕ is expanded in a series ascending according to powers of r,

$$= -\frac{Ca^3}{4\pi}\int_0^\infty \int_0^{2\pi} \frac{(2h^2 - q'^2)\{h\cos\theta + q'\sin\theta\cos(\omega - \omega')\}q'dq'd\omega'}{(h^2 + q'^2)^4}$$

$$= -\tfrac{1}{2}Ca^3h\cos\theta\int_0^\infty \frac{(2h^2 - q'^2)q'dq'}{(h^2 + q'^2)^4} = -\frac{Ca^3\cos\theta}{8h^3} \ldots\ldots\ldots\ldots(17).$$

In order now to determine the motion reflected from the plane and again from the sphere, we must suppose the several points of the sphere to be moved with a normal velocity

$Ca^3\cos\theta./8h^3$, or, which is the same, we must suppose the whole sphere to be moved towards the plane with a velocity $Ca^3/8h^3$. Hence the value of ϕ corresponding to this motion will be given by the equation

$$\phi = -\frac{Ca^6\cos\theta}{16h^3r^2} \quad\quad\quad\quad (18).$$

For points at a great distance from the centre of the sphere, the motion which is twice reflected will be very small compared with that which is but once reflected. For points close to the sphere however, with which alone we are concerned, those motions will be of the same order of magnitude, and if we take account of the one we must take account of the other.

Putting $q = r\sin\theta$, $z = h - r\cos\theta$ in (16), expanding, and retaining the two most important terms, we have

$$\phi = C\left(K - \frac{a^3r\cos\theta}{8h^3}\right) \quad\quad\quad (19),$$

K being a constant, the value of which is not required, and the second term being evidently found by multiplying the quantity at the second side of (17) by r. Adding together the parts of ϕ given by equations (15), (18) and (19), putting $r = a$, replacing C by dC/dt, and taking for h the value which it has in equilibrium, just as in the case of the oscillating cylinder in Article 8, we have for the small motion of the sphere

$$\frac{d\phi}{dt} = K\frac{dC}{dt} - \frac{a}{2}\left(1 + \frac{3a^3}{8h^3}\right)\frac{dC}{dt}\cos\theta.$$

The resultant of the part of the pressure due to the first term is zero: that due to the second term is greater than if the plane were removed in the ratio of $1 + 3a^3/8h^3$ to 1. Consequently, if we neglect quantities of the order a^4/h^4, the effect of the inertia of the fluid is, to add a mass equal to $(1 + 3a^3/8h^3).\frac{1}{2}\mu$ to that of the sphere, without increasing the moment of inertia of the latter about its diameter. The effect therefore of a large spherical case is eight times as great as that of a tangent plane to the case, perpendicular to the direction of the motion of the ball.

The effect of a distant rigid plane parallel to the direction of motion of an oscillating sphere might be calculated in the same manner, but as the method is sufficiently explained by the first case, it will be well to employ the artifice before alluded to, an artifice which is frequently employed in this subject. It consists in supposing an exactly symmetrical motion to take place on the opposite side of a rigid plane, by which means we may evidently conceive the plane removed.

Let the sphere be oscillating in the direction of the axis of x, the oscillations in this case, as in the last, being so small that they may be taken as rectilinear in calculating the motion of the fluid; and instead of a rigid plane conceive an equal sphere to exist at an equal distance on the opposite side of the plane xy, moving in the same direction and with the same velocity as the actual sphere. Let r, θ, ω, be the polar co-ordinates of any particle measured from the centre of the sphere, θ being the angle between r and a line drawn through the centre parallel to the axis of x; and ω the angle which the plane passing through these lines makes with the plane xz. Let r', θ', ω', be the corresponding quantities symmetrically measured from the centre of the imaginary sphere.

If the fluid were infinite we should have for the motion corresponding to that of the given sphere

$$\phi = - \frac{Ca^3 \cos \theta}{2r^2} \quad\dots\dots\dots\dots\dots\dots(20).$$

The motion reflected from the plane is evidently the same as that corresponding to the motion of the imaginary sphere in an infinite mass of fluid, for which we have

$$\phi = - \frac{Ca^3 \cos \theta'}{2r'^2} \quad\dots\dots\dots\dots\dots\dots(21).$$

Now $r' \cos \theta' = r \cos \theta$, $r' \sin \theta' \sin \omega' = r \sin \theta \sin \omega$,

$$r' \sin \theta' \cos \omega' + r \sin \theta \cos \omega = 2h;$$

whence $r'^2 = r^2 + 4h^2 - 4hr \sin \theta \cos \omega$,

and equation (21) is reduced to

$$\phi = - \frac{Ca^3 r \cos \theta}{2 \left\{ r^2 + 4h^2 - 4hr \sin \theta \cos \omega \right\}^{\frac{3}{2}}} \cdot$$

Retaining only the terms of the order $a^3 r/h^3$ or r^4/h^3, so as to get the value of $d\phi/dr$ to the order a^3/h^3, the above equation is reduced to

$$\phi = - \frac{Ca^3 r \cos \theta}{16h^3} \quad\dots\dots\dots\dots\dots\dots(22),$$

and the value of $d\phi/dr$ when $r = a$ is, to the required degree of approximation,

$$- \frac{Ca^3 \cos \theta}{16h^3}.$$

For the value of ϕ corresponding to the motion of the imaginary sphere reflected from the real sphere, we shall therefore have

$$\phi = - \frac{Ca^6 \cos \theta}{32h^3 r^2} \dots\dots\dots\dots\dots\dots\dots(23).$$

Adding together the values of ϕ given by (20), (22) and (23), putting $r = a$, and replacing C by dC/dt, we have, to the requisite degree of approximation,

$$\frac{d\phi}{dt} = - \frac{a}{2} \left(1 + \frac{3}{16} \frac{a^3}{h^3} \right) \frac{dC}{dt} . \cos \theta.$$

Hence in this case the motion of the sphere will be the same as if an additional mass equal to $(1 + 3a^3/16h^3) . \frac{1}{2}\mu$ were collected at its centre. The effect therefore of a distant rigid plane which is parallel to the direction of the motion of a ball pendulum will be half that of a plane at the same distance, and perpendicular to that direction. It would seem from Poisson's words at page 562 of the eleventh volume of the *Mémoires de l'Académie*, that he supposed the effect in the former case to depend on a higher order of small quantities than that in the latter.

If the ball oscillate in a direction inclined to the plane, the motion may be easily deduced from that in the two cases just given, by means of the principle of superposition.

11. The values of ϕ which have been given for the motion of translation of a sphere and cylinder do not require us to suppose that either the velocity, or the distance to which the centre of the sphere or axis of the cylinder has been moved, is small, provided the same particles remain in contact with the surface. The same indeed is true of the values corresponding to a motion of translation combined with a motion of contraction or expansion which is the same in all directions, but varies in any manner with the time. The value of ϕ corresponding to a motion of translation of the cylinder is $- Ca^2 \cos \theta . r^{-1}$, C being the velocity of the axis, and θ being measured from a line drawn in the direction of its motion. The whole resultant of the part of the pressure due to the square of the velocity is zero, since the velocity at the point whose co-ordinates are r, θ, is the same as that at

S.   4

the point whose co-ordinates are r and $\pi - \theta$. To find the resultant of the part depending on $d\phi/dt$, it will be necessary to express ϕ by means of co-ordinates referred to axes fixed in space. Let Ox, Oy, be rectangular axes passing through the centre of any section of the cylinder, ϖ the angle which the direction of motion of the axis makes with Ox, θ' the inclination of any radius vector to Ox; then

$$\phi = - \frac{Ca^2}{r^2} \left(r \cos \theta' \cos \varpi + r \sin \theta' \sin \varpi \right)$$

$$= - \frac{a^2 \left(C'x + C''y \right)}{x^2 + y^2},$$

putting C' and C'' for the resolved parts of the velocity C along the axes of x and y respectively. Taking now axes Ax', Ay', parallel to the former and fixed in space, putting α and β for the co-ordinates of O, differentiating ϕ with respect to t, and replacing $d\alpha/dt$ by C', and $d\beta/dt$ by C'', and then supposing α and β to vanish, we have

$$\frac{d\phi}{dt} = \frac{a^2 C^2}{x^2 + y^2} - \frac{2a^2 \left(C'x + C''y \right)^2}{(x^2 + y^2)^2} - \frac{a^2 \left(x \dfrac{dC'}{dt} + y \dfrac{dC''}{dt} \right)}{x^2 + y^2}.$$

The resultant of the part of the pressure due to the first two terms is zero, since the pressure at the point (x, y) depending on these terms is the same as that at the point $(-x, -y)$. It will be easily found that the resultant of the whole pressure parallel to x, and acting in the negative direction, on a length l of the cylinder, is equal to $\pi \rho l a^2 . dC'/dt$, and that parallel to y equal to $\pi \rho l a^2 . dC''/dt$. The resultant of these two will be $\pi \rho l a^2 F$, where F is the effective force of a point in the axis of the cylinder, and will act in a direction opposite to that of F. Hence the only effect of the motion of the fluid will be, to increase the mass of the cylinder by that of the fluid displaced. In a similar manner it may be proved that, when a solid sphere moves in any manner in an infinite fluid, the only effect of the motion of the fluid is to increase the mass of the sphere by half that of the fluid displaced.

A similar result may be proved to be true for any solid symmetrical with respect to two planes at right angles to each other, and moving in the direction of the line of their intersection in an infinitely extended fluid, the solid and fluid having been at first at rest. Let the planes of symmetry be taken for the planes of xy and xz, the origin being fixed in the body: then it is evident

that the resultant of the pressure on the solid due to the motion
will be in the direction of the axis of x, and that there will be
no resultant couple. Let C be the velocity of the solid at any
time; then the value of ϕ at that time will be of the form
$C\psi\,(x, y, z)$, where C alone contains t (Art. 4), and the velocity
of the particle whose co-ordinates are x, y, z, being proportional
to C, the *vis viva* of the solid and the fluid together will be
proportional to C^2. Now if no forces act on the fluid and solid,
except the pressure of the fluid, this *vis viva* must be constant*;
therefore C must be constant; therefore the resultant of the fluid
pressure on the solid must be zero. If now C be a function of t
we shall have

$$p = - \rho\psi\,(x, y, z)\,\frac{dC}{dt} + p',$$

p' being the pressure when C is constant. Since therefore the
resultant of the fluid pressure varies for the same solid and fluid
as dC/dt the effective force, and for different fluids varies as ρ,
the effect of the inertia of the fluid will be, to increase the mass
of the solid by n times that of the fluid displaced, n depending
only on the particular solid considered.

Let us consider two such solids, similar to each other, and
having the co-ordinate planes similarly situated, and moving with
the same velocities. Let the linear dimensions of the second
be greater than those of the first in the ratio of m to 1. Let

* If an incompressible fluid which is homogeneous or heterogeneous, and contains in it any number of rigid bodies, be in motion, the rigid bodies being also
in motion, if the rigid bodies are perfectly smooth, and no contacts are formed or
broken among them, and if no forces act except the pressure of the fluid, the
principle of *vis viva* gives

$$\frac{d\Sigma mv^2}{dt} = 2\iint p\,\nu dS \dots\dots\dots(a),$$

where v is the whole velocity of the mass m, and the sign Σ extends over the whole
fluid and the rigid bodies spoken of, and where dS is an element of the surface
which bounds the whole, p, the pressure about the element dS, and ν the normal
velocity of the particles in that element, reckoned positive when tending into the
fluid, and where the sign $\iint$ extends to all points of the bounding surface. To apply
equation (a) to the case of motion at present considered, let us first confine ourselves to a spherical portion of the fluid, whose radius is r, and whose centre is near
the solid, so that dS refers to the surface of this portion. Let us now suppose r to
become infinite: then the second side of (a) will vanish, provided p, remain finite,
and ν decrease in a higher ratio than r^{-2}. Both of these will be true, (Art. 9); for
ν will vary ultimately as r^{-3}, since there is no alteration of volume. Hence if the
sign Σ extend to infinity, we shall have Σmv^2 constant.

u, v, w, be the velocities, parallel to the axes, of the particle (x, y, z) in the fluid about the first; then shall the corresponding velocities at the point (mx, my, mz) in the fluid about the second be also u, v, w. For

$$udmx + vdmy + wdmz = m \left(udx + vdy + wdz\right) \ldots\ldots(24),$$

and is therefore an exact differential, since $udx + vdy + wdz$ is one : also the normal at the point (x, y, z) in the first surface will be inclined to the axes at the same angles as the normal at the point (mx, my, mz) of the second surface is inclined to its axes, and therefore the normal velocities of the two surfaces at these points are the same; and the velocities of the fluid at these two points parallel to the axes being also the same, it follows that the normal velocity of each point of the second surface is equal to that of the fluid in contact with it. Lastly, the motion about the first solid being supposed to vanish at an infinite distance from it, that about the second will vanish also. Hence the supposition made with respect to the motion of the fluid about the second surface is correct. Now putting ϕ for $\int(udx + vdy + wdz)$ for the fluid in the first case, the corresponding integral for the fluid in the second case will be $m\phi$, if the constant be properly chosen, as follows from equation (24). Consequently the value of that part of the expression for the pressure, on which the resistance depends, will be m times as great for any point in the second case as it is for the corresponding point in the first. Also, each element of the surface of the second solid will be m^2 times as great as the corresponding element of the surface of the first. Hence the whole resistance on the second solid will be m^3 times as great as that on the first, and therefore the quantity n depends only on the *form*, and not on the *size* of the solid.

When forces act on the fluid, it will only be necessary to add the corresponding pressure. Hence when a sphere descends from rest in a fluid by the action of gravity, the motion will be the same as if a moving force equal to that of the sphere *minus* that of the fluid displaced acted on a mass equal to that of the sphere *plus* half that of the fluid displaced. For a cylinder which is so long that we may suppose the length infinite, descending horizontally, every thing will be the same, except that the mass to be moved will be equal to that of the cylinder *plus* the whole of the fluid displaced. In these cases, as well as in that of any solid which is symmetrical with respect to two vertical planes at right angles to each other, the motion will be uniformly accelerated, and similar solids of the same material will descend with equal velocities. These results are utterly opposed even to the commonest observation, which shews that large solids descend much more rapidly than small ones of the same shape and material, and that the velocity of a body falling in a fluid (such as water), does not sensibly increase after a little time. It becomes then of importance in the theory of resistances to enquire what may be the cause of this discrepancy between theory and observation. The following are the only ways of accounting for it which suggest themselves to me.

First. It has been supposed that the same particles remain in contact with the solid throughout the motion. It must be remembered that we suppose the ultimate molecules of fluids (if such exist), to be so close that their distance is quite insensible, a supposition of the truth of which there can be hardly any doubt. Consequently we reason on a fluid as if it were infinitely divisible. Now if the motion which takes place in the cases of the sphere and cylinder be examined, supposing for simplicity their motions to be rectilinear, it will be found that a particle in contact with the surface of either moves along that surface with a velocity which at last becomes infinitely small, and that it does not reach the end of the sphere or cylinder from which the whole is moving until after an infinite time, while any particle not in contact with the surface is at last left behind. It seems difficult to conceive of what other kind the motion can be, without supposing a line (or rather surface) of particles to make an abrupt turn. If it should be said that the particles may come off in tangents, it must be remembered that this sort of motion is included in the condition which has been assumed with respect to the surface.

Secondly. The discrepancy alluded to might be supposed to arise from the friction of the fluid against the surface of the solid. But, for the reason mentioned in the beginning of this paper, this explanation does not appear to me satisfactory.

Thirdly. It appears to me very probable that the *spreading out* motion of the fluid, which is supposed to take place behind the middle of the sphere or cylinder, though dynamically possible, nay, the *only* motion dynamically possible when the conditions

which have been supposed are accurately satisfied, is unstable;
so that the slightest cause produces a disturbance in the fluid,
which accumulates as the solid moves on, till the motion is quite
changed. Common observation seems to shew that, when a solid
moves rapidly through a fluid at some distance below the surface,
it leaves behind it a succession of eddies in the fluid. When the
solid has attained its terminal velocity, the product of the resistance, or rather the mean resistance, and any space through which the
solid moves, will be equal to half the *vis viva* of the corresponding
portion of its *tail of eddies,* so that the resistance will be measured
by the *vis viva* in the length of two units of that tail. So far
therefore as the resistance which a ship experiences depends
on the disturbance of the water which is independent of its
elevation or depression, that ship which leaves the least wake
ought, according to this view, to be *cæteris paribus* the best sailer.
The resistance on a ship differs from that on a solid in motion
immersed in a fluid in the circumstance, that part of the resistance is employed in producing a wave.

Fourthly. The discrepancy alluded to may be due to the
mutual friction, or imperfect fluidity of the fluid.

12. *Motion about an elliptic cylinder of small eccentricity* *.

The value of ϕ, which has been deduced (Art. 8), for the
motion of the fluid about a circular cylinder, is found on the
supposition that for each value of r there exists, or may be

[* This particular problem, so far at least as concerns motion of translation,
is of little interest in itself, because Green (see *Transactions of the Royal Society
of Edinburgh,* Vol. XIII. p. 54, or p. 315 of his collected works) has determined the
motion of a fluid about an ellipsoid moving in any manner with a motion of translation only; and the ellipsoid includes of course as a particular case an elliptic
cylinder of any eccentricity. The problem in the text will however serve as an
example of the mode of proceeding in the case of a cylinder of any kind differing
little from a circular cylinder.

In the case of such a cylinder, supposed to be free from abrupt changes of form,
it might safely be assumed that the expression for ϕ which applies to the fluid
beyond the greatest radius vector of any point of the surface might also be used
for some distance within, as explained in the text. By starting with this assumption,
which would be verified in the end, the process of solution would of course be
shortened. We should simply have to take the expression (31′), form the expression
(26′) for the velocity normal to the surface, putting $r = c\,(1 + \epsilon \cos 2\theta)$, and expanding as far as the first power of ϵ, and equate the result to the expression (26). We
should thus determine the arbitrary constants in (31′), which would complete the
solution of the problem.]

supposed to exist, a real and finite value of ϕ. This will be true, in any case of motion in two dimensions where $u\,dx + v\,dy$ is an exact differential, *for those values of* r *for which the fluid is not interrupted*, but will be true for values of r for which it is interrupted by solids only when it is possible to replace those solids at any instant by masses of fluid, without affecting the motion of the fluid exterior to them, those masses moving in such a manner that the motion of the whole fluid might have been produced instantaneously by impact. In some cases such a substitution could be made, while in others it probably could not. In any case however we may try whether the expansion given by equation (3) will enable us to get a result, and if it will, we need be in no fear that it is wrong (Art. 2). The same remarks will apply to the question of the possibility of the expansion of ϕ in the series of Laplace's coefficients given in equation (10), for values of r for which the fluid is interrupted. They will also apply to such a question as that of finding the permanent temperature of the earth due to the solar heat, the earth being supposed to be a homogeneous oblate spheroid, and the points of the surface being supposed to be kept up to constant temperatures, given by observation, depending on the latitude.

In cases of fluid motion such as those mentioned, the motion may be determined by conceiving the whole mass of fluid divided into two or more portions, taking the most general value of ϕ for each portion, this value being in general expressed in a different manner for the different portions, then limiting the general value of ϕ for each portion so as to satisfy the conditions with respect to the surfaces of solids belonging to that portion, and lastly introducing the condition that the velocity and direction of motion of each pair of contiguous particles in any two of the portions are the same. The question first proposed will afford an example of this method of solution.

Let an elliptic cylinder be moving with a velocity C, in the direction of the major axis of a section of it made by a plane perpendicular to its axis. The motion being supposed to be in two dimensions, it will be sufficient to consider only this section. Let

$$r = c\,(1 + \epsilon \cos 2\theta)$$

be the approximate equation to the ellipse so formed, the centre being the pole, and powers of ϵ above the first being neglected. Let a circle be described about the same centre, and having a radius γ equal to $(1 + k) c$, k being $\not< \epsilon$, and being a small quantity of the order ϵ. Let the portions of fluid within and without the radius γ be considered separately, and putting

$$r = c + z,$$

let the value of ϕ corresponding to the former portion be

$$P + Qz + Rz^2,$$

P, Q and R being functions of θ, and the term in z^2 being retained, in order to get the value of $d\phi/dr$ true to the order ϵ, while the terms in z^3, &c. are omitted. Substituting this value of ϕ in equation (2), and equating to zero coefficients of different powers of z, we have

$$R = -\frac{Q}{2c} - \frac{1}{2c^2}\frac{d^2P}{d\theta^2},$$

which is the only condition to be satisfied, since the other equations would only determine the coefficients of z^3, &c. in terms of the preceding ones. We have then

$$\phi = P + Qz - \frac{1}{2c}\left(Q + \frac{1}{c}\frac{d^2P}{d\theta^2}\right)z^2 \ldots\ldots\ldots(25).$$

Now if ξ be the angle between the normal at any point of the ellipse, and the major axis, we have

$$\xi = \theta + 2\epsilon \sin 2\theta,$$

and the velocity of the ellipse resolved along the normal

$$= C \cos \xi = C (1 - \epsilon) \cos \theta + C\epsilon \cos 3\theta \ldots\ldots(26).$$

The velocity of the fluid at the same point resolved along the normal is

$$\frac{d\phi}{dr} + 2\epsilon \sin 2\theta \frac{d\phi}{rd\theta}\ldots\ldots\ldots\ldots(26'),$$

or $\quad \dfrac{d\phi}{dz} + \dfrac{2\epsilon}{c} \sin 2\theta \dfrac{d\phi}{d\theta} \ldots\ldots\ldots\ldots(27).$

Let P and Q be expanded in series of cosines of θ and its multiples, so that

$$P = \Sigma_0^\infty P_n \cos n\theta, \quad Q = \Sigma_0^\infty Q_n \cos n\theta,$$

there being no sines in the expansions of P and Q, since the motion is symmetrical with respect to the major axis; then

$$\phi = \Sigma_0^\infty \left\{ P_n + Q_n z - \frac{1}{2c}\left(Q_n - \frac{n^2}{c}P_n\right) z^2 \right\} \cos n\theta \ldots(28);$$

$$\frac{d\phi}{dz} = \Sigma_0^\infty \left\{ Q_n - \frac{1}{c}\left(Q_n - \frac{n^2}{c}P_n\right) z \right\} \cos n\theta \ldots\ldots\ldots(29);$$

$$\frac{1}{c+z}\frac{d\phi}{d\theta} = -\Sigma_0^\infty n \left\{ \frac{P_n}{c} + \left(\frac{Q_n}{c} - \frac{P_n}{c^2}\right) z \right\} \sin n\theta \ldots\ldots\ldots(30).$$

For a point in the ellipse, $z = c\epsilon \cos 2\theta$, whence from (27), (29) and (30), we find that the normal velocity of the fluid

$$= \Sigma_0^\infty \left\{ Q_n \cos n\theta + \frac{\epsilon}{2}\left[n(n-2)\frac{P_n}{c} - Q_n \right] \cos (n-2)\theta \right.$$
$$\left. + \frac{\epsilon}{2}\left[n(n+2)\frac{P_n}{c} - Q_n \right] \cos (n+2)\theta \right\},$$

which is the same thing as

$$\Sigma_0^\infty \left\{ \frac{\epsilon}{2}\left[n(n-2)\frac{P_{n-2}}{c} - Q_{n-2} \right] + Q_n \right.$$
$$\left. + \frac{\epsilon}{2}\left[n(n+2)\frac{P_{n+2}}{c} - Q_{n+2} \right] \right\} \cos n\theta \ldots(31),$$

if we suppose P and Q to be zero when affected with a negative suffix. This expression will have to be equated to the value of $C\cos\xi$ given by equation (26).

For the part of the fluid without the radius γ we have

$$\phi = A_0 \log r + \Sigma_1^\infty \frac{A_n}{r_n} \cos n\theta* \ldots\ldots\ldots(31'),$$

since there will be no sines in the expression for ϕ, because the motion is symmetrical with respect to the major axis, and no positive powers of r, because the velocity vanishes at an infinite distance.

From the above value of ϕ we have, for the points at a distance γ from the centre,

---

* The first term of this expression is accurately equal to zero, since there is no expansion or contraction of the solid (Art. 8). I have however retained it, in order to render the solution of the problem in the present article independent of the proposition referred to.

$$\frac{d\phi}{dr} = \frac{A_0}{\gamma} - \Sigma_1^\infty \frac{nA_n}{\gamma^{n+1}} \cos n\theta,$$

$$\frac{d\phi}{rd\theta} = -\Sigma_1^\infty \frac{nA_n}{\gamma^{n+1}} \sin n\theta.$$

Equating the above expressions to the velocities along and perpendicular to the radius vector given by equations (29) and (30), when z is put $= kc$, and then equating coefficients of corresponding sines and cosines, we have

$$(1-k)\,Q_n + kn^2 \frac{P_n}{c} = -\frac{nA_n}{\gamma^{n+1}} \quad\ldots\ldots\ldots\ldots (32),$$

$$(1-k)\frac{P_n}{c} + kQ_n = \frac{A_n}{\gamma^{n+1}} \quad\ldots\ldots\ldots\ldots (33),$$

when $n > 0$, and equating constant terms we have

$$(1-k)\,Q_0 = \frac{A_0}{\gamma},$$

from which equation with (32) and (33) we have, putting

$$\gamma = (1+k)\,c,$$

$$\frac{P_n}{c} = \frac{A_n}{c^{n+1}}, \quad Q_n = -\frac{nA_n}{c^{n+1}}, \quad \text{when } n > 0, \quad \text{and} \quad Q_0 = \frac{A_0}{c}.$$

Substituting these values in the expression (31), it becomes

$$\Sigma_0^\infty \left\{ \frac{\epsilon}{2}(n+1)(n-2)\frac{A_{n-2}}{c^{n-1}} - \frac{nA_n}{c^{n+1}} + \frac{\epsilon}{2}(n+1)(n+2)\frac{A_{n+2}}{c^{n+3}} \right\} \cos n\theta$$

$$+ \frac{A_0}{c} - \frac{\epsilon A_0}{2c} \cos 2\theta.$$

In the case of a circular cylinder the quantities A_0, A_2, A_3, &c. are each zero. In the present case therefore they are small quantities depending on ϵ. Hence, neglecting quantities of the order ϵ^2 in the above expression, it becomes

$$\frac{A_0}{c} + \frac{2\epsilon A_1}{c^2} \cos 3\theta - \Sigma_1^\infty \frac{nA_n}{c^{n+1}} \cos n\theta,$$

which must be equal to $C\{(1-\epsilon)\cos\theta + \epsilon \cos 3\theta\}$. Equating coefficients of corresponding cosines, we have

$$A_1 = -C(1-\epsilon)\,c^2,$$
$$A_3 = -C\,\epsilon c^4,$$

and the other quantities A_0, A_2, &c. are of an order higher than ϵ.

Hence, for the part of the fluid which lies without the radius γ, we have

$$\phi = -\, C \left\{ (1 - \epsilon) \frac{c^2}{r} \cos \theta + \frac{\epsilon c^4}{r^3} \cos 3\theta \right\} \quad \ldots\ldots\ldots(34),$$

and for the part which lies between that radius and the ellipse we have from (28)

$$\phi = -\, Cc \left\{ (1 - \epsilon) \cos \theta + \epsilon \cos 3\theta \right\} + C \left\{ (1 - \epsilon) \cos \theta + 3\epsilon \cos 3\theta \right\} z$$

$$-\, \frac{C}{c} \cos \theta z^2 \quad \ldots\ldots\ldots\ldots\ldots\ldots\ldots(35).$$

The value of ϕ given by equation (35) may be deduced from that given by equation (34) by putting $r = c + z$, and expanding as far as to z^2. In the case of the elliptic cylinder then it appears that the same value of ϕ serves for the part of the fluid without, and the part within the radius γ. If the cylinder be moving with a velocity C' in the direction of the minor axis of a section, the value of ϕ will be found from that given by equation (34) by changing the sign of ϵ, putting C' for C, and supposing θ to be measured from the minor axis.

If the cylinder revolve round its axis with an angular velocity ω, the normal velocity of the surface at any point will be $2\omega\epsilon c \sin 2\theta$. Since ϵ^2 is neglected, we may suppose this normal velocity to take place on the surface of a circular cylinder whose radius is c; whence (Art. 8) the corresponding value of ϕ will be

$$-\, \frac{\omega \epsilon c^4}{r^2} \sin 2\theta.$$

If we suppose all these motions to take place together, we have only (Art. 5) to add together the values of ϕ corresponding to each. If we suppose the motion very small, so as to neglect the square of the velocity, we need only retain the terms depending on $d\omega/dt$, dC/dt and dC'/dt, in the value of $d\phi/dt$, and we may calculate the pressure due to each separately. The resultant of the pressure due to the term $d\omega/dt$ will evidently be zero, on account of the symmetry of the corresponding motion, while the resultant couple will be of the order ϵ^2, since the pressure on any point of the surface, and the perpendicular from the centre on the normal at that point, are each of the order ϵ. The pressure due to the term dC/dt will evidently have a resultant in the direction of the major axis of a section of the cylinder; and it will be easily proved that the resultant pressure on a length l of the cylinder is $\pi\rho c^2 l (1-2\epsilon) dC/dt$. That due to the term dC'/dt will be $\pi\rho c^2 l (1+2\epsilon) dC'/dt$, acting along the minor axis. If the cylinder be constrained to oscillate so that its axis oscillates in a direction making an angle α with the major axis, and if C'' be its velocity, which is supposed to be very small, the resultant pressures along the major and minor axes will be

$$\mu (1-2\epsilon) \cos \alpha \frac{dC''}{dt} \text{ and } \mu (1+2\epsilon) \sin \alpha \frac{dC''}{dt}$$

respectively, where μ is the mass of the fluid displaced. Resolving these pressures in the direction of the motion, the resolved part will be $\mu (1-2\epsilon \cos 2\alpha) dC''/dt$, or $\mu (1-\frac{1}{2}e^2 \cos 2\alpha) dC''/dt$, e being the eccentricity; so that the effect of the inertia of the fluid will be, to increase the mass of the solid by a mass equal to $\mu (1-\frac{1}{2}e^2 \cos 2\alpha)$, which must be supposed to be collected at the axis.

A similar method of calculation would apply to any given solid differing little either from a circular cylinder or from a sphere. In the latter case it would be necessary to use expansions in series of Laplace's coefficients, instead of expansions in series of sines and cosines.

13. *Motion of fluid in a closed box whose interior is of the form of a rectangular parallelepiped.*

The motion being supposed to begin from rest, the motion at any time may be supposed to have been produced by impact (Art. 4). The motion of the box at any instant may be resolved into a motion of translation and three motions of rotation about three axes parallel to the edges, and passing through the centre of gravity of the fluid, and the part of ϕ due to each of these motions may be calculated separately. Considering any one of the motions of rotation, we shall see that the normal velocity of each face in consequence of it will ultimately be the same as if that face revolved round an axis passing through its centre, and that the latter motion would not alter the volume of the fluid. Consequently, in calculating the part of ϕ due to any one of the angular velocities, we may calculate separately the part due to the motion of each face.

Let the origin be in a corner of the box, the axes coinciding with its edges. Let a, b, c, be these edges, U, V, W, the velocities, parallel to the axes, of the centre of gravity of the interior of the box, ω', ω'', ω'''; the angular velocities of the box about axes through this point parallel to those of x, y, z. Let us first consider the part of ϕ due to the motion of the face xz in consequence of the angular velocity ω'''.

The value of ϕ corresponding to this motion must satisfy the equation

$$\frac{d^2\phi}{dx^2} + \frac{d^2\phi}{dy^2} = 0 \dots\dots\dots\dots\dots(36),$$

with the conditions

$$\frac{d\phi}{dx} = 0, \text{ when } x = 0 \text{ or } a \dots\dots\dots(37),$$

$$\frac{d\phi}{dy} = 0, \text{ when } y = b \dots\dots\dots\dots(38),$$

$$\frac{d\phi}{dy} = \omega'''(x - \tfrac{1}{2}a), \text{ when } y = 0 \dots\dots(39),$$

within limits corresponding to those of the box.

Now, for a given value of y, the value of ϕ between $x = 0$ and $x = a$ can be expanded in a convergent series of cosines of $\pi x/a$ and its multiples; and, since (37) is satisfied, the series by which $d\phi/dx$ will be expressed will also hold good for the limiting values of x, and will be convergent. The general value of ϕ then will be of the form $\Sigma_0^\infty Y_n \cos n\pi x/a$. Substituting in (36), and equating coefficients of corresponding cosines, which may be done, since any function of x can be expanded in but one such series of cosines between the limits 0 and a, we find that the general value of Y_n is $Ce^{n\pi y/a} + C'e^{-n\pi y/a}$, or, changing the constants,

$$Y_n = A_n \left(e^{n\pi(b-y)/a} + e^{-n\pi(b-y)/a}\right) + B_n \left(e^{n\pi y/a} + e^{-n\pi y/a}\right),$$

when $n > 0$, and for $n = 0$,

$$Y_0 = A_0 y + B_0.$$

From the condition (38) we have

$$A_0 + \pi a^{-1} \Sigma_1^\infty n B_n \left(e^{n\pi b/a} - e^{-n\pi b/a}\right) \cos n\pi x/a = 0 :$$

whence $A_0 = 0$, $B_n = 0$, and, omitting B_0,

$$\phi = \Sigma_1^\infty A_n \left(e^{n\pi(b-y)/a} + e^{-n\pi(b-y)/a}\right) \cos n\pi x/a.$$

From the condition (39), we have

$$- \pi a^{-1} \Sigma_1^\infty n A_n \left(e^{n\pi b/a} - e^{-n\pi b/a} \right) \cos n\pi x/a = \omega'''\left(x - \tfrac{1}{2}a\right).$$

Determining the coefficients in the usual manner, we have

$$A_n = \frac{2a^2 \omega'''}{n^3 \pi^3} \left\{ 1 - (-1)^n \right\} \div \left(e^{n\pi b/a} - e^{-n\pi b/a} \right);$$

whence

$$\phi = \frac{4a^2 \omega'''}{\pi^3} \Sigma_0 \frac{1}{n^3} \frac{e^{n\pi(b-y)/a} + e^{-n\pi(b-y)/a}}{e^{n\pi b/a} - e^{-n\pi b/a}} \cos n\pi x/a,$$

putting Σ_0, for shortness, to denote the sum corresponding to *odd* integral values of n from 1 to ∞.

It is evident that the value of ϕ corresponding to the motion of the opposite face in consequence of the angular velocity ω''' will be found from that just given by putting $b - y$ for y, and changing the sign of ω'''; whence the value corresponding to the motion of these two faces in consequence of ω''' will be

$$\frac{4\omega''' a^2}{\pi^3} \Sigma_0 \frac{1}{n^3} \frac{\left(e^{n\pi b/a} - 1\right) e^{-n\pi y/a} + \left(e^{-n\pi b/a} - 1\right) e^{n\pi y/a}}{e^{n\pi b/a} - e^{-n\pi b/a}} \cos n\pi x/a.$$

Let this expression be denoted by $\omega''' \psi(x, a, y, b)$. It is evident that the part of ϕ due to the motion of the two faces parallel to the plane yz will be got by interchanging x and y, a and b, and changing the sign of ω''' in the last expression, and will therefore be $- \omega''' \psi(y, b, x, a)$. The parts of ϕ corresponding to the angular velocities ω', ω'', will be got by interchanging the requisite quantities. Also the part of ϕ due to the velocities U, V, W, will be $Ux + Vy + Wz$ (Art. 7), and therefore we have for the complete value of ϕ

$$Ux + Vy + Wz + \omega''' \left\{ \psi(x, a, y, b) - \psi(y, b, x, a) \right\} + \omega' \left\{ \psi(y, b, z, c) \right.$$
$$\left. - \psi(z, c, y, b) \right\} + \omega'' \left\{ \psi(z, c, x, a) - \psi(x, a, z, c) \right\}.$$

According to Art. 7 we may consider separately the motion of translation of the box and fluid, and the motion of rotation about the centre of gravity of the latter; and the whole pressure will be compounded of the pressures due to each. The pressures at the several points of the box due to the motion of translation will have a single resultant, which will be the same as if the mass of the fluid were collected at its centre of gravity. Those due to the

motion of rotation will have a single resultant couple, to calculate
which we have

$$\phi = \omega''' \{\psi\,(x,\, a,\, y,\, b) - \psi\,(y,\, b,\, x,\, a)\} + \&\text{c}.$$

Since for the motion of rotation there is no resultant force,
we may find the resultant couple of the pressures round *any*
origin, that for instance which has been chosen. If now we
suppose the motion very small, so as to neglect the square of
the velocity, we may find $d\phi/dt$ as if the axes were fixed in space.
We have then for the motion of rotation

$$p = -\rho\,\frac{d\omega'''}{dt}\,\{\psi\,(x,\, a,\, y,\, b) - \psi\,(y,\, b,\, x,\, a)\} - \&\text{c}.$$

Hence we may calculate separately the couples due to each of
the quantities $d\omega'''/dt$, $d\omega'/dt$ and $d\omega''/dt$. It is evident from the
symmetry of the motion that that due to $d\omega'''/dt$ will act round
the axis of z, and that the pressures on the two faces perpendicular
to that axis will have resultants which are equal and opposite.
Also, since $\psi\,(a,\, a,\, y,\, b) = -\psi\,(0,\, a,\, y,\, b)$ and $\psi\,(x,\, a,\, b,\, b,) = -\psi$
$(x,\, a,\, 0,\, b)$, it will be seen that the couples due to the pressures
on the faces perpendicular to the axes of x and y will be twice
as great respectively as those due to the pressures on the planes
yz and xz. The pressure on the element $dydz$ of the plane yz will
be $p_{x=0}\,dydz$, and the moment of this pressure round the axis of z,
reckoned positive when it tends to turn the box from x to y,
will be

$$-\rho\,\frac{d\omega'''}{dt}\,y\,\{\psi\,(0,\, a,\, y,\, b) - \psi\,(y,\, b,\, 0,\, a)\}\,dydz.$$

Substituting the values of the functions, integrating from $y = 0$ to
$y = b$, and from $z = 0$ to $z = c$, replacing $\Sigma_0\,1/n^5$ by its value
$\pi^4/96$, and reducing the other terms, it will be found that the couple
due to the pressure on the plane yz is

$$\frac{\rho a^3 bc}{24}\,\frac{d\omega'''}{dt} - \frac{8\rho a^4 c}{\pi^5}\,\frac{d\omega'''}{dt}\,\Sigma_0\,\frac{1}{n^5}\frac{1 - e^{-n\pi b/a}}{1 + e^{-n\pi b/a}}$$

$$-\frac{8\rho b^4 c}{\pi^5}\,\frac{d\omega'''}{dt}\,\Sigma_0\,\frac{1}{n^5}\frac{1 - e^{-n\pi a/b}}{1 + e^{-n\pi a/b}}\,.$$

We shall get the couple due to the pressure on the plane xz
by interchanging a and b, changing the sign of ω''', and measuring
the couple in the opposite direction, or, which is the same, by
merely interchanging a and b. Adding together these two couples

and doubling their sum we shall find that the couple due to $d\omega'''/dt$ is $-C\,d\omega'''/dt$, where

$$C = \frac{32\rho c}{\pi^5}\,\Sigma_0\,\frac{1}{n^5}\left\{a^4\frac{1-e^{-n\pi b/a}}{1+e^{-n\pi b/a}} + b^4\frac{1-e^{-n\pi a/b}}{1+e^{-n\pi a/b}}\right\}$$

$$-\frac{\rho abc}{12}\,(a^2+b^2)\ldots\ldots\ldots(40).$$

Similarly, the couple due to $d\omega'/dt$ will be $-A\,d\omega'/dt$, tending to turn the box from y to z, and that due to $d\omega''/dt$ will be $-B\,d\omega''/dt$, tending to turn the box from z to x, where A and B are derived from C by interchanging the requisite quantities. Hence, considering the motions both of translation and rotation of the box, we see that the small motions of the box will take place as if the fluid were replaced by a solid having the same mass, centre of gravity, and principal axes, and having A, B and C for its principal moments. This will be true whether forces act on the fluid or not, provided that if there are any they are of the kind mentioned in Art. 1.

Putting $A_{,}$, $B_{,}$, $C_{,}$ for the principal moments of inertia of the solidified fluid, we have

$$C_{,} = \frac{\rho abc}{12}\,(a^2+b^2).$$

Taking the ratio of C to $C_{,}$, replacing each term such as

$$\frac{1-e^{-n\pi b/a}}{1+e^{-n\pi b/a}}\text{ by }1-\frac{2e^{-n\pi b/a}}{1+e^{-n\pi b/a}}\,,\text{ putting for }\frac{384}{\pi^5}\Sigma_0\frac{1}{n^5}$$

its approximate value $1{\cdot}260497$, and for $384/\pi^5$ its approximate value $1{\cdot}254821$, and employing subsidiary angles, we have

$$\frac{C}{C_{,}} = 1{\cdot}260497\,\frac{a^4+b^4}{ab\,(a^2+b^2)} - 1{\cdot}254821\left\{\frac{a^3}{b\,(a^2+b^2)}\,\Sigma_0\frac{1}{n^5}\text{ versin }2\theta_n\right.$$

$$\left.+\frac{b^3}{a\,(a^2+b^2)}\,\Sigma_0\frac{1}{n^5}\text{ versin }2\theta'_n\right\}-1^*,$$

where $\qquad\tan\theta_n = e^{-n\pi b/2a},\ \tan\theta'_n = e^{-n\pi a/2b},$
so that
$$L\tan\theta_n = 10 - k\,nb/a,\quad L\tan\theta'_n = 10 - k\,na/b,$$
where $\qquad\qquad k = {\cdot}6821882.$

* [It will be shewn further on, in a supplement to this paper, that either of these two infinite series may be expressed by means of the other, so that we shall have only one of the infinite series to calculate in any case, for which we may choose the more rapidly convergent.]

The numerical calculation of this ratio is very easy, on account of the great rapidity with which the series contained in it converge, both on account of the coefficients, and on account of the rapid diminution of the angles θ_n and θ'_n. The values of $A/A_{,}$ and $B/B_{,}$ will be derived from that of $C/C_{,}$ by putting c for a in the first case, and c for b in the second. The calculation of the small motions of the box will thus be reduced to a question of ordinary rigid dynamics*.

When one of the quantities a, b, becomes infinitely great compared with the other, the ratio $C/C_{,}$ becomes 1, as will be seen from equation (40). This result might have been expected. When $a = b$ the value of $C/C_{,}$ is ·156537†.

The experiment of the box appears capable of great variety as well as accuracy. We may take boxes in which the edges have

* [Corresponding to the two simple cases of steady motion referred to in the foot-note to p. 7, are two in which the motion of the fluid within a box of simple form can be expressed in finite terms, the box and the fluid being initially at rest, and the box being then moved about its axis.

The first is that in which the box is of the form of a right prism, having for its base an equilateral triangle. If as before a be the perpendicular from the centre of the triangle on one side, and θ be measured from this perpendicular, we shall have

$$\phi = -\frac{\omega}{6a} r^3 \sin 3\theta;$$

and by performing the integrations we shall find that if k be the radius of gyration of what we may call the *equivalent solid*, that is, the solid, of the same mass as the fluid, by which the fluid may be replaced without affecting the motion of the box under given forces,

$$k^2 = \tfrac{2}{5}a^2;$$

and as a is the radius of gyration for the fluid supposed solidified, the moment of inertia of the equivalent solid is two-fifths of that of the solidified fluid.

The other is that of a box of the form of a right elliptic prism. In this case ϕ is of the form $cr^2 \sin 2\theta$, θ being measured from the major axis; and determining c so as to suit an ellipse of which a and b are the semiaxes, we find

$$\phi = \frac{\omega(a^2 - b^2)}{2(a^2 + b^2)} r^2 \sin 2\theta.$$

k having the same meaning as before, it will be found that

$$k^2 = \tfrac{1}{4} \frac{(a^2 - b^2)^2}{a^2 + b^2},$$

so that the ratio of the moment of inertia of the equivalent solid to that of the solidified fluid is that of $(a^2 - b^2)^2$ to $(a^2 + b^2)^2$.]

† [A passage containing a proposal to compare this result with experiment is here omitted, as the experiment is described, in the form in which it was actually carried out, in the supplement before referred to.]

S. 5

various ratios to each other, and may make the same box oscillate in various positions.

14. *Initial motion in a rectangular box, the several points of the surface of which are moved with given velocities, consistent with the condition that the volume of the fluid is not altered.*

Employing the same notation as in the last case, let $F(x, y)$ be the given normal velocity at any point of the face in the plane xy. Let $\int_0^a \int_0^b F(x, y)\, dx dy = Wab$, and let

$$F(x, y) = f(x, y) + W:$$

then, since the normal motion of the above face due to the function $f(x, y)$ does not alter the volume of the fluid, we may consider separately the part of ϕ due to this quantity. For this part we have

$$\frac{d^2\phi}{dx^2} + \frac{d^2\phi}{dy^2} + \frac{d^2\phi}{dz^2} = 0 \dots\dots\dots\dots(41),$$

with the conditions

$$\frac{d\phi}{dx} = 0, \text{ when } x = 0 \text{ or } a\dots\dots\dots(42),$$

$$\frac{d\phi}{dy} = 0, \text{ when } y = 0 \text{ or } b\dots\dots\dots(43),$$

$$\frac{d\phi}{dz} = 0, \text{ when } z = c\dots\dots\dots\dots(44),$$

$$\frac{d\phi}{dz} = f(x, y), \text{ when } z = 0 \dots\dots\dots(45),$$

within limits corresponding to those of the box.

For a given value of z the value of ϕ from $x = 0$ to $x = a$ and from $y = 0$ to $y = b$ may be expanded in a series of the form

$$\Sigma_0^\infty \Sigma_0'^\infty P_{m,n} \cos m\pi x/a \, . \, \cos n\pi y/b,$$

the sign Σ referring to m, and Σ' to n: and since the values of ϕ, $d\phi/dx$ and $d\phi/dy$ do not alter abruptly, and equations (42) and (43) are satisfied, it follows that the series by which ϕ, $d\phi/dx$ and $d\phi/dy$ are expressed are convergent, and hold good for the limiting values of x and y. Substituting the value of ϕ just given in (41), equating to zero coefficients of corresponding cosines, and introducing the condition (44), we have, omitting the constant, or supposing $A_{0,0}=0$,

$$\phi = \Sigma_0^\infty \Sigma_0'^\infty A_{m,n}\{e^{p\pi(c-z)/c}+e^{-p\pi(c-z)/c}\}\cos m\pi x/a \,.\, \cos n\pi y/b,$$

where
$$\frac{p^2}{c^2}=\frac{m^2}{a^2}+\frac{n^2}{b^2}.$$

Determining the coefficients such as $A_{m,n}$ from the condition (45) in the usual manner we have, m and n being > 0,

$$A_{m,n}=-\frac{4c}{\pi pab}\,(e^{\pi p}-e^{-p\pi})^{-1}\int_0^a\int_0^b f(x,y)\cos m\pi x/a.\cos n\pi y/b.dxdy,$$

$$A_{0,n}=-\frac{2}{\pi na}\,(e^{n\pi c/b}-e^{-n\pi c/b})^{-1}\int_0^a\int_0^b f(x,y)\cos n\pi y/b.dxdy*,$$

with a similar expression for $A_{m,0}$, whence the value of ϕ corresponding to $f(x,y)$ is known. In a similar manner we may find the values corresponding to the similar functions belonging to each of the other faces. If W' be the quantity corresponding to W for the face opposite to the plane xy, and U, U', correspond to W, W', for the faces perpendicular to the axis of x, and if V, V', be the corresponding quantities for y, there remains only to be found the part of ϕ due to these six quantities. Since U, U', are the velocities parallel to the axis of x of the faces perpendicular to that axis, and so for V, V', &c., the motion corresponding to these six quantities may be resolved into three motions of translation parallel to the three axes, the velocities being U, V and W, and that motion which is due to the motions of the faces opposite to the planes yz, zx, xy, moving with velocities $U'-U$, $V'-V$, $W'-W$, parallel to the axes of x, y, z, respectively. The condition that the volume of the fluid remains the same requires that

$$\frac{1}{a}(U'-U)+\frac{1}{b}(V'-V)+\frac{1}{c}(W'-W)=0.$$

It will be found that the velocities

$$u=\frac{x}{a}(U'-U),\quad v=\frac{y}{b}(V'-V),\quad w=\frac{z}{c}(W'-W),$$

satisfy all the requisite conditions. Hence the part of ϕ due to

* The function $f(x,y)$ in these integrals may be replaced by $F(x,y)$, since $\int_0^a\int_0^b W\cos n\pi y/b\,.\,\cos n\pi x/a\,.\,dxdy=0$, unless $m=n=0$.

the six quantities U, U', V, V', W, W', is

$$Ux + Vy + Wz + (U'-U)\frac{x^2}{2a} + (V'-V)\frac{y^2}{2b} + (W'-W)\frac{z^2}{2c}.$$

This quantity, added to the six others which have already been given, gives the value of ϕ which contains the complete solution of the problem.

The case of motion which has just been given seems at first sight to be an imaginary one, capable of no practical application. It may however be applied to the determination of the small motion of a ball pendulum oscillating in a case in the form of a rectangular parallelepiped, the dimensions of the case being great compared with the radius of the ball. For this purpose it will be necessary to calculate the motion of the ball reflected from the case, by means of the formulæ just given, and then the motion again reflected from the sphere, exactly as has been done in the case of a rigid plane, Art. 10. In the present instance however the result contains definite integrals, the numerical calculation of which would be very troublesome.

[From the *Cambridge Mathematical Journal*, Vol. IV. p. 28. (*Nov.* 1843).]

ON THE MOTION OF A PISTON AND OF THE AIR IN A CYLINDER.

WHEN a piston is in motion in a cylinder which also contains air, if the motion of the piston be not very rapid, so that its velocity is inconsiderable compared with the velocity of propagation of sound, the motions of the air may be divided into two classes, the one consisting of those which depend directly on the motion of the piston, the other, of those which are propagated with the velocity of sound, and depend on the initial state of the air, or on a breach of continuity in the motion of the piston. If we suppose the initial velocity and condensation of the air in each section of the cylinder to be given, and also the initial velocity of the piston, both kinds of motion will in general take place, and the solution of the problem will be complicated. If, however, we restrict ourselves to motions of the first class, the approximate solution, though rather long, will be simple. In this case we must suppose the inital velocity and condensation of the air not to be given arbitrarily, but to be connected, according to a certain law which is yet to be found, with the motion of the piston. The problem as so simplified may perhaps be of some interest, as affording an example of the application of the partial differential equations of fluid motion, without requiring the employment of that kind of analysis which is necessary in most questions of that sort. It is, moreover, that motion of the air which it is proposed to consider, which principally affects the motion of the piston.

Conceive an air-tight piston to move in a cylinder which is closed at one end, and contains a mass of air between the closed end and the piston. For more simplicity, suppose the rest of the cylinder to contain no air. Let a point in the closed end be taken for origin, and let x be measured along the cylinder. Let x_1 be the abscissa of the piston; a the initial value of x_1; u the velocity parallel to x of any particle of air whose abscissa is x; p the pressure, ρ the density about that particle; Π the initial mean pressure; p_1 the value of p when $x = x_1$; X, a function of x, the accelerating force acting on the air; then for the motion of the air we have

$$\left. \begin{aligned} \frac{1}{\rho}\frac{dp}{dx} &= X - \frac{du}{dt} - u\frac{du}{dx}, \\ \frac{d\rho}{dt} + \frac{d\rho u}{dx} &= 0, \end{aligned} \right\} \quad \dots\dots\dots\dots (1),$$

and
$$p = k\rho,$$

neglecting the variation of temperature.

We have also the conditions

$$u = 0 \quad \text{when} \quad x = 0 \dots\dots\dots\dots(2);$$

$$u = \frac{dx_1}{dt} \quad \text{when} \quad x = x_1 \dots\dots\dots\dots(3),$$

for positive values of t, and

$$\int_0^a p\,dx = \Pi a \quad \text{when} \quad t = 0 \dots\dots\dots(4).$$

Eliminating ρ from equations (1), we have

$$\frac{1}{p}\frac{dp}{dx} = \frac{1}{k}\left(X - \frac{du}{dt} - u\frac{du}{dx} \right)\dots\dots\dots(5);$$

$$\frac{dp}{dt} + \frac{dpu}{dx} = 0 \dots\dots\dots\dots(6).$$

Now, k being very large, for a first approximation let $\frac{1}{k}$ be neglected; then, integrating (5),

$$p = \phi(t).$$

Substituting in (6), and integrating,

$$u = \psi(t) - \frac{\phi'(t)}{\phi(t)}\,x.$$

The conditions (2) and (3) give

$$\psi(t) = 0; \quad \frac{\phi'(t)}{\phi(t)} = -\frac{1}{x_1}\frac{dx_1}{dt};$$

whence
$$\phi(t) = \frac{C}{x_1}.$$

Substituting in (4) the value of p when $t = 0$, we have

$$\int_0^a \frac{C}{a}dx = C = \Pi a;$$

whence
$$p = \Pi\frac{a}{x_1}; \quad u = \frac{x}{x_1}\frac{dx_1}{dt}.$$

Let now, for a second approximation,

$$p = \Pi\frac{a}{x_1} + p'; \quad u = \frac{x}{x_1}\frac{dx_1}{dt} + u';$$

so that p' and u' are small quantities of the order $1/k$; then, substituting these values in (5) and (6), remembering that the quantities which are not small must destroy each other, and retaining only small quantities of the first order, we have

$$\frac{dp'}{dx} = \frac{\Pi a}{kx_1}\left(X - \frac{x}{x_1}\frac{d^2x_1}{dt^2}\right)\dots\dots\dots(7);$$

$$\frac{dp'}{dt} + \frac{1}{x_1}\frac{dx_1}{dt}\frac{dp'x}{dx} + \Pi\frac{a}{x_1}\frac{du'}{dx} = 0\dots\dots\dots(8);$$

and the conditions (2), (3) and (4) give

$$u' = 0 \text{ when } x = 0, \text{ or } x = x_1, \text{ and } t \text{ is positive} \dots(9);$$

$$\int_0^a p'dx = 0 \text{ when } t = 0\dots\dots\dots\dots(10).$$

Integrating (7), we have

$$p' = \frac{\Pi a}{kx_1}\left(\int_0^x Xdx - \frac{x^2}{2x_1}\frac{d^2x_1}{dt^2}\right) + \omega(t)\dots\dots(11).$$

Substituting the values of p' and of its differential coefficients in (8), and integrating, we obtain

$$u' = \frac{x^3}{6kx_1{}^2}\frac{d}{dt}\left(x_1\frac{d^2x_1}{dt^2}\right) - \frac{1}{kx_1}\frac{dx_1}{dt}\int_0^x Xxdx - \frac{x}{\Pi a}\frac{d}{dt}\{x_1\omega(t)\} + \zeta(t)$$
$$\dots\dots\dots\dots(12).$$

The conditions (9) give $\zeta(t) = 0$;

$$\frac{1}{6k}\frac{d}{dt}\left(x_1\frac{d^2x_1}{dt^2}\right) - \frac{1}{kx_1^2}\frac{dx_1}{dt}\int_0^{x_1}Xx\,dx - \frac{1}{\Pi a}\frac{d}{dt}\{x_1\omega(t)\} = 0;$$

and integrating, we get

$$x_1\omega(t) = \frac{\Pi ax_1}{6k}\frac{d^2x_1}{dt^2} - \frac{\Pi a}{k}\int_a^{x_1}\left(\int_0^{x_1}Xx\,dx\right)\frac{dx_1}{x_1^2} + C\ldots(13).$$

Putting f for the initial value of d^2x_1/dt^2 we have, from (10) and (11),

$$\frac{\Pi}{k}\left(\int_0^a dx\int_0^x X\,dx - \frac{fa^2}{6}\right) + \omega(0)\,a = 0;$$

and substituting the value of $\omega(0)$ given by this equation in (13), after having made $t = 0$, $x_1 = a$, $d^2x_1/dt^2 = f$ in the latter, we have

$$C = -\frac{\Pi}{k}\int_0^a dx\int_0^x X\,dx.$$

Substituting this value of C in that of $\omega(t)$, and substituting in (11) and (12), and then substituting the values of p' and u' in those of p and u, we have

$$p = \Pi\,\frac{a}{x_1} + \frac{\Pi a}{kx_1}\left(\int_0^x X\,dx - \frac{x^2}{2x_1}\frac{d^2x_1}{dt^2}\right)$$

$$+ \frac{\Pi a}{6k}\frac{d^2x_1}{dt^2} - \frac{\Pi a}{kx_1}\int_a^{x_1}\left(\int_0^{x_1}Xx\,dx\right)\frac{dx_1}{x_1^2} - \frac{\Pi}{kx_1}\int_0^a\left(\int_0^x X\,dx\right)dx *$$

$$\ldots\ldots\ldots\ldots\ldots\ldots(14);$$

$$u = \frac{x}{x_1}\frac{dx_1}{dt} - \frac{x}{6k}\left(1 - \frac{x^2}{x_1^2}\right)\frac{d}{dt}\left(x_1\frac{d^2x_1}{dt^2}\right)$$

$$+ \frac{1}{kx_1}\frac{dx_1}{dt}\left\{\frac{x}{x_1}\int_0^{x_1}Xx\,dx - \int_0^x Xx\,dx\right\}\ldots\ldots(15).$$

Let A be the area of a section of the cylinder, and let $\Pi Aa/k = \mu$, so that μ is the mass of the air; then we have

$$p_1A = \Pi A\,\frac{a}{x_1} - \frac{\mu}{3}\frac{d^2x_1}{dt^2}$$

$$+ \frac{\mu}{x_1}\int_0^{x_1}X\,dx - \frac{\mu}{x_1}\int_a^{x_1}\left(\int_0^{x_1}Xx\,dx\right)\frac{dx_1}{x_1^2} - \frac{\mu}{ax_1}\int_0^a dx\int_0^x X\,dx.$$

[* It is best at once to get rid of the double integrals by integration by parts, which simplifies the expression, converting the last two terms into

$$-\frac{\Pi a}{kx_1^2}\int_0^{x_1}(x_1 - x)\,Xdx.\,]$$

If there were no motion, the term $- \frac{1}{3}\mu \ d^2x_1/dt^2$ would disappear. But in that case the value of p_1A, the pressure on the piston, might be deduced immediately from the equation of equilibrium of an elastic fluid

$$\frac{1}{p}\frac{dp}{dx} = \frac{X}{k}.$$

Integrating this equation, determining the constant by the condition that $\int_0^{x_1} p\,dx = \Pi a$, multiplying by A, and putting $x = x_1$, we have, neglecting $1/k^2$,

$$p_1A = \Pi A \frac{a}{x_1} + \frac{\mu}{x_1}\int_0^{x_1} X\,dx - \frac{\mu}{x_1^2}\int_0^{x_1}\left(\int_0^x X\,dx\right)dx.$$

Comparing this expression with the above, when the second term of the latter is left out, we have

$$\int_a^{x_1}\left(\int_0^{x_1} Xx\,dx\right)\frac{dx_1}{x_1^2} + \frac{1}{a}\int_0^a dx\int_0^x X\,dx = \frac{1}{x_1}\int_0^{x_1} dx\int_0^x X\,dx,$$

a formula which may also be proved directly. We have then

$$p_1A = \Pi A \frac{a}{x_1} - \frac{\mu}{3}\frac{d^2x_1}{dt^2} + \mu\frac{d}{dx_1}\left(\frac{1}{x_1}\int_0^{x_1} dx_1\int_0^{x_1} X\,dx\right).$$

The first term would be the value of the pressure on the piston if the air had no inertia and were acted on by no external forces; the second term is that due to the *inertia* of the air; the last term is that due to the external forces, and in the case of gravity expresses the effect of the *weight* of the air. If M be the mass of the piston, P the accelerating force parallel to x acting on it, not including the pressure of the air, its equation of motion is

$$\left(M + \frac{\mu}{3}\right)\frac{d^2x_1}{dt^2} = MP + \Pi A \frac{a}{x_1} + \mu\frac{d}{dx_1}\left(\frac{1}{x_1}\int_0^{x_1} dx_1\int_0^{x_1} X\,dx\right)\dots(16).$$

Hence the effect of the inertia of the air is to increase the mass of the piston by one third of that of the air, without increasing the moving force acting on it. If we could integrate equation (16) twice, we should determine the arbitrary constants by means of the initial values of x_1 and dx_1/dt, and thus get x_1 in terms of t: then, substituting in (14) and (15), we should obtain p and u as functions of x and t.

If the cylinder be vertical and smooth and turned upwards, we have $P = X = -g$; and if, moreover, the motion be very small, putting $x_1 = a + y$, and neglecting y^2, we have

$$\left(M + \frac{\mu}{3}\right) \frac{d^2 y}{dt^2} + \frac{\Pi A}{a} y = \Pi A - \left(M + \frac{\mu}{2}\right) g.$$

The term at the second side of this equation is by hypothesis small, and if we suppose the mean value of x to be taken for a, it is zero. On this supposition $\Pi A = \left(M + \frac{\mu}{2}\right) g$, and the time of a small oscillation will be $2\pi \sqrt{\dfrac{M + \dfrac{\mu}{3}}{M + \dfrac{\mu}{2}}} \cdot \dfrac{a}{g}$, which becomes,

since μ^2 is neglected throughout, $2\pi \left(1 - \dfrac{\mu}{12M}\right) \sqrt{\dfrac{a}{g}}$.

The reader who wishes to see the complete solution of the problem, in the case where no forces act on the air, and the air and piston are at first at rest, may consult a paper of Lagrange's with additions made by Poisson in the *Journal de l'École Polytechnique*. T. XIII. (21e *Cah.*) p. 187.

[From the *Transactions of the Cambridge Philosophical Society*,
Vol. VIII. p. 287.]

On the Theories of the Internal Friction of Fluids
in Motion, and of the Equilibrium and Motion of
Elastic Solids.

[Read April 14, 1845.]

The equations of Fluid Motion commonly employed depend
upon the fundamental hypothesis that the mutual action of two
adjacent elements of the fluid is normal to the surface which
separates them. From this assumption the equality of pressure
in all directions is easily deduced, and then the equations of
motion are formed according to D'Alembert's principle. This
appears to me the most natural light in which to view the subject; for the two principles of the absence of tangential action,
and of the equality of pressure in all directions ought not to be
assumed as independent hypotheses, as is sometimes done, inasmuch as the latter is a necessary consequence of the former*
The equations of motion so formed are very complicated, but yet
they admit of solution in some instances, especially in the case
of small oscillations. The results of the theory agree on the
whole with observation, so far as the time of oscillation is concerned. But there is a whole class of motions of which the
common theory takes no cognizance whatever, namely, those
which depend on the tangential action called into play by the
sliding of one portion of a fluid along another, or of a fluid along
the surface of a solid, or of a different fluid, that action in fact
which performs the same part with fluids that friction does with
solids.

* This may be easily shewn by the consideration of a tetrahedron of the fluid,
as in Art. 4.

Thus, when a ball pendulum oscillates in an indefinitely extended fluid, the common theory gives the arc of oscillation
constant. Observation however shews that it diminishes very
rapidly in the case of a liquid, and diminishes, but less rapidly,
in the case of an elastic fluid. It has indeed been attempted to
explain this diminution by supposing a friction to act on the ball,
and this hypothesis may be approximately true, but the imperfection of the theory is shewn from the circumstance that no
account is taken of the equal and opposite friction of the ball on
the fluid.

Again, suppose that water is flowing down a straight aqueduct
of uniform slope, what will be the discharge corresponding to
a given slope, and a given form of the bed ? Of what magnitude
must an aqueduct be, in order to supply a given place with
a given quantity of water? Of what form must it be, in order
to ensure a given supply of water with the least expense of
materials in the construction ? These, and similar questions are
wholly out of the reach of the common theory of Fluid Motion,
since they entirely depend on the laws of the transmission of that
tangential action which in it is wholly neglected. In fact, according to the common theory the water ought to flow on with
uniformly accelerated velocity; for even the supposition of a
certain friction against the bed would be of no avail, for such
friction could not be transmitted through the mass. The practical
importance of such questions as those above mentioned has made
them the object of numerous experiments, from which empirical
formulæ have been constructed. But such formulæ, although
fulfilling well enough the purposes for which they were constructed, can hardly be considered as affording us any material
insight into the laws of nature; nor will they enable us to pass
from the consideration of the phenomena from which they were
derived to that of others of a different class, although depending
on the same causes.

In reflecting on the principles according to which the motion
of a fluid ought to be calculated when account is taken of the
tangential force, and consequently the pressure not supposed the
same in all directions, I was led to construct the theory explained
in the first section of this paper, or at least the main part of it,
which consists of equations (13), and of the principles on which

they are formed. I afterwards found that Poisson had written a memoir on the same subject, and on referring to it I found that he had arrived at the same equations. The method which he employed was however so different from mine that I feel justified in laying the latter before this Society *. The leading principles of my theory will be found in the hypotheses of Art. 1, and in Art. 3.

The second section forms a digression from the main object of this paper, and at first sight may appear to have little connexion with it. In this section I have, I think, succeeded in shewing that Lagrange's proof of an important theorem in the ordinary theory of Hydrodynamics is untenable. The theorem to which I refer is the one of which the object is to shew that $udx+vdy+wdz$, (using the common notation,) is always an exact differential when it is so at one instant. I have mentioned the principles of M. Cauchy's proof, a proof, I think, liable to no sort of objection. I have also given a new proof of the theorem, which would have served to establish it had M. Cauchy not been so fortunate as to obtain three first integrals of the general equations of motion. As it is, this proof may possibly be not altogether useless.

Poisson, in the memoir to which I have referred, begins with establishing, according to his theory, the equations of equilibrium and motion of elastic solids, and makes the equations of motion of fluids depend on this theory. On reading his memoir, I was led to apply to the theory of elastic solids principles precisely analogous to those which I have employed in the case of fluids. The formation of the equations, according to these principles, forms the subject of Sect. III.

The equations at which I have thus arrived contain two arbitrary constants, whereas Poisson's equations contain but one. In Sect. IV. I have explained the principles of Poisson's theories of elastic solids, and of the motion of fluids, and pointed out what appear to me serious objections against the truth of one of the hypotheses which he employs in the former. This theory seems to be very generally received, and in consequence it is usual to deduce the measure of the cubical compressibility of elastic solids from that of their extensibility, when formed into rods or wires,

---

* The same equations have also been obtained by Navier in the case of an incompressible fluid (*Mém. de l'Académie*, t. VI. p. 389), but his principles differ from mine still more than do Poisson's.

or from some quantity of the same nature. If the views which I have explained in this section be correct, the cubical compressibility deduced in this manner is too great, much too great in the case of the softer substances, and even the softer metals. The equations of Sect. III. have, I find, been already obtained by M. Cauchy in his *Exercises Mathématiques*, except that he has not considered the effect of the heat developed by sudden compression. The method which I have employed is different from his, although in some respects it much resembles it.

The equations of motion of elastic solids given in Sect. III. are the same as those to which different authors have been led, as being the equations of motion of the luminiferous ether in vacuum. It may seem strange that the same equations should have been arrived at for cases so different; and I believe this has appeared to some a serious objection to the employment of those equations in the case of light. I think the reflections which I have made at the end of Sect. IV., where I have examined the consequences of the law of continuity, a law which seems to pervade nature, may tend to remove the difficulty.

SECTION I.

Explanation of the Theory of Fluid Motion proposed. Formation of the Differential Equations. Application of these Equations to a few simple cases.

1.   Before entering on the explanation of this theory, it will be necessary to define, or fix the precise meaning of a few terms which I shall have occasion to employ.

In the first place, the expression "the velocity of a fluid at any particular point" will require some notice. If we suppose a fluid to be made up of ultimate molecules, it is easy to see that these molecules must, in general, move among one another in an irregular manner, through spaces comparable with the distances between them, when the fluid is in motion. But since there is no doubt that the distance between two adjacent molecules is quite insensible, we may neglect the irregular part of the velocity, compared with the common velocity with which all the molecules in the neighbourhood of the one considered are moving. Or, we may consider the mean velocity of the molecules in the neighbourhood of the one considered, apart from the velocity due to the irregular motion. It is this regular velocity which I shall understand by the *velocity of a fluid at any point*, and I shall accordingly regard it as varying continuously with the co-ordinates of the point.

Let P be any material point in the fluid, and consider the instantaneous motion of a very small element E of the fluid about P. This motion is compounded of a motion of translation, the same as that of P, and of the motion of the several points of E relatively to P. If we conceive a velocity equal and opposite to that of P impressed on the whole element, the remaining velocities form what I shall call the *relative velocities* of the points of the fluid about P; and the motion expressed by these velocities is what I shall call the *relative motion* in the neighbourhood of P.

It is an undoubted result of observation that the molecular forces, whether in solids, liquids, or gases, are forces of enormous intensity, but which are sensible at only insensible distances. Let E' be a very small element of the fluid circumscribing E, and of a thickness greater than the distance to which the molecular forces are sensible. The forces acting on the element E are the external forces, and the pressures arising from the molecular action of E'. If the molecules of E were in positions in which they could remain at rest if E were acted on by no external force and the molecules of E' were held in their actual positions, they would be in what I shall call a state of *relative equilibrium*. Of course they may be far from being in a state of actual equilibrium. Thus, an element of fluid at the top of a wave may be sensibly in a state of relative equilibrium, although far removed from its position of equilibrium. Now, in consequence of the intensity of the molecular forces, the pressures arising from the molecular action on E will be very great compared with the external moving forces acting on E. Consequently the state of relative equilibrium, or of relative motion, of the molecules of E will not be sensibly affected by the external forces acting on E. But the pressures in different directions about the point P depend on that state of relative equilibrium or motion, and consequently will not be sensibly affected by the external moving forces acting on E. For the same reason they will not be sensibly affected by any motion of rotation common to all the points of E; and it is a direct consequence of the second law of motion, that they will

not be affected by any motion of translation common to the whole
element. If the molecules of E were in a state of relative equilibrium, the pressure would be equal in all directions about P,
as in the case of fluids at rest. Hence I shall assume the following principle :—

*That the difference between the pressure on a plane in a given
direction passing through any point* P *of a fluid in motion and the
pressure which would exist in all directions about* P *if the fluid in
its neighbourhood were in a state of relative equilibrium depends
only on the relative motion of the fluid immediately about* P ; *and
that the relative motion due to any motion of rotation may be elimi-
nated without affecting the differences of the pressures above mentioned.*

Let us see how far this principle will lead us when it is
carried out.

2. It will be necessary now to examine the nature of the
most general instantaneous motion of an element of a fluid.
The proposition in this article is however purely geometrical, and
may be thus enunciated :—" Supposing space, or any portion of
space, to be filled with an infinite number of points which move
in any continuous manner, retaining their identity, to examine
the nature of the instantaneous motion of any elementary portion
of these points."

Let u, v, w be the resolved parts, parallel to the rectangular
axes, Ox, Oy, Oz, of the velocity of the point P, whose co-ordinates
at the instant considered are x, y, z. Then the relative velocities
at the point P', whose co-ordinates are $x + x'$, $y + y'$, $z + z'$, will be

$$\frac{du}{dx} x' + \frac{du}{dy} y' + \frac{du}{dz} z' \text{ parallel to } x,$$

$$\frac{dv}{dx} x' + \frac{dv}{dy} y' + \frac{dv}{dz} z' \quad\ldots\ldots\ldots\ldots y,$$

$$\frac{dw}{dx} x' + \frac{dw}{dy} y' + \frac{dw}{dz} z' \quad\ldots\ldots\ldots\ldots z,$$

neglecting squares and products of x', y', z'. Let these velocities
be compounded of those due to the angular velocities ω', ω'', ω'''
about the axes of x, y, z, and of the velocities U, V, W parallel

to x, y, z. The linear velocities due to the angular velocities being $\omega''z' - \omega'''y'$, $\omega'''x' - \omega'z'$, $\omega'y' - \omega''x'$ parallel to the axes of x, y, z, we shall therefore have

$$U = \frac{du}{dx}x' + \left(\frac{du}{dy} + \omega'''\right)y' + \left(\frac{du}{dz} - \omega''\right)z',$$

$$V = \left(\frac{dv}{dx} - \omega'''\right)x' + \frac{dv}{dy}y' + \left(\frac{dv}{dz} + \omega'\right)z',$$

$$W = \left(\frac{dw}{dx} + \omega''\right)x' + \left(\frac{dw}{dy} - \omega'\right)y' + \frac{dw}{dz}z'.$$

Since ω', ω'', ω''' are arbitrary, let them be so assumed that

$$\frac{dU}{dy'} = \frac{dV}{dx'}, \quad \frac{dV}{dz'} = \frac{dW}{dy'}, \quad \frac{dW}{dx'} = \frac{dU}{dz'},$$

which gives

$$\omega' = \tfrac{1}{2}\left(\frac{dw}{dy} - \frac{dv}{dz}\right), \quad \omega'' = \tfrac{1}{2}\left(\frac{du}{dz} - \frac{dw}{dx}\right), \quad \omega''' = \tfrac{1}{2}\left(\frac{dv}{dx} - \frac{du}{dy}\right), \quad ...(1),$$

$$U = \frac{du}{dx}x' + \tfrac{1}{2}\left(\frac{du}{dy} + \frac{dv}{dx}\right)y' + \tfrac{1}{2}\left(\frac{du}{dz} + \frac{dw}{dx}\right)z', \Big]$$
$$V = \tfrac{1}{2}\left(\frac{dv}{dx} + \frac{du}{dy}\right)x' + \frac{dv}{dy}y' + \tfrac{1}{2}\left(\frac{dv}{dz} + \frac{dw}{dy}\right)z', \Big\} \quad ......(2).$$
$$W = \tfrac{1}{2}\left(\frac{dw}{dx} + \frac{du}{dz}\right)x' + \tfrac{1}{2}\left(\frac{dw}{dy} + \frac{dv}{dz}\right)y' + \frac{dw}{dz}z', \Big]$$

The quantities ω', ω'', ω''' are what I shall call the *angular velocities of the fluid* at the point considered. This is evidently an allowable definition, since, in the particular case in which the element considered moves as a solid might do, these quantities coincide with the angular velocities considered in rigid dynamics. A further reason for this definition will appear in Sect. III.

Let us now investigate whether it is possible to determine x', y', z' so that, considering only the relative velocities U, V, W, the line joining the points P, P' shall have no angular motion. The conditions to be satisfied, in order that this may be the case, are evidently that the increments of the relative co-ordinates x', y', z' of the second point shall be ultimately proportional to those co-ordinates. If e be the rate of extension of the line joining the two points considered, we shall therefore have

$$\begin{aligned} Fx' + hy' + gz' &= ex', \Big] \\ hx' + Gy' + fz' &= ey', \Big\} \quad ......(3); \\ gx' + fy' + Hz' &= ez', \Big] \end{aligned}$$

6

where

$$F = \frac{du}{dx}, \quad G = \frac{dv}{dy}, \quad H = \frac{dw}{dz}, \quad 2f = \frac{dv}{dz} + \frac{dw}{dy},$$

$$2g = \frac{dw}{dx} + \frac{du}{dz}, \quad 2h = \frac{du}{dy} + \frac{dv}{dx}.$$

If we eliminate from equations (3) the two ratios which exist between the three quantities x', y', z', we get the well known cubic equation

$$(e-F)(e-G)(e-H) - f^2(e-F) - g^2(e-G) - h^2(e-H) - 2fgh = 0 \ldots (4),$$

which occurs in the investigation of the principal axes of a rigid body, and in various others. As in these investigations, it may be shewn that there are in general three directions, at right angles to each other, in which the point P' may be situated so as to satisfy the required conditions. If two of the roots of (4) are equal, there is one such direction corresponding to the third root, and an infinite number of others situated in a plane perpendicular to the former; and if the three roots of (4) are equal, a line drawn in any direction will satisfy the required conditions.

The three directions which have just been determined I shall call *axes of extension*. They will in general vary from one point to another, and from one instant of time to another. If we denote the three roots of (4) by e', e'', e''', and if we take new rectangular axes $Ox_{,}$, $Oy_{,}$, $Oz_{,}$, parallel to the axes of extension, and denote by $u_{,}$, $U_{,}$, &c. the quantities referred to these axes corresponding to u, U, &c., equations (3) must be satisfied by $y_{,}' = 0$, $z_{,}' = 0$, $e = e'$, by $x_{,}' = 0$, $z_{,}' = 0$, $e = e''$, and by $x_{,}' = 0$, $y_{,}' = 0$, $e = e'''$, which requires that $f_{,} = 0$, $g_{,} = 0$, $h_{,} = 0$, and we have

$$e' = F_{,} = \frac{du_{,}}{dx_{,}}, \quad e'' = G_{,} = \frac{dv_{,}}{dy_{,}}, \quad e'' = H_{,} = \frac{dw_{,}}{dz_{,}}.$$

The values of $U_{,}$, $V_{,}$, $W_{,}$ which correspond to the residual motion after the elimination of the motion of rotation corresponding to ω', ω'' and ω''', are

$$U_{,} = e'x_{,}', \quad V_{,} = e''y_{,}', \quad W_{,} = e'''z_{,}'.$$

The angular velocity of which ω', ω'', ω''' are the components is independent of the arbitrary directions of the co-ordinate axes: the same is true of the directions of the axes of extension, and of the values of the roots of equation (4). This might be proved in various ways; perhaps the following is the simplest. The conditions by which ω', ω'', ω''' are determined are those which express that the relative velocities U, V, W, which remain after eliminating a certain angular velocity, are such that $U dx' + V dy' + W dz'$ is ultimately an exact differential, that is to say when squares and products of x', y' and z' are neglected. It appears moreover from the solution that there is only one way in which these conditions can be satisfied for a given system of co-ordinate axes. Let us take new rectangular axes, Ox, Oy, Oz, and let U, V, W be the resolved parts along these axes of the velocities U, V, W, and x', y', z', the relative co-ordinates of P'; then

$$U = \mathrm{U} \cos x\mathrm{x} + \mathrm{V} \cos x\mathrm{y} + \mathrm{W} \cos x\mathrm{z},$$

$$dx' = \cos x\mathrm{x}\, d\mathrm{x}' + \cos x\mathrm{y}\, d\mathrm{y}' + \cos x\mathrm{z}\, d\mathrm{z}', \ \&c.;$$

whence, taking account of the well known relations between the cosines involved in these equations, we easily find

$$U dx' + V dy' + W dz' = \mathrm{U}\, d\mathrm{x}' + \mathrm{V}\, d\mathrm{y}' + \mathrm{W}\, d\mathrm{z}'.$$

It appears therefore that the relative velocities U, V, W, which remain after eliminating a certain angular velocity, are such that $\mathrm{U}\, d\mathrm{x}' + \mathrm{V}\, d\mathrm{y}' + \mathrm{W}\, d\mathrm{z}'$ is ultimately an exact differential. Hence the values of U, V, W are the same as would have been obtained from equations (2) applied directly to the new axes, whence the truth of the proposition enunciated at the head of this paragraph is manifest.

The motion corresponding to the velocities $U_{,}$, $V_{,}$, $W_{,}$ may be further decomposed into a motion of dilatation, positive or negative, which is alike in all directions, and two motions which I shall call *motions of shifting*, each of the latter being in two dimensions, and not affecting the density. For let δ be the rate of linear extension corresponding to a uniform dilatation; let $\sigma x_{,}' - \sigma y_{,}'$ be the velocities parallel to $x_{,}, y_{,}$, corresponding to a motion of shifting parallel to the plane $x_{,}y_{,}$, and let $\sigma' x_{,}', - \sigma' z_{,}'$ be the velocities parallel to $x_{,}, z_{,}$, corresponding to a similar motion of shifting parallel to the plane $x_{,}z_{,}$. The velocities parallel to $x_{,}, y_{,}, z_{,}$ respectively corresponding to the quantities δ, σ and σ' will be $(\delta + \sigma + \sigma')x_{,}'$, $(\delta - \sigma)y_{,}'$, $(\delta - \sigma')z_{,}'$, and equating these to $U_{,}$, $V_{,}$, $W_{,}$ we shall get

$$\delta = \tfrac{1}{3}(e' + e'' + e'''), \quad \sigma = \tfrac{1}{3}(e' + e''' - 2e''), \quad \sigma' = \tfrac{1}{3}(e' + e'' - 2e''').$$

Hence the most general instantaneous motion of an elementary portion of a fluid is compounded of a motion of translation, a motion of rotation, a motion of uniform dilatation, and two motions of shifting of the kind just mentioned.

3. Having determined the nature of the most general instantaneous motion of an element of a fluid, we are now prepared to consider the normal pressures and tangential forces called into play by the relative displacements of the particles. Let p be the pressure which would exist about the point P if the neighbouring molecules were in a state of relative equilibrium: let $p + p_{,}$ be the normal pressure, and $t_{,}$ the tangential action, both referred to a unit of surface, on a plane passing through P and having a given direction. By the hypotheses of Art. 1. the quantities $p_{,}$, $t_{,}$ will be independent of the angular velocities ω', ω'', ω''', depending only on the residual relative velocities U, V, W, or, which comes to the same, on e', e'' and e''', or on σ, σ' and δ. Since this residual motion is symmetrical with respect to the axes of extension, it follows that if the plane considered is perpendicular to any one of these axes the tangential action on it is zero, since there is no reason why it should act in one direction rather than in the opposite; for by the hypotheses of Art. 1 the change of density and temperature about the point P is to be neglected, the constitution of the fluid being ultimately uniform about that point. Denoting then by $p + p'$, $p + p''$, $p + p'''$ the pressures on planes perpendicular to the axes of $x_{,}$, $y_{,}$, $z_{,}$, we must have

$$p' = \phi(e', e'', e'''), \quad p'' = \phi(e'', e''', e'), \quad p''' = \phi(e''', e', e''),$$

$\phi(e', e'', e''')$ denoting a function of e', e'' and e''' which is symmetrical with respect to the two latter quantities. The question is now to determine, on whatever may seem the most probable hypothesis, the form of the function ϕ.

Let us first take the simpler case in which there is no dilatation, and only one motion of shifting, or in which $e'' = -e'$, $e''' = 0$, and let us consider what would take place if the fluid consisted of smooth molecules acting on each other by actual contact. On this supposition, it is clear, considering the magnitude of the pressures acting on the molecules compared with their masses, that they would be sensibly in a position of relative equilibrium, except when the equilibrium of any one of them became impossible from the displacement of the adjoining ones, in which case the molecule in question would start into a new position of equilibrium. This start would cause a corresponding displacement in the molecules immediately about the one which started, and this disturbance would be propagated immediately in all directions, the nature of the displacement however being different in different directions, and would soon become insensible. During the continuance of this disturbance, the pressure on a small plane drawn through the element considered would not be the same in all directions, nor normal to the plane: or, which comes to the same, we may suppose a uniform normal pressure p to act, together with a normal pressure $p_{,,}$, and a tangential force $t_{,,}$, $p_{,,}$ and $t_{,,}$ being forces of great intensity and short duration, that is being of the nature of impulsive forces. As the number of molecules comprised in the element considered has been supposed extremely great, we may take a time τ so short that all summations with respect to such intervals of time may be replaced without sensible error by integrations, and yet so long that a very great number of starts shall take place in it. Consequently we have only to consider the average effect of such starts, and moreover we may without sensible error replace the impulsive forces such as $p_{,,}$ and $t_{,,}$, which succeed one another with great rapidity, by continuous forces. For planes perpendicular to the axes of extension these continuous forces will be the normal pressures p', p'', p'''.

Let us now consider a motion of shifting differing from the former only in having e' increased in the ratio of m to 1. Then, if we suppose each start completed before the starts which would be sensibly affected by it are begun, it is clear that the same series of starts will take place in the second case as in the first, but at intervals of time which are less in the ratio of 1 to m. Consequently the continuous pressures by which the impulsive actions due to these starts may be replaced must be increased in the ratio of m to 1. Hence the pressures p', p'', p''' must be proportional to e', or we must have

$$p' = Ce', \quad p'' = C'e', \quad p''' = C''e'.$$

It is natural to suppose that these formulæ hold good for negative as well as positive values of e'. Assuming this to be true, let the sign of e' be changed. This comes to interchanging x and y, and consequently p''' must remain the same, and p' and p'' must be interchanged. We must therefore have $C'' = 0$, $C' = -C$. Putting then $C = -2\mu$ we have,

$$p' = -2\mu e', \quad p'' = 2\mu e', \quad p''' = 0.$$

It has hitherto been supposed that the molecules of a fluid are in actual contact. We have every reason to suppose that this is not the case. But precisely the same reasoning will apply if they are separated by intervals as great as we please compared with their magnitudes, provided only we suppose the force of restitution called into play by a small displacement of *any one* molecule to be very great.

Let us now take the case of two motions of shifting which co-exist, and let us suppose $e' = \sigma + \sigma'$, $e'' = -\sigma$, $e''' = -\sigma'$. Let the small time τ be divided into $2n$ equal portions, and let us suppose that in the first interval a shifting motion corresponding to $e' = 2\sigma$, $e'' = -2\sigma$ takes place parallel to the plane $x y_{,}$, and that in the second interval a shifting motion corresponding to $e' = 2\sigma'$, $e''' = -2\sigma'$ takes place parallel to the plane $x z_{,}$, and so on alternately. On this supposition it is clear that if we suppose the time $\tau/2n$ to be extremely small, the continuous forces by which the effect of the starts may be replaced will be $p' = -2\mu(\sigma + \sigma')$, $p'' = 2\mu\sigma$, $p''' = 2\mu\sigma'$. By supposing n indefinitely increased, we might make the motion considered approach as near as we please to that in which the two motions of shifting coexist; but we are not at liberty to do so, for in order to apply the above reasoning we must suppose the time $\tau/2n$ to be so large that the average effect of the starts which occur in it may be taken. Consequently it must be taken as an additional assumption, and not a matter of absolute demonstration, that the effects of the two motions of shifting are superimposed.

Hence if $\delta = 0$, *i.e.* if $e' + e'' + e''' = 0$, we shall have in general

$$p' = -2\mu e', \quad p'' = -2\mu e'', \quad p''' = -2\mu e''' \dots\dots\dots(5).$$

It was by this hypothesis of starts that I first arrived at these equations, and the differential equations of motion which result from them. On reading Poisson's memoir however, to which I shall have occasion to refer in Section IV., I was led to reflect that however intense we may suppose the molecular forces to be, and however near we may suppose the molecules to be to their positions of relative equilibrium, we are not therefore at liberty to suppose them *in* those positions, and consequently not at liberty to suppose the pressure equal in all directions in the intervals of time between the starts. In fact, by supposing the molecular forces indefinitely increased, retaining the same ratios to each other, we may suppose the displacements of the molecules from their positions of relative equilibrium indefinitely diminished, but on the other hand the force of restitution called into action by a given displacement is indefinitely increased in the same proportion. But be these displacements what they may, we know that the forces of restitution make equilibrium with forces equal and opposite to the effective forces ; and in calculating the effective forces we may neglect the above displacements, or suppose the molecules to move in the paths in which they would move if the shifting motion took place with indefinite slowness. Let us first consider a single motion of shifting, or one for which $e'' = -e'$, $e''' = 0$, and let p_i and t_i denote the same quantities as before. If we now suppose e' increased in the ratio of m to 1, all the effective forces will be increased in that ratio, and consequently p_i and t_i will be increased in the same ratio. We may deduce the values of p' p'', and p''' just as before, and then pass by the same reasoning to the case of two motions of shifting which coexist, only that in this case the reasoning will be demonstrative, since we *may* suppose the time $\tau/2n$ indefinitely diminished. If we suppose the state of things considered in this paragraph to exist along with the motions of starting already considered, it is easy to see that the expressions for p', p'' and p''' will still retain the same form.

There remains yet to be considered the effect of the dilatation. Let us first suppose the dilatation to exist without any shifting : then it is easily seen that the relative motion of the fluid at the point considered is the same in all directions. Consequently the only effect which such a dilatation could have would be to introduce a normal pressure p_i, alike in all directions, in addition to that due to the action of the molecules supposed to be in a state of relative equilibrium. Now the pressure p_i could only arise from the aggregate of the molecular actions called into play by the displacements of the molecules from their positions of relative equilibrium; but since these displacements take place, on an average, indifferently in all directions, it follows that the actions of which p_i is composed neutralize each other, so that $p_i = 0$. The same conclusion might be drawn from the hypothesis of starts, supposing, as it is natural to do, that each start calls into action as much increase of pressure in some directions as diminution of pressure in others.

If the motion of uniform dilatation coexists with two motions of shifting, I shall suppose, for the same reason as before, that the effects of these different motions are superimposed. Hence subtracting δ from each of the three quantities e', e'' and e''', and putting the remainders in the place of e', e'' and e''' in equations (5), we have

$$p' = \tfrac{2}{3}\mu(e'' + e''' - 2e'), \quad p'' = \tfrac{2}{3}\mu(e''' + e' - 2e''),$$
$$p''' = \tfrac{2}{3}\mu(e' + e'' - 2e''') \dots\dots\dots\dots(6).$$

If we had started with assuming $\phi(e', e'', e''')$ to be a linear function of e', e'' and e''', avoiding all speculation as to the molecular constitution of a fluid, we should have had at once $p' = Ce' + C'(e'' + e''')$, since p' is symmetrical with respect to e'' and e'''; or, changing the constants, $p' = \tfrac{2}{3}\mu(e'' + e''' - 2e') + \kappa(e' + e'' + e''')$. The expressions for p'' and p''' would be obtained by interchanging the requisite quantities. Of course we may at once put $\kappa = 0$ if we assume that in the case of a uniform motion of dilatation the pressure at any instant depends only on the actual density and temperature at that instant, and not on the rate at which the former changes with the time. In most cases to which it would be interesting to apply the theory of the friction of fluids the density of the fluid is either constant, or may without sensible error be regarded as constant, or else changes slowly with the time. In the first two cases the results would be the same, and in the third case nearly the same, whether κ were equal to zero or not. Consequently, if theory and experiment should in such cases agree, the experiments must not be regarded as confirming that part of the theory which relates to supposing κ to be equal to zero.

4. It will be easy now to determine the oblique pressure, or resultant of the normal pressure and tangential action, on any plane. Let us first consider a plane drawn through the point P parallel to the plane yz. Let $Ox_{,}$ make with the axes of x, y, z angles whose cosines are l', m', n'; let l'', m'', n'' be the same for $Oy_{,}$, and l''', m''', n''' the same for $Oz_{,}$. Let P_1 be the pressure, and (xty), (xtz) the resolved parts, parallel to y, z respectively, of the tangential force on the plane considered, all referred to a unit of surface, (xty) being reckoned positive when the part of the fluid towards $-x$ urges that towards $+x$ in the positive direction of y, and similarly for (xtz). Consider the portion of the fluid comprised within a tetrahedron having its vertex in the point P, its base parallel to the plane yz, and its three sides parallel to the

planes $x_{,}y_{,}$, $y_{,}z_{,}$, $z_{,}x_{,}$ respectively.   Let A be the area of the base, and therefore $l'A$, $l''A$, $l'''A$ the areas of the faces perpendicular to the axes of $x_{,}$, $y_{,}$, $z_{,}$.   By D'Alembert's principle, the pressures and tangential actions on the faces of this tetrahedron, the moving forces arising from the external attractions, not including the molecular forces, and forces equal and opposite to the effective moving forces will be in equilibrium, and therefore the sums of the resolved parts of these forces in the directions of x, y and z will each be zero.   Suppose now the dimensions of the tetrahedron indefinitely diminished, then the resolved parts of the external, and of the effective moving forces will vary ultimately as the cubes, and those of the pressures and tangential forces on the sides as the squares of homologous lines.   Dividing therefore the three equations arising from equating to zero the resolved parts of the above forces by A, and taking the limit, we have

$$P_1 = \Sigma l'^2\,(p+p'), \qquad (xty) = \Sigma l'm'\,(p+p'), \qquad (xtz) = \Sigma l'n'\,(p+p'),$$

the sign Σ denoting the sum obtained by taking the quantities corresponding to the three axes of extension in succession.   Putting for p', p'' and p''' their values given by (6), putting $e'+e''+e'''=3\delta$, and observing that $\Sigma l'^2 = 1$, $\Sigma l'm' = 0$, $\Sigma l'n' = 0$, the above equations become

$$P_1 = p - 2\mu\Sigma l'^2 e' + 2\mu\delta, \qquad (xty) = -2\mu\Sigma l'm'e', \qquad (xtz) = -2\mu\Sigma l'n'e'.$$

The method of determining the pressure on any plane from the pressures on three planes at right angles to each other, which has just been given, has already been employed by MM. Cauchy and Poisson.

The most direct way of obtaining the values of $\Sigma l'^2 e'$ &c. would be to express l', m' and n' in terms of e' by any two of equations (3), in which x', y', z' are proportional to l', m', n', together with the equation $l'^2 + m'^2 + n'^2 = 1$, and then to express the resulting symmetrical function of the roots of the cubic equation (4) in terms of the coefficients.   But this method would be excessively laborious, and need not be resorted to.   For after eliminating the angular motion of the element of fluid considered the remaining velocities are $e'x'_{,}$, $e''y'_{,}$, $e'''z'_{,}$, parallel to the axes of $x_{,}$, $y_{,}$, $z_{,}$.   The sum of the resolved parts of these parallel to the axis of x is $l'e'x'_{,} + l''e''y'_{,} + l'''e'''z'_{,}$.   Putting for $x'_{,}$, $y'_{,}$, $z'_{,}$ their values $l'x' + m'y' + n'z'$ &c., the above sum becomes

$$x'\Sigma l'^2 e' + y'\Sigma l'm'e' + z'\Sigma l'n'e' \,;$$

but this sum is the same thing as the velocity U in equation (2), and therefore we have

$$\Sigma l'^2 e' = \frac{du}{dx}, \quad \Sigma l'm'e' = \tfrac{1}{2}\left(\frac{du}{dy} + \frac{dv}{dx}\right), \quad \Sigma l'n'e' = \tfrac{1}{2}\left(\frac{du}{dz} + \frac{dw}{dx}\right).$$

It may also be very easily proved directly that the value of 3δ, the rate of cubical dilatation, satisfies the equation

$$3\delta = \frac{du}{dx} + \frac{dv}{dy} + \frac{dw}{dz} \quad\dots\dots\dots\dots\dots (7).$$

Let P_2, (ytz), (ytx) be the quantities referring to the axis of y, and P_3, (ztx), (zty) those referring to the axis of z, which correspond to P_1 &c. referring to the axis of x. Then we see that $(ytz) = (zty)$, $(ztx) = (xtz)$, $(xty) = (ytx)$. Denoting these three quantities by T_1, T_2, T_3, and making the requisite substitutions and interchanges, we have

$$\left.\begin{aligned}
P_1 &= p - 2\mu\left(\frac{du}{dx} - \delta\right), \\[2mm]
P_2 &= p - 2\mu\left(\frac{dv}{dy} - \delta\right), \\[2mm]
P_3 &= p - 2\mu\left(\frac{dw}{dz} - \delta\right), \\[2mm]
T_1 &= -\mu\left(\frac{dv}{dz} + \frac{dw}{dy}\right), \\[2mm]
T_2 &= -\mu\left(\frac{dw}{dx} + \frac{du}{dz}\right), \\[2mm]
T_3 &= -\mu\left(\frac{du}{dy} + \frac{dv}{dx}\right),
\end{aligned}\right\} \quad\dots\dots\dots (8).$$

It may also be useful to know the components, parallel to x, y, z, of the oblique pressure on a plane passing through the point P, and having a given direction. Let l, m, n be the cosines of the angles which a normal to the given plane makes with the axes of x, y, z; let P, Q, R be the components, referred to a unit of surface, of the oblique pressure on this plane, P, Q, R being reckoned positive when the part of the fluid in which is situated the normal to which l, m and n refer is urged by the other part in the positive directions of x, y, z, when l, m and n are positive. Then considering as before a tetrahedron of which the base is parallel to the given plane, the vertex in the point P, and the sides parallel to the co-ordinate planes, we shall have

$$\left.\begin{array}{c} P = lP_1 + mT_3 + nT_2, \\ Q = lT_3 + mP_2 + nT_1, \\ R = lT_2 + mT_1 + nP_3, \end{array}\right\} \quad \dots\dots\dots\dots (9).$$

In the simple case of a sliding motion for which $u = 0$, $v = f(x)$, $w = 0$, the only forces, besides the pressure p, which act on planes parallel to the co-ordinate planes are the two tangential forces T_3, the value of which in this case is $-\mu \, dv/dx$. In this case it is easy to shew that the axes of extension are, one of them parallel to Oz, and the two others in a plane parallel to xy, and inclined at angles of $45°$ to Ox. We see also that it is necessary to suppose μ to be positive, since otherwise the tendency of the forces would be to increase the relative motion of the parts of the fluid, and the equilibrium of the fluid would be unstable.

5.   Having found the pressures about the point P on planes parallel to the co-ordinate planes, it will be easy to form the equations of motion. Let X, Y, Z be the resolved parts, parallel to the axes, of the external force, not including the molecular force ; let ρ be the density, t the time. Consider an elementary parallelepiped of the fluid, formed by planes parallel to the co-ordinate planes, and drawn through the point (x, y, z) and the point $(x + \Delta x, \ y + \Delta y, \ z + \Delta z)$. The mass of this element will be ultimately $\rho \Delta x \Delta y \Delta z$, and the moving force parallel to x arising from the external forces will be ultimately $\rho X \Delta x \Delta y \Delta z$; the effective moving force parallel to x will be ultimately $\rho \, Du/Dt \, . \, \Delta x \Delta y \Delta z$, where D is used, as it will be in the rest of this paper, to denote differentiation in which the independent variables are t and three parameters of the particle considered, (such for instance as its initial cordinates,) and not t, x, y, z. It is easy also to shew that the moving force acting on the element considered arising from the oblique pressures on the faces is ultimately

$$\left(\frac{dP}{dx} + \frac{dT_3}{dy} + \frac{dT_2}{dz}\right) \Delta x \, \Delta y \, \Delta z,$$

acting in the negative direction. Hence we have by D'Alembert's principle

$$\rho \left(\frac{Du}{Dt} - X\right) + \frac{dP_1}{dx} + \frac{dT_3}{dy} + \frac{dT_2}{dz} = 0, \ \&c. \ \dots\dots(10),$$

in which equations we must put for Du/Dt its value

$$\frac{du}{dt} + u\frac{du}{dx} + v\frac{du}{dy} + w\frac{du}{dz},$$

and similarly for Dv/dt and Dw/dt. In considering the general equations of motion it will be needless to write down more than one, since the other two may be at once derived from it by interchanging the requisite quantities. The equations (10), the ordinary equation of continuity, as it is called,

$$\frac{d\rho}{dt} + \frac{d\rho u}{dx} + \frac{d\rho v}{dy} + \frac{d\rho w}{dz} = 0 \quad \dots\dots\dots\dots(11),$$

which expresses the condition that there is no generation or destruction of mass in the interior of a fluid, the equation connecting p and ρ, or in the case of an incompressible fluid the equivalent equation $D\rho/Dt = 0$, and the equation for the propagation of heat, if we choose to take account of that propagation, are the only equations to be satisfied at every point of the interior of the fluid mass.

As it is quite useless to consider cases of the utmost degree of generality, I shall suppose the fluid to be homogeneous, and of a uniform temperature throughout, except in so far as the temperature may be raised by sudden compression in the case of small vibrations. Hence in equations (10) μ may be supposed to be constant as far as regards the temperature; for, in the case of small vibrations, the terms introduced by supposing it to vary with the temperature would involve the square of the velocity, which is supposed to be neglected. If we suppose μ to be independent of the pressure also, and substitute in (10) the values of P_1 &c. given by (8), the former equations become

$$\rho\left(\frac{Du}{Dt} - X\right) + \frac{dp}{dx} - \mu\left(\frac{d^2u}{dx^2} + \frac{d^2u}{dy^2} + \frac{d^2u}{dz^2}\right)$$

$$-\frac{\mu}{3}\frac{d}{dx}\left(\frac{du}{dx} + \frac{dv}{dy} + \frac{dw}{dz}\right) = 0, \&c. \dots\dots (12).$$

Let us now consider in what cases it is allowable to suppose μ to be independent of the pressure. It has been concluded by Dubuat, from his experiments on the motion of water in pipes and canals, that the total retardation of the velocity due to friction is not increased by increasing the pressure. The total retardation depends, partly on the friction of the water against the sides of the pipe or canal, and partly on the mutual friction, or tangential action, of the different portions of the water. Now if these two parts of the whole retardation were separately variable with p, it is very unlikely that they should when combined give a result independent of p. The amount of the internal friction of the water depends on the value of μ. I shall therefore suppose that for water, and by analogy for other incompressible fluids, μ is independent of the pressure. On this supposition, we have from equations (11) and (12)

$$\rho \left(\frac{Du}{Dt} - X \right) + \frac{dp}{dx} - \mu \left(\frac{d^2u}{dx^2} + \frac{d^2u}{dy^2} + \frac{d^2u}{dz^2} \right) = 0, \&c....(13),$$

$$\frac{du}{dx} + \frac{dv}{dy} + \frac{dw}{dz} = 0.$$

These equations are applicable to the determination of the motion of water in pipes and canals, to the calculation of the effect of friction on the motions of tides and waves, and such questions.

If the motion is very small, so that we may neglect the square of the velocity, we may put $Du/Dt = du/dt$, &c. in equations (13). The equations thus simplified are applicable to the determination of the motion of a pendulum oscillating in water, or of that of a vessel filled with water and made to oscillate. They are also applicable to the determination of the motion of a pendulum oscillating in air, for in this case we may, with hardly any error, neglect the compressibility of the air.

The case of the small vibrations by which sound is propagated in a fluid, whether a liquid or a gas, is another in which $d\mu/dp$ may be neglected. For in the case of a liquid reasons have been shewn for supposing μ to be independent of p, and in the case of a gas we may neglect $d\mu/dp$, if we neglect the small change in the value of μ, arising from the small variation of pressure due to the forces X, Y, Z.

6. Besides the equations which must hold good at any point in the interior of the mass, it will be necessary to form also the equations which must be satisfied at its boundaries. Let M be a point in the boundary of the fluid. Let a normal to the surface at M, drawn on the outside of the fluid, make with the axes angles whose cosines are l, m, n. Let P', Q', R' be the components

of the pressure of the fluid about M on the solid or fluid with which it is in contact, these quantities being reckoned positive when the fluid considered presses the solid or fluid outside it in the positive directions of x, y, z, supposing l, m and n positive. Let S be a very small element of the surface about M, which will be ultimately plane, S' a plane parallel and equal to S, and directly opposite to it, taken within the fluid. Let the distance between S and S' be supposed to vanish in the limit compared with the breadth of S, a supposition which may be made if we neglect the effect of the curvature of the surface at M; and let us consider the forces acting on the element of fluid comprised between S and S', and the motion of this element. If we suppose equations (8) to hold good to within an insensible distance from the surface of the fluid, we shall evidently have forces ultimately equal to PS, QS, RS, (P, Q and R being given by equations (9),) acting on the inner side of the element in the positive directions of the axes, and forces ultimately equal to $P'S$, $Q'S$, $R'S$ acting on the outer side in the negative directions. The moving forces arising from the external forces acting on the element, and the effective moving forces will vanish in the limit compared with the forces PS, &c.: the same will be true of the pressures acting about the edge of the element, if we neglect capillary attraction, and all forces of the same nature. Hence, taking the limit, we shall have

$$P' = P, \quad Q' = Q, \quad R' = R.$$

The method of proceeding will be different according as the bounding surface considered is a free surface, the surface of a solid, or the surface of separation of two fluids, and it will be necessary to consider these cases separately. Of course the surface of a liquid exposed to the air is really the surface of separation of two fluids, but it may in many cases be regarded as a free surface if we neglect the inertia of the air: it may always be so regarded if we neglect the friction of the air as well as its inertia.

Let us first take the case of a free surface exposed to a pressure Π, which is supposed to be the same at all points, but may vary with the time; and let $L = 0$ be the equation to the surface. In this case we shall have $P' = l\Pi$, $Q' = m\Pi$, $R' = n\Pi$; and putting for P, Q, R their values given by (9), and for P_1 &c. their values given by (8), and observing that in this case $\delta = 0$, we shall have

$$l\left(\Pi - p\right) + \mu \left\{ 2l\,\frac{du}{dx} + m\left(\frac{du}{dy} + \frac{dv}{dx}\right) + n\left(\frac{du}{dz} + \frac{dw}{dx}\right) \right\} = 0,\ \&\text{c}.\ \ldots(14),$$

in which equations l, m, n will have to be replaced by dL/dx, dL/dy, dL/dz, to which they are proportional.

If we choose to take account of capillary attraction, we have only to diminish the pressure Π by the quantity $H\left(\dfrac{1}{r_1} + \dfrac{1}{r_2}\right)$, where H is a positive constant depending on the nature of the fluid, and r_1, r_2, are the principal radii of curvature at the point considered, reckoned positive when the fluid is concave outwards. Equations (14) with the ordinary equation

$$\frac{dL}{dt} + u\,\frac{dL}{dx} + v\,\frac{dL}{dy} + w\,\frac{dL}{dz} = 0 \ldots\ldots\ldots\ldots\ldots(15),$$

are the conditions to be satisfied for points at the free surface. Equations (14) are for such points what the three equations of motion are for internal points, and (15) is for the former what (11) is for the latter, expressing in fact that there is no generation or destruction of fluid at the free surface.

The equations (14) admit of being differently expressed, in a way which may sometimes be useful. If we suppose the origin to be in the point considered, and the axis of z to be the external normal to the surface, we have $l = m = 0$, $n = 1$, and the equations become

$$\frac{dw}{dx} + \frac{du}{dz} = 0,\quad \frac{dw}{dy} + \frac{dv}{dz} = 0,\quad \Pi - p + 2\mu\,\frac{dw}{dz} = 0 \ldots\ldots\ldots(16).$$

The relative velocity parallel to z of a point $(x', y', 0)$ in the free surface, indefinitely near the origin, is $dw/dx\,.\,x' + dw/dy\,.\,y'$: hence we see that dw/dx, dw/dy are the angular velocities, reckoned from x to z and from y to z respectively, of an element of the free surface. Subtracting the linear velocities due to these angular velocities from the relative velocities of the point (x', y', z'), and calling the remaining relative velocities U, V, W, we shall have

$$U = \frac{du}{dx}\, x' + \frac{du}{dy}\, y' + \left(\frac{du}{dz} + \frac{dw}{dx}\right) z',$$

$$V = \frac{dv}{dx}\, x' + \frac{dv}{dy}\, y' + \left(\frac{dv}{dz} + \frac{dw}{dy}\right) z',$$

$$W = \frac{dw}{dz}\, z'.$$

Hence we see that the first two of equations (16) express the conditions that $dU/dz' = 0$ and $dV/dz' = 0$, which are evidently the
conditions to be satisfied in order that there may be no sliding
motion in a direction parallel to the free surface. It would be
easy to prove that these are the conditions to be satisfied in order
that the axis of z may be an axis of extension.

The next case to consider is that of a fluid in contact with a
solid. The condition which first occurred to me to assume for
this case was, that the film of fluid immediately in contact with
the solid did not move relatively to the surface of the solid. I
was led to try this condition from the following considerations.
According to the hypotheses adopted, if there was a very large
relative motion of the fluid particles immediately about any imaginary surface dividing the fluid, the tangential forces called into
action would be very large, so that the amount of relative motion
would be rapidly diminished. Passing to the limit, we might suppose that if at any instant the velocities altered discontinuously
in passing across any imaginary surface, the tangential force called
into action would immediately destroy the finite relative motion
of particles indefinitely close to each other, so as to render the
motion continuous; and from analogy the same might be supposed
to be true for the surface of junction of a fluid and solid. But
having calculated, according to the conditions which I have mentioned, the discharge of long straight circular pipes and rectangular
canals, and compared the resulting formulæ with some of the
experiments of Bossut and Dubuat, I found that the formulæ did
not at all agree with experiment. I then tried Poisson's conditions
in the case of a circular pipe, but with no better success. In fact,
it appears from experiment that the tangential force varies nearly
as the square of the velocity with which the fluid flows past the
surface of a solid, at least when the velocity is not very small. It
appears however from experiments on pendulums that the total

friction varies as the first power of the velocity, and consequently we may suppose that Poisson's conditions, which include as a particular case those which I first tried, hold good for very small velocities. I proceed therefore to deduce these conditions in a manner conformable with the views explained in this paper.

First, suppose the solid at rest, and let $L = 0$ be the equation to its surface. Let M' be a point within the fluid, at an insensible distance h from M. Let ϖ be the pressure which would exist about M if there were no motion of the particles in its neighbourhood, and let $p_{\prime}$ be the additional normal pressure, and $t_{\prime}$ the tangential force, due to the relative velocities of the particles, both with respect to one another and with respect to the surface of the solid. If the motion is so slow that the starts take place independently of each other, on the hypothesis of starts, or that the molecules are very nearly in their positions of relative equilibrium, and if we suppose as before that the effects of different relative velocities are superimposed, it is easy to shew that $p_{\prime}$ and $t_{\prime}$ are linear functions of u, v, w and their differential coefficients with respect to x, y and z; u, v, &c. denoting here the velocities of the fluid about the point M', in the expressions for which however the co-ordinates of M may be used for those of M', since h is neglected. Now the relative velocities about the points M and M' depending on du/dx, &c. are comparable with $du/dx \cdot h$, while those depending on u, v and w are comparable with these quantities, and therefore in considering the action of the fluid on the solid it is only necessary to consider the quantities u, v and w. Now since, neglecting h, the velocity at M' is tangential to the surface at M, u, v, and w are the components of a certain velocity V tangential to the surface. The pressure $p_{\prime}$ must be zero; for changing the signs u, v, and w the circumstances concerned in its production remain the same, whereas its analytical expression changes sign. The tangential force at M will be in the direction of V, and proportional to it, and consequently its components along the axes of x, y, z will be proportional to u, v, w. Reckoning the tangential force positive when, l, m, and n being positive, the solid is urged in the positive directions of x, y, z, the resolved parts of the tangential force will therefore be νu, νv, νw, where ν must evidently be positive, since the effect of the forces must be to check the relative motion of the fluid and solid. The normal pressure of the fluid on the solid being equal to ϖ, its components will be evidently $l\varpi$, $m\varpi$, $n\varpi$.

Suppose now the solid to be in motion, and let u', v', w' be the resolved parts of the velocity of the point M of the solid, and ω', ω'', ω''' the angular velocities of the solid. By hypothesis, the forces by which the pressure at any point differs from the normal pressure due to the action of the molecules supposed to be in a state of relative equilibrium about that point are independent of any velocity of translation or rotation. Supposing then linear and angular velocities equal and opposite to those of the solid impressed both on the solid and on the fluid, the former will be for an instant at rest, and we have only to treat the resulting velocities of the fluid as in the first case. Hence $P' = l\varpi + \nu(u - u')$, &c.; and in the equations (8) we may employ the actual velocities u, v, w, since the pressures P, Q, R are independent of any motion of translation and rotation common to the whole fluid. Hence the equations $P' = P$, &c. gives us

$$l(\varpi - p) + \nu(u - u')$$

$$+ \mu \left\{ 2l \left(\frac{du}{dx} - \delta \right) + m \left(\frac{du}{dy} + \frac{dv}{dx} \right) + n \left(\frac{du}{dz} + \frac{dw}{dx} \right) \right\} = 0, \&c., \quad \ldots \ldots (17),$$

which three equations with (15) are those which must be satisfied at the surface of a solid, together with the equation $L = 0$. It will be observed that in the case of a free surface the pressures P', Q', R' are given, whereas in the case of the surface of a solid they are known only by the solution of the problem. But on the other hand the form of the surface of the solid is given, whereas the form of the free surface is known only by the solution of the problem.

Dubuat found by experiment that when the mean velocity of water flowing through a pipe is less than about one inch in a second, the water near the inner surface of the pipe is at rest. If these experiments may be trusted, the conditions to be satisfied in the case of small velocities are those which first occurred to me, and which are included in those just given by supposing $\nu = \infty$.

I have said that when the velocity is not very small the tangential force called into action by the sliding of water over the inner surface of a pipe varies nearly as the square of the velocity. This fact appears to admit of a natural explanation. When a current of water flows past an obstacle, it produces a resistance varying nearly as the square of the velocity. Now even if the inner surface of a pipe is polished we may suppose that little irregularities exist, forming so many obstacles to the current. Each little protuberance will experience a resistance varying nearly as the square of the velocity, from whence there will result a tangential action of the fluid on the surface of the pipe, which will vary nearly as the square of the velocity; and the same will be true of the equal and opposite reaction of the pipe on the fluid. The tangential force due to this cause will be combined with that by which the fluid close to the pipe is kept at rest when the velocity is sufficiently small*.

[* Except in the case of capillary tubes, or, in case the tube be somewhat wider, of excessively slow motions, the main part of the resistance depends upon the formation of eddies. This much appears clear; but the precise way in which the eddies act is less evident. The explanation in the text gives probably the correct account of what takes place in the case of a river flowing over a rough stony bed; but in the case of a pipe of fairly smooth interior surface the minute protuberances would be too small to produce much resistance of the same kind as that contemplated in the paragraph beginning near the foot of p. 53.

What actually happens appears to be this. The rolling motion of the fluid belonging to the eddies is continually bringing the more swiftly moving fluid which is found nearer to the centre of the pipe close to the surface. And in consequence the gliding or shifting motion of the fluid in the immediate neighbourhood of the surface in such places is very greatly increased, and with it the tangential pressure.

Thus while in some respects these two classes of resistances are similar, in others they are materially different. As typical examples of the two classes we may take, for the first, that of a polished sphere of glass of some size descending by its weight in deep water; for the second, that of a very long circular glass pipe down which water is flowing. In both cases alike eddies are produced, and the eddies once produced ultimately die away. In both cases alike the internal friction of the fluid, and the friction between the fluid and the solid, are intimately connected with the formation of eddies, and it is by friction that the eddies die away, and the kinetic energy of the mass is converted into molecular kinetic energy, that is, heat. But in the first case the resistance depends mainly on the difference of the pressure p in front and rear, the resultant of the other forces of which the expressions are given in equations (8) being comparatively insignificant, while in the second case it is these latter pressures that we are concerned with, the resultant of the pressure p in the direction of the axis of the tube being practically nil, even though the polish of the surface be not mathematically perfect.

Hence if, the motion being what it actually is, the fluidity of the fluid were suddenly to become perfect, the immediate effect on the resistance in the first case would be insignificant, while in the second case the resistance would practically vanish. Of course if the fluidity were to $remain$ perfect, the motion after some time would be very different from what it had been before; but that is not a point under consideration.

Some questions connected with the effect of friction in altering the motion of a nearly perfect fluid will be considered further on in discussing the case of motion given in Art. 55 of a paper *On the Critical Values of the Sums of Periodic Series*.]

7—2

There remains to be considered the case of two fluids having a common surface. Let u', v', w', μ', δ' denote the quantities belonging to the second fluid corresponding to u, &c. belonging to the first. Together with the two equations $L = 0$ and (15) we shall have in this case the equation derived from (15) by putting u', v', w' for u, v, w; or, which comes to the same, we shall have the two former equations with

$$l(u - u') + m(v - v') + n(w - w') = 0 \dots\dots\dots\dots(18).$$

If we consider the principles on which equations (17) were formed to be applicable to the present case, we shall have six more equations to be satisfied, namely (17), and the three equations derived from (17) by interchanging the quantities referring to the two fluids, and changing the signs l, m, n. These equations give the value of ϖ, and leave five equations of condition. If we must suppose $\nu = \infty$, as appears most probable, the six equations above mentioned must be replaced by the six $u' = u$, $v' = v$, $w' = w$, and

$$lp - \mu f(u, v, w) = lp' - \mu' f(u', v', w'), \text{ &c.,}$$

$f(u, v, w)$ denoting the coefficient of μ in the first of equations (17). We have here six equations of condition instead of five, but then the equation (18) becomes identical.

7.   The most interesting questions connected with this subject require for their solution a knowledge of the conditions which must be satisfied at the surface of a solid in contact with the fluid, which, except perhaps in case of very small motions, are unknown. It may be well however to give some applications of the preceding equations which are independent of these conditions. Let us then in the first place consider in what manner the transmission of sound in a fluid is affected by the tangential action. To take the simplest case, suppose that no forces act on the fluid, so that the pressure and density are constant in the state of equilibrium, and conceive a series of plane waves to be propagated in the direction of the axis of x, so that $u = f(x, t)$, $v = 0$, $w = 0$. Let $p_{,}$ be the pressure, and $\rho_{,}$ the density of the fluid when it is in equilibrium, and put $p = p_{,} + p'$. Then we have from equations (11) and (12), omitting the square of the disturbance,

$$\frac{1}{\rho_{,}}\frac{d\rho}{dt} + \frac{du}{dx} = 0, \quad \rho_{,}\frac{du}{dt} + \frac{dp'}{dx} - \frac{4}{3}\mu\frac{d^2u}{dx^2} = 0 \dots\dots\dots\dots(19),$$

Let $A\Delta\rho$ be the increment of pressure due to a very small increment $\Delta\rho$ of density, the temperature being unaltered, and let m
be the ratio of the specific heat of the fluid when the pressure is
constant to its specific heat when the volume is constant; then
the relation between p' and ρ will be

$$p' = mA\,(\rho - \rho_{\prime})\dots\dots\dots\dots\dots\dots(20).$$

Eliminating p' and ρ from (19) and (20) we get

$$\frac{d^2u}{dt^2} - mA\frac{d^2u}{dx^2} - \frac{4\mu}{3\rho_{\prime}}\frac{d^3u}{dt\,dx^2} = 0.$$

To obtain a particular solution of this equation, let

$$u = \phi\,(t)\cos\frac{2\pi x}{\lambda} + \psi\,(t)\sin\frac{2\pi x}{\lambda}.$$

Substituting in the above equation, we see that $\phi\,(t)$ and $\psi\,(t)$
must satisfy the same equation, namely,

$$\phi''\,(t) + \frac{4\pi^2}{\lambda^2}\,mA\phi\,(t) + \frac{16\pi^2\mu}{3\lambda^2\rho_{\prime}}\,\phi'\,(t) = 0,$$

the integral of which is

$$\phi\,(t) = \epsilon^{-ct}\left(C\cos\frac{2\pi bt}{\lambda} + C'\sin\frac{2\pi bt}{\lambda}\right),$$

where

$$c = \frac{8\pi^2\mu}{3\lambda^2\rho_{\prime}}, \quad b^2 = mA - \frac{16\pi^2\mu^2}{9\lambda^2\rho_{\prime}^2},$$

C and C' being arbitrary constants. Taking the same expression
with different arbitrary constants for $\psi\,(t)$, replacing products of
sines and cosines by sums and differences, and combining the
resulting sines and cosines two and two, we see that the resulting
value of u represents two series of waves propagated in opposite
directions. Considering only those waves which are propagated
in the positive direction of x, we have

$$u = C_1\epsilon^{-ct}\cos\left\{\frac{2\pi}{\lambda}\,(bt - x) + C_2\right\}\dots\dots\dots\dots(21).$$

We see then that the effect of the tangential force is to make
the intensity of the sound diminish as the time increases, and to
render the velocity of propagation less than what it would otherwise be. Both effects are greater for high, than for low notes;
but the former depends on the first power of μ, while the latter
depends only on μ^2. It appears from the experiments of M. Biot,
made on empty water pipes in Paris, that the velocity of propaga-

tion of sound is sensibly the same whatever be its pitch. Hence it is necessary to suppose that for air $\mu^2/\lambda^2\rho_{,}^{2}$ is insensible compared with A or $p_{,}/\rho_{,}$. I am not aware of any similar experiments made on water, but the ratio of $(\mu/\lambda\rho_{,})^2$ to A would probably be insensible for water also. The diminution of intensity as the time increases is, in the case of plane waves, due *entirely* to friction; but as we do not possess any means of measuring the intensity of sound the theory cannot be tested, nor the numerical value of μ determined, in this way.

The velocity of sound in air, deduced from the note given by a known tube, is sensibly less than that determined by direct observation. Poisson thought that this might be due to the retardation of the air by friction against the sides of the tube. But from the above investigation it seems unlikely that the effect produced by that cause would be sensible.

The equation (21) may be considered as expressing in all cases the effect of friction; for we may represent an arbitrary disturbance of the medium as the aggregate of series of plane waves propagated in all directions.

8. Let us now consider the motion of a mass of uniform inelastic fluid comprised between two cylinders having a common axis, the cylinders revolving uniformly about their axis, and the fluid being supposed to have attained its permanent state of motion. Let the axis of the cylinders be taken for that of z, and let q be the actual velocity of any particle, so that $u = -q \sin\theta$, $v = q \cos\theta$, $w = 0$, r and θ being polar co-ordinates in a plane parallel to xy.

Observing that

$$\frac{d^2f}{dx^2} + \frac{d^2f}{dy^2} = \frac{d^2f}{dr^2} + \frac{1}{r}\frac{df}{dr} + \frac{1}{r^2}\frac{d^2f}{d\theta^2},$$

where f is any function of x and y, and that $dp/d\theta = 0$, we have from equations (13), supposing after differentiation that the axis of x coincides with the radius vector of the point considered, and omitting the forces, and the part of the pressure due to them,

$$\frac{dp}{dr} - \rho\,\frac{q^2}{r} = 0,$$

$$\frac{d^2q}{dr^2} + \frac{1}{r}\frac{dq}{dr} - \frac{q}{r^2} = 0, \quad \dots\dots\dots\dots\dots (22),$$

and the equation of continuity is satisfied identically.

AND THE EQUILIBRIUM AND MOTION OF ELASTIC SOLIDS. 103

The integral of (22) is

$$q = \frac{C}{r} + C'r.$$

If a is the radius of the inner, and b that of the outer cylinder, and if q_1, q_2 are the velocities of points close to these cylinders respectively, we must have $q = q_1$ when $r = a$, and $q = q_2$ when $r = b$, whence

$$q = \frac{1}{b^2 - a^2} \left\{ (bq_1 - aq_2) \frac{ab}{r} + (bq_2 - aq_1) r \right\} \quad \ldots\ldots(23).$$

If the fluid is infinitely extended, $b = \infty$, and

$$\frac{q}{q_1} = \frac{a}{r}.$$

These cases of motion were considered by Newton (*Principia*, Lib. II. Prop. 51). The hypothesis which I have made agrees in this case with his, but he arrives at the result that the velocity is constant, not, that it varies inversely as the distance. This arises from his having taken, as the condition of their being no acceleration or retardation of the motion of an annulus, that the force tending to turn it in one direction must be equal to that tending to turn it in the opposite, whereas the true condition is that the moment of the force tending to turn it one way must be equal to the moment of the force tending to turn it the other. Of course, making this alteration, it is easy to arrive at the above result by Newton's reasoning. The error just mentioned vitiates the result of Prop. 52. It may be shewn from the general equations that in this case a permanent motion in annuli is impossible, and that, whatever may be the law of friction between the solid sphere and the fluid. Hence it appears that it is necessary to suppose that the particles move in planes passing through the axis of rotation, while they at the same time move round it. In fact, it is easy to see that from the excess of centrifugal force in the neighbourhood of the equator of the revolving sphere the particles in that part will recede from the sphere, and approach it again in the neighbourhood of the poles, and this circulating motion will be combined with a motion about the axis. If however we leave the centrifugal force out of consideration, as Newton has done, the motion in annuli becomes possible, but the solution is different from Newton's, as might have been expected.

The case of motion considered in this article may perhaps admit of being compared with experiment, without knowing the conditions which must be satisfied at the surface of a solid. A hollow, and a solid cylinder might be so mounted as to admit of being turned with different uniform angular velocities round their common axis, which is supposed to be vertical. If both cylinders are turned, they ought to be turned in opposite directions, if only one, it ought to be the outer one; for if the inner were made to revolve too fast, the fluid near it would have a tendency to fly outwards in consequence of the centrifugal force, and eddies would be produced. As long as the angular velocities are not great, so that the surface of the liquid is very nearly plane, it is not of much importance that the fluid is there terminated; for the conditions which must be satisfied at a free surface are satisfied for any section of the fluid made by a horizontal plane, so long as the motion about that section is supposed to be the same as it would be if the cylinders were infinite. The principal difficulty would probably be to measure accurately the time of revolution, and distance from the axis, of the different annuli. This would probably be best done by observing motes in the fluid. It might be possible also to discover in this way the conditions to be satisfied at the surface of the cylinders; or at least a law might be suggested, which could be afterwards compared more accurately with experiment by means of the discharge of pipes and canals.

If the rotations of the cylinders are in opposite directions, there will be a certain distance from the axis at which the fluid will not revolve at all. Writing $-q_1$ for q_1 in equation (23), we have for this distance $\sqrt{\dfrac{ab\,(bq_1 + aq_2)}{bq_2 + aq_1}}$.

9. Although the discharge of a liquid through a long straight pipe or canal, under given circumstances, cannot be calculated without knowing the conditions to be satisfied at the surface of contact of the fluid and solid, it may be well to go a certain way towards the solution.

Let the axis of z be parallel to the generating lines of the pipe or canal, and inclined at an angle α to the horizon; let the plane yz be vertical, and let y and z be measured downwards.

The motion being uniform, we shall have $u = 0$, $v = 0$, $w = f(x, y)$, and we have from equations (13)

$$\frac{dp}{dx} = 0, \quad \frac{dp}{dy} = g\rho \cos\alpha, \quad \frac{dp}{dz} = g\rho \sin\alpha + \mu \left(\frac{d^2w}{dx^2} + \frac{d^2w}{dy^2} \right).$$

In the case of a canal $dp/dz = 0$; and the calculation of the motion in a pipe may always be reduced to that of the motion in the same pipe when dp/dz is supposed to be zero, as may be shewn by reasoning similar to Dubuat's. Moreover the motion in a canal is a particular case of the motion in a pipe. For consider a pipe for which $dp/dz = 0$, and which is divided symmetrically by the plane xz. From the symmetry of the motion, it is clear that we must have $dw/dy = 0$ when $z = 0$; but this is precisely the condition which would have to be satisfied if the fluid had a free surface coinciding with the plane xz; hence we may suppose the upper half of the fluid removed, without affecting the motion of the rest, and thus we pass to the case of a canal. Hence it is the same thing to determine the motion in a canal, as to determine that in the pipe formed by completing the canal symmetrically with respect to the surface of the fluid.

We have then, to determine the motion, the equation

$$\frac{d^2w}{dx^2} + \frac{d^2w}{dy^2} + \frac{g\rho \sin\alpha}{\mu} = 0.$$

In the case of a rectangular pipe, it would not be difficult to express the value of w at any point in terms of its values at the several points of the perimeter of a section of the pipe. In the case of a cylindrical pipe the solution is extremely easy: for if we take the axis of the pipe for that of z, and take polar co-ordinates r, θ in a plane parallel to xy, and observe that $dw/d\theta = 0$, since the motion is supposed to be symmetrical with respect to the axis, the above equation becomes

$$\frac{d^2w}{dr^2} + \frac{1}{r} \frac{dw}{dr} + \frac{g\rho \sin\alpha}{\mu} = 0.$$

Let a be the radius of the pipe, and U the velocity of the fluid close to the surface; then, integrating the above equation, and determining the arbitrary constants by the conditions that w shall be finite when $r = 0$, and $w = U$ when $r = a$, we have

$$w = \frac{g\rho \sin\alpha}{4\mu} (a^2 - r^2) + U.$$

SECTION II.

Objections to Lagrange's proof of the theorem that if udx+vdy+wdz
is an exact differential at any one instant it is always so, the
pressure being supposed equal in all directions. Principles of
M. Cauchy's proof. A new proof of the theorem. A physical
interpretation of the circumstance of the above expression
being an exact differential.

10. The proof of this theorem given by Lagrange depends
on the legitimacy of supposing u, v and w capable of expansion
according to positive integral powers of t, for a sufficiently small
finite value of t. It is clear that the expansion cannot contain
negative powers of t, since u, v and w are supposed to be finite
when $t = 0$; but it may be objected to Lagrange's proof that there
are functions of t of which the expansion contains fractional
powers of t, and that we do not know but that u, v and w may
be such functions. This objection has been considered by Mr
Power*, who has shewn that the theorem is true if we suppose
u, v and w capable of expansion according to any powers of t.
Still the proof remains unsatisfactory, in fact inconclusive, for
these are functions of t, (for instance, e^{-1/t^2}, $t \log t$,) which do not
admit of expansion according to powers of t, integral or fractional,
and we do not know but that u, v and w may be functions of this
nature. I do not here mention the proof which Poisson has
given of the theorem in his *Traité de Mécanique*, because it
appears to me liable to an objection to which I shall presently
have occasion to refer : in fact, Poisson himself did not think the
theorem generally true.

It is remarkable that Mr Power's proof, if it were legitimate,
would establish the theorem even when account is taken of the
variation of pressure in different directions, according to the
theory explained in Section I., if we suppose that $d\mu/dp = 0$. To
shew this we have only got to treat equations (12) as Mr Power
has treated the three equations of fluid motion formed on the
ordinary hypothesis. Yet in this case the theorem is evidently
untrue. Thus, conceive a mass of fluid which is bounded by
a solid plane coinciding with the plane yz, and which extends

* *Cambridge Philosophical Transactions*, Vol. VII. (Part 3) p. 455.

infinitely in every direction on the positive side of the axis of x, and suppose the fluid at first to be at rest. Suppose now the solid plane to be moved in any manner parallel to the axis of y; then, unless the solid plane exerts no tangential force on the fluid, (and we may suppose that it does exert some,) it is clear that at a given time we shall have $u = 0$, $v = f(x)$, $w = 0$, and therefore $u\,dx + v\,dy + w\,dz$ will not be an exact differential. It will be interesting then to examine in this case the nature of the function of t which expresses the value of v.

Supposing X, Y, Z to be zero in equations (12), and observing that in the case considered we have $dp/dy = 0$, we get

$$\frac{dv}{dt} = \frac{\mu}{\rho}\,\frac{d^2v}{dx^2} \dotfill (24).$$

Differentiating this equation $n - 1$ times with respect to t, we easily get

$$\frac{d^n v}{dt^n} = \left(\frac{\mu}{\rho}\right)^n \frac{d^{2n}v}{dx^{2n}}:$$

but when $t = 0$, $v = 0$ when $x > 0$, and therefore for a given value of x all the differential coefficients of v with respect to t are zero. Hence for indefinitely small values of t the value of v at a given point increases more slowly than if it varied ultimately as any power of t, however great; hence v cannot be expanded in a series according to powers of t. This result is independent of the condition to be satisfied at the surface of the solid plane.

I think what has just been proved shews clearly that Lagrange's proof of the theorem considered, even with Mr Power's improvement of it, is inadmissible.

11. The theorem is however true, and a proof of it has been given by M. Cauchy*, which appears to me perfectly free from objection, and which is very simple in principle, although it depends on rather long equations. M. Cauchy first eliminates p from the three equations of motion by means of the conditions that $d^2p/dx\,dy = d^2p/dy\,dx$, &c., he then changes the independent variables from x, y, z, t to a, b, c, t, where a, b, c are the initial

---

* *Mémoire sur la Théorie des Ondes*, in the first volume of the *Mémoires des savans Étrangers*. M. Cauchy has not had occasion to enunciate the theorem, but it is contained in his equations (16). This equation may be obtained in the same manner in the more general case in which p is supposed to be a function of ρ.

co-ordinates of the particles. The three transformed equations admit each of being once integrated with respect to t; and determining the arbitrary functions of a, b, c by the initial values of u, v and w, the three integrals have the form

$$\omega_0' = F\omega' + G\omega'' + H\omega''', \&c.,$$

ω', ω'' and ω''' denoting here the same as in Art. 2, and ω_0', &c. denoting the initial values of ω', &c. for the same particle. Solving the above equations with respect to ω', ω'' and ω''', the resulting equations are

$$\omega' = \frac{1}{S}\left(\frac{dx}{da}\omega_0' + \frac{dx}{db}\omega_0'' + \frac{dx}{dc}\omega_0'''\right), \&c.$$

where S is a function of the differential coefficients of x, y and z with respect to a, b and c, which by the condition of continuity is shewn to be equal to ρ_0/ρ, ρ_0 being the initial density about the particle whose density at the time considered is ρ. Since dx/da, &c. are finite, (for to suppose them infinite would be equivalent to supposing a discontinuity to exist in the fluid,) it follows at once from the preceding equations that if $\omega_0' = 0$, $\omega_0'' = 0$, $\omega_0''' = 0$, that is if $u_0 da + v_0 db + w_0 dc$ be an exact differential, either for the whole fluid or for any portion of it, then shall $\omega' = 0$, $\omega'' = 0$, $\omega''' = 0$, i.e. $u dx + v dy + w dz$ will be an exact differential, at any subsequent time, either for the whole mass or for the above portion of it.

12.   It is not from seeing the smallest flaw in M. Cauchy's proof that I propose a new one, but because it is well to view the subject in different lights, and because the proof which I am about to give does not require such long equations. It will be necessary in the first place to prove the following lemma.

LEMMA.   If ω_1, ω_2,...ω_n are n functions of t, which satisfy the n differential equations

$$\left.\begin{aligned}\frac{d\omega_1}{dt} &= P_1\omega_1 + Q_1\omega_2 \ldots + V_1\omega_n, \\ &\cdots\cdots\cdots\cdots\cdots\cdots\cdots\cdots\cdots\cdots \\ \frac{d\omega_n}{dt} &= P_n\omega_1 + Q_n\omega_2 \ldots + V_n\omega_n, \end{aligned}\right\} \ldots\ldots\ldots\ldots(25),$$

where P_1, Q_1...V_n may be functions of t, ω_1...ω_n, and if when $\omega_1 = 0$, $\omega_2 = 0$...$\omega_n = 0$, none of the quantities P_1, ...V_n is infinite for any value of t from 0 to T, and if $\omega_1...\omega_n$ are each zero when $t = 0$, then shall each of these quantities remain zero for all values of t from 0 to T.

DEMONSTRATION. Let τ be a finite value of t, then by hypothesis τ may be taken so small that the values of $\omega_1...\omega_n$ are sufficiently small to exclude all values which might render any one of the quantities $P_1...V_n$ infinite. Let L be a superior limit to the numerical values of the several quantities $P_1...V_n$ for all values of t from 0 to τ; then it is evident that $\omega_1...\omega_n$ cannot increase faster than if they satisfied the equations

$$\left.\begin{array}{l} \dfrac{d\omega_1}{dt} = L(\omega_1 + \omega_2 ... + \omega_n), \\ .... \\ \dfrac{d\omega_n}{dt} = L(\omega_1 + \omega_2 ... + \omega_n), \end{array}\right\} ....(26),$$

vanishing in this case also when $t = 0$. But if $\omega_1 + \omega_2... + \omega_n = \Omega$, we have by adding together the above equations

$$\frac{d\Omega}{dt} = nL\Omega:$$

if now Ω be not equal to zero, dividing this equation by Ω and integrating, we have

$$\Omega = Ce^{nLt};$$

but no value of C different from zero will allow Ω to vanish when $t = 0$, whereas by hypothesis it does vanish; hence $\Omega = 0$; but Ω is the sum of n quantities which evidently cannot be negative, and therefore each of these must be zero. Since then $\omega_1...\omega_n$ would have to be equal to zero for all values of t from 0 to τ even if they satisfied equations (26), they must à fortiori be equal to zero in the actual case, since they satisfy equations (25). Hence there is no value of t from 0 to T at which any one of the quantities $\omega_1...\omega_n$ can begin to differ from zero, and therefore these quantities must remain equal to zero for all values of t from 0 to T.

This lemma might be extended to the case in which $n = \infty$, with certain restrictions as to the convergency of the series. We may also, instead of the integers 1, 2...n, have a continuous variable α which varies from 0 to a, so that ω is a function of the independent variables α and t, satisfying the differential equation

$$\frac{d\omega}{dt} = \int_0^a \psi\left(\alpha,\, \omega,\, t\right)\omega d\alpha,$$

where $\psi(\alpha,\, 0,\, t)$ does not become infinite for any value of α from 0 to a combined with any value of t from 0 to T. It may be shewn, just as before, that if $\omega = 0$ when $t = 0$ for all values of α from 0 to a, then must $\omega = 0$ for all values of t from 0 to T. The proposition might be further extended to the case in which $a = \infty$, with a certain restriction as to the convergency of the integral, but equations (25) are already more general than I shall have occasion to employ.

It appears to me to be sometimes assumed as a principle that two variables, functions of another, t, are proved to be equal for all values of t when it is shewn that they are equal for a certain value of t, and that whenever they are equal for the same value of t their increments for the same increment of t are *ultimately* equal. But according to this principle, if two curves could be shewn always to touch when they meet they must always coincide, a conclusion manifestly false. I confess I cannot see that Newton in his *Principia*, Lib. I., Prop. 40, has proved more than that if the velocities of the two bodies are equal at equal distances, the increments of those velocities for equal increments of the distances are ultimately equal: at least something additional seems required to put the proof quite out of the reach of objection. Again it is usual to speak of the condition, that the motion of a particle of fluid in contact with the surface of a solid at rest is tangential to the surface, as the same thing as the condition that the particle shall always remain in contact with the surface. That it is the same thing might be shewn by means of the lemma in this article, supposing the motion continuous; but independently of proof I do not see why a particle should not move in a curve not coinciding with the surface, but touching it where it meets it. The same remark will apply to the condition that a particle which at one instant lies in a free surface, or is in contact with a solid, shall ultimately lie in the free surface, or be in contact with the solid, at the consecutive instant. I refer here to the more general case in which the solid is at rest or in motion. For similar reasons Poisson's proof of the Hydrodynamical theorem which forms the principal subject of this section has always appeared to me unsatisfactory, in fact far less satisfactory than Lagrange's. I may add that Poisson's proof, as well as Lagrange's, would apply to the case in which friction is taken into account, in which case the theorem is not true.

13. Supposing ρ to be a function of p, $1/f'(p)$, the ordinary equations of Hydrodynamics are

$$\frac{df(p)}{dx} = X - \frac{Du}{Dt}, \quad \frac{df(p)}{dy} = Y - \frac{Dv}{Dt}, \quad \frac{df(p)}{dz} = Z - \frac{Dw}{Dt} \ldots(27).$$

The forces X, Y, Z will here be supposed to be such that $Xdx + Ydy + Zdz$ is an exact differential, this being the case for any forces emanating from centres, and varying as any functions of the distances. Differentiating the first equations (27) with respect to y, and the second with respect to x, subtracting, putting for Du/Dt and Dv/Dt their values, adding and subtracting $du/dz \,.\, dv/dz$, and employing the notation of Art. 2, we obtain

$$\frac{D\omega'''}{Dt} = \frac{du}{dz}\omega' + \frac{dv}{dz}\omega'' - \left(\frac{du}{dx} + \frac{dv}{dy}\right)\omega'''\ldots\ldots\ldots(28).$$

By treating the first and third, and then the second and third of equations (27) in the same manner, we should obtain two more equations, which may be got at once from that which has just been found by interchanging the requisite quantities. Now for points in the interior of the mass the differential coefficients du/dz, &c. will not be infinite, on account of the continuity of the motion, and therefore the three equations just obtained are a particular case of equations (25). If then $udx + vdy + wdz$ is an exact differential for any portion of the fluid when $t = 0$, that is, if ω', ω'' and ω''' are each zero when $t = 0$, it follows from the lemma of the last article that ω', ω'' and ω''' will be zero for any value of t, and therefore $udx + vdy + wdz$ will always remain an exact differential. It will be observed that it is for the same portion of fluid, not for the fluid occupying the same portion of space, that this is true, since equations (28), &c. contain the differential coefficients $D\omega'/Dt$, &c., and not $d\omega'/dt$, &c.

14. The circumstance of $udx + vdy + wdz$ being an exact differential admits of a physical interpretation which may be

noticed, as it is well to view a subject of this nature in different lights.

Conceive an indefinitely small element of a fluid in motion to become suddenly solidified, and the fluid about it to be suddenly destroyed ; let the form of the element be so taken that the resulting solid shall be that which is the simplest with respect to rotatory motion, namely, that which has its three principal moments about axes passing through the centre of gravity equal to each other, and therefore every axis passing through that point a principal axis, and let us enquire what will be the linear and angular motion of this element just after solidification.

By the instantaneous solidification, velocities will be suddenly generated or destroyed in the different portions of the element, and a set of mutual impulsive forces will be called into action. Let x, y, z be the co-ordinates of the centre of gravity G of the element at the instant of solidification, $x + x'$, $y + y'$, $z + z'$ those of any other point in it. Let u, v, w be the velocities of G along the three axes just before solidification, u', v', w' the relative velocities of the point whose relative co-ordinates are x', y', z'. Let $\bar{u}$, $\bar{v}$, $\bar{w}$ be the velocities of G, $u_{,}$, $v_{,}$, $w_{,}$ the relative velocities of the point above mentioned, and ω', ω'', ω''' the angular velocities just after solidification. Since all the impulsive forces are internal, we have

$$\bar{u} = u, \quad \bar{v} = v, \quad \bar{w} = w.$$

We have also, by the principle of conservation of areas,

$$\Sigma m \left\{ y' \left(w_{,} - w' \right) - z' \left(v_{,} - v' \right) \right\} = 0, \text{ &c.,}$$

m denoting an element of the mass of the element considered. But $u_{,} = \omega'' z' - \omega''' y'$, u' is ultimately equal to

$$\frac{du}{dx} x' + \frac{du}{dy} y' + \frac{du}{dz} z',$$

and similar expressions hold good for the other quantities. Substituting in the above equations, and observing that

$$\Sigma m y' z' = \Sigma m' z' x' = \Sigma m x' y' = 0, \text{ and } \Sigma m x'^2 = \Sigma m y'^2 = \Sigma m z'^2,$$

we have $\omega' = \frac{1}{2} \left(\dfrac{dw}{dy} - \dfrac{dv}{dz} \right)$, &c.

We see then that an indefinitely small element of the fluid, of which the three principal moments about the centre of gravity are equal, if suddenly solidified and detached from the rest of the fluid will begin to move with a motion simply of translation, which may however vanish, or a motion of translation combined with one of rotation, according as $u\,dx + v\,dy + w\,dz$ is, or is not an exact differential, and in the latter case the angular velocities will be the same as in Art. 2.

The principle which forms the subject of this section might be proved, at least in the case of a homogeneous incompressible fluid, by considering the change in the motion of a spherical element of the fluid in the indefinitely small time dt. This method of proving the principle would shew distinctly its intimate connexion with the hypothesis of normal pressure, or the equivalent hypothesis of the equality of pressure in all directions, since the proof depends on the impossibility of an angular velocity being generated in the element in the indefinitely small time dt by the pressure of the surrounding fluid, inasmuch as the direction of the pressure at any point of the surface ultimately passes through the centre of the sphere. The proof I speak of is however less simple than the one already given, and would lead me too far from my subject.

SECTION III.

Application of a method analogous to that of Sect. I. to the determination of the equations of equilibrium and motion of elastic solids.

15. All solid bodies are more or less elastic, as is shewn by the capability they possess of transmitting sound, and vibratory motions in general. The solids considered in this section are supposed to be homogeneous and uncrystallized, so that when in their natural state the average arrangement of their particles is the same at one point as at another, and the same in one direction as in another. The natural state will be taken to be that in which no forces act on them, from which it may be shewn that the pressure in the interior is zero at all points and in all directions, neglecting the small pressure depending on attractions of the nature of capillary attraction.

Let x, y, z be the co-ordinates of any point P in the solid considered when in its natural state, α, β, γ the increments of those

S.   8

co-ordinates at the time considered, whether the body be in a state
of constrained equilibrium or of motion. It will be supposed that
a, β and γ are so small that their squares and products may be
neglected. All the theorems proved in Art. 2 with reference to
linear and angular velocities will be true here with reference to
linear and angular displacements, since these two sets of quantities
are resolved according to the same laws, as long as the angular
displacements are supposed to be very small. Thus, the most
general displacement of a very small element of the solid consists
of a displacement of translation, an angular displacement, and three
displacements of extension in the direction of three rectangular
axes, which may be called in this case, with more propriety than in
the former, *axes of extension.* The three displacements of extension
may be resolved into two displacements of shifting, each in two
dimensions, and a displacement of uniform dilatation, positive or
negative. The pressures about the element considered will depend
on the displacements of extension only; there may also, in the
case of motion, be a small part depending on the relative velocities,
but this part may be neglected, unless we have occasion to consider
the effect of the internal friction in causing the vibrations of solid
bodies to subside. It has been shewn (Art. 7) that the effect of
this cause is insensible in the case of sound propagated through
air; and there is no reason to suppose it greater in the case of
solids than in the case of fluids, but rather the contrary. The
capability which solids possess of being put into a state of isochronous vibration shews that the pressures called into action by small
displacements depend on homogeneous functions of those displacements of one dimension. I shall suppose moreover, according to
the general principle of the superposition of small quantities, that
the pressures due to different displacements are superimposed, and
consequently that the pressures are linear functions of the displacements. Since squares of a, β and γ are neglected, these
pressures may be referred to a unit of surface in the natural state
or after displacement indifferently, and a pressure which is normal
to any surface after displacement may be regarded as normal to
the original position of that surface. Let $-A\delta$ be the pressure
corresponding to a uniform linear dilatation δ when the solid is in
equilibrium, and suppose that it becomes $-mA\delta$, in consequence
of the heat developed, when the solid is in a state of rapid vibration. Suppose also that a displacement of shifting parallel to

the plane xy, for which $\alpha = kx$, $\beta = -ky$, $\gamma = 0$, calls into action a pressure $-Bk$ on a plane perpendicular to the axis of x, and a pressure Bk on a plane perpendicular to that of y; the pressures on these planes being equal and of opposite signs, that on a plane perpendicular to the axis of z being zero, and the tangential forces on those planes being zero, for the same reasons as in Sect. I. It may also be shewn as before that it is necessary to suppose B positive, in order that the equilibrium of the solid medium may be stable, and it is easy to see that the same must be the case with A for the same reason.

It is clear that we shall obtain the expressions for the pressures from those already found for the case of a fluid by merely putting α, β, γ, B for u, v, w, μ, and $-A\delta$ or $-mA\delta$ for p, according as we are considering the case of equilibrium or of vibratory motion, the body being in the latter case supposed to be constrained only in so far as depends on the motion.

For the case of equilibrium then we have from equations (8)

$$P_1 = -A\delta - 2B\left(\frac{d\alpha}{dx} - \delta\right), \quad T_1 = -B\left(\frac{d\beta}{dz} + \frac{d\gamma}{dy}\right), \ \&c. \ldots(29),$$

δ being here $= \frac{1}{3}\left(\frac{d\alpha}{dx} + \frac{d\beta}{dy} + \frac{d\gamma}{dz}\right)$; and the equations of equilibrium will be obtained from (12) by putting $Du/Dt = 0$, $p = -A\delta$, making the same substitution as before for u, v, w and μ. We have therefore, for the equations of equilibrium,

$$\rho X + \frac{1}{3}(A + B)\frac{d}{dx}\left(\frac{d\alpha}{dx} + \frac{d\beta}{dy} + \frac{d\gamma}{dz}\right)$$
$$+ B\left(\frac{d^2\alpha}{dx^2} + \frac{d^2\alpha}{dy^2} + \frac{d^2\alpha}{dz^2}\right) = 0, \ \&c. \ldots\ldots\ldots(30).$$

In the case of a vibratory motion, when the body is in its natural state except so far as depends on the motion, we have from equations (8)

$$P_1 = -mA\delta - 2B\left(\frac{d\alpha}{dx} - \delta\right), \quad T_1 = -B\left(\frac{d\beta}{dz} + \frac{d\gamma}{dy}\right), \ \&c. \ \ldots(31),$$

and the equations of motion will be derived from (12) as before, only Du/Dt &c. must be replaced by $d^2\alpha/dt^2$ &c., and X, Y, Z put equal to zero. The equations of motion, then, are

$$\rho \frac{d^2\alpha}{dt^2} = \tfrac{1}{3}(mA + B) \, \frac{d}{dx}\left(\frac{d\alpha}{dx} + \frac{d\beta}{dy} + \frac{d\gamma}{dz}\right)$$

$$+ B\left(\frac{d^2\alpha}{dx^2} + \frac{d^2\alpha}{dy^2} + \frac{d^2\alpha}{dz^2}\right), \text{ &c.} \ldots\ldots\ldots (32).$$

16. The conditions to be satisfied at the surface of the solid
may be easily deduced from the analogous conditions in the case
of a fluid with a free surface, only it will be necessary to replace
the normal pressure Π by an oblique pressure, of which the components will be denoted by X_1, Y_1, Z_1. We have then, making
the necessary changes in the quantities involved in (14),

$$X_1 + lA\delta + B\left\{2l\frac{d\alpha}{dx} + m\left(\frac{d\alpha}{dy} + \frac{d\beta}{dx}\right) + n\left(\frac{d\alpha}{dz} + \frac{d\gamma}{dx}\right)\right\} = 0, \text{ &c.,}$$

for the case of equilibrium, and for the case of motion such as that
just considered it will only be necessary to replace A by mA in
these equations. If we measure the angles of which l, m, n are
the cosines from the external normal, the forces X_1, Y_1, Z_1 must be
reckoned positive when, l, m and n being positive, the surface of
the solid is urged in the negative directions of x, y, z, and in other
cases the signs must be taken conformably.

If the solid considered is in a state of constraint when at rest,
and is moreover put into a state of vibration, the pressures and
displacements due to these two causes must be calculated separately
and added together. If m were equal to 1, they could be calculated together from the same equations.

SECTION IV.

*Principles of Poisson's theory of elastic solids, and of the oblique
pressures existing in fluids in motion. Objections to one of his
hypotheses. Reflections on the constitution, and equations of
motion of the luminiferous ether in vacuum.*

17. In the twentieth *Cahier* of the *Journal de l'École Polytechnique* may be found a memoir by Poisson, entitled *Mémoire sur les
Équations générales de l'Équilibre et du Mouvement des Corps
solides élastiques et des Fluides*, which contains the substance of
two memoirs presented by him to the Academy, brought together
with some additions. In this memoir the author treats principally

of the equations of equilibrium and motion of elastic solids, of the equations of equilibrium of fluids, with reference especially to capillary attraction, and of the equations of motion of fluids, supposing the pressure not to be equal in all directions.

It is supposed by Poisson that all bodies, whether solid or fluid, are composed of ultimate molecules, separated from each other by vacant spaces. In the cases of an uncrystallized solid in its natural state, and of a fluid in equilibrium, he supposes that the molecules are arranged irregularly, and that the average arrangement is the same in all directions. These molecules he supposes to act on each other with forces, of which the main part is a force in the direction of the line joining the centres of gravity, and varying as some function of the distance of these points, and the remainder a secondary force, or it may be two secondary forces, depending on the molecules not being mathematical points. He supposes that it is on these secondary forces that the solidity of solid bodies depends. He supposes however that in calculating the pressures these secondary forces may be neglected, partly because they become insensible at much smaller distances than the main part of the forces, and partly because they act, on the average, alike in all directions. He supposes that the molecular force decreases very rapidly as the distance increases, yet not so rapidly but that the sphere in which the molecular action is sensible contains an immense number of molecules. He supposes consequently that in estimating the resultant force of a hemisphere of the medium on a molecule in the centre of its base the action of the neighbouring molecules, which are situated irregularly, may be neglected compared with the action of those more remote, of which the average may be taken. The consequence of this supposition of course is that the total action is normal to the base of the hemisphere, and sensibly the same for one molecule as for an adjacent one.

The rest of the reasoning by which Poisson establishes the equations of motion and equilibrium of elastic solids is purely mathematical, sufficient data having been already assumed. It might appear that the reasoning in Art. 16 of his memoir, by which the expression for N is simplified, required the fresh hypothesis of a symmetrical arrangement of the molecules; but it really does not, being admissible according to the principle of averages.

Taking for the natural state of the body that in which the pressure is zero, the equations at which Poisson arrives contain only *one* unknown constant k, whereas the equations of Sect. III. of this paper contain *two*, A or mA and B. This difference depends on the assumption made by Poisson that the irregular part of the force exerted by a hemisphere of the medium on a molecule in the centre of its base may be neglected in comparison with the whole force. As a result of this hypothesis, Poisson finds that the change in direction, and the proportionate change in length, of a line joining two molecules are continuous functions of the co-ordinates of one of the molecules and the angles which determine the direction of the line; whereas in Sect. III., if we adopt the hypothesis of ultimate molecules at all, it is allowable to suppose that these quantities vary irregularly in passing from one pair of molecules to an adjacent pair. Of course the equations of Sect. III. ought to reduce themselves to Poisson's equations for a particular relation between A and B. Neglecting the heat developed by compression, as Poisson has done, and therefore putting $m = 1$, this relation is $A = 5B$.

18.   Poisson's theory of fluid motion is as follows. The time t is supposed to be divided into a number n of equal parts, each equal to τ. In the first of these the fluid is supposed to be displaced as an elastic solid would be, according to Poisson's previous theory, and therefore the pressures are given by the same equations. If the causes producing the displacement were now to cease, the fluid would re-arrange itself, so that the average arrangement about each point should be the same in all directions after a very short time. During this time, the pressures would have altered, in an unknown manner, from those corresponding to a displaced solid to a normal pressure equal to $p + Dp/Dt \cdot \tau$, the pressures during the alteration involving an unknown function of the time elapsed since the end of the interval τ. Another displacement and another re-arrangement may now be supposed to to take place, and so on. But since these very small relative motions will take place independently of each other, we may suppose each displacement to begin at the expiration of the time during which the preceding one is supposed to remain, and we may suppose each re-arrangement to be going on during the succeeding displacements. Supposing now n to become infinite, we pass to

the case in which the fluid is supposed to be continually beginning to be displaced as a solid would, and continually re-arranging itself so as to make the average arrangement about each point the same in all directions.

Poisson's equations (9), page 152, which are applicable to the motion of a liquid, or of an elastic fluid in which the change of density is small, agree with equations (12) of this paper. For the quantity ψt is the pressure p which would exist at any instant if the motion were then to cease, and the increment, $\dfrac{d\psi t}{dt}\,\tau$ or $\dfrac{Dp}{Dt}\,\tau$, of this quantity in the very small time τ will depend only on the increment, $\dfrac{d\chi t}{dt}\,\tau$ or $\dfrac{D\rho}{Dt}\,\tau$, of the density χt or ρ. Consequently the value of $\dfrac{d\psi t}{dt}\,\tau$ will be the same as if the density of the particle considered passed from χt to $\chi t + \dfrac{d\chi t}{dt}\,\tau$ in the time τ by a uniform motion of dilatation. I suppose that according to Poisson's views such a motion would not require a re-arrangement of the molecules, since the pressure remains equal in all directions. On this supposition we shall get the value of $\dfrac{d\psi t}{dt}$ from that of $R_i' - K$ in the equations of page 140 by putting

$$\frac{du}{dx} = \frac{dv}{dy} = \frac{dw}{dz} = -\frac{1}{3\chi t}\frac{d\chi t}{dt}.$$

We have therefore

$$\alpha\frac{d\psi t}{dt} = \frac{\alpha}{3}\,(K - 5k)\,\frac{d\chi t}{\chi t\, dt}.$$

Putting now for $\beta + \beta'$ its value $2\alpha k$, and for $\dfrac{1}{\chi t}\dfrac{d\chi t}{dt}$ its value given by equation (2), the expression for ϖ, page 152, becomes

$$\varpi = p + \frac{\alpha}{3}\,(K + k)\left(\frac{du}{dx} + \frac{dv}{dy} + \frac{dw}{dz}\right).$$

Observing that $\alpha(K + k) = \beta$, this value of ϖ reduces Poisson's equations (9) to the equations (12) of this paper.

Poisson himself has not made this reduction of his equations, nor any equivalent one, so that his equations, as he has left them, involve two arbitrary constants. The reduction of these two to one depends on the assumption that a uniform expansion of any particle does not require a re-arrangement of the molecules, as it leaves the pressure still equal in all directions. If we do not make this assumption, but retain the two arbitrary constants, the equations will be the same as those which would be obtained by the method of this paper, supposing the quantity κ of Art. 3 not to be zero.

19. There is one hypothesis made in the common theory of elastic solids, the truth of which appears to me very questionable. That hypothesis is the one to which I have already alluded in Art. 17, respecting the legitimacy of neglecting the irregular part of the action of the molecules in the immediate neighbourhood of the one considered, in comparison with the total action of those more remote, which is regular. It is from this hypothesis that it follows as a result that the molecules are not displaced among one another in an irregular manner, in consequence of the directive action of neighbouring molecules. Now it is obvious that the molecules of a fluid admit of being displaced among one another with great readiness. The molecules of solids, or of most solids at any rate, must admit of new arrangements, for most solids admit of being bent, permanently, without being broken. Are we then to suppose that when a solid is constrained it has no *tendency* to relieve itself from the state of constraint, in consequence of its molecules *tending towards* new relative positions, provided the amount of constraint be very small? It appears to me to be much more natural to suppose à *priori* that there should be some such tendency.

In the case of a uniform dilatation or contraction of a particle, a re-arrangement of its molecules would be of little or no avail towards relieving it from constraint, and therefore it is natural to suppose that in this case there is little or no tendency towards such a re-arrangement. It is quite otherwise, however, in the case of what I have called a displacement of shifting. Consequently B will be less than if there were no tendency to a re-arrangement. On the hypothesis mentioned in this article, of which the absence of such tendency is a consequence, I have said that a relation has been found between A and B, namely $A = 5B$. It is natural then to expect to find the ratio of A to B greater than 5, approaching more nearly to 5 as the solid considered is more hard and brittle, but differing materially from 5 for the softer solids, especially such as India rubber, or, to take an extreme case, jelly. According to this view the relation $A = 5B$ belongs only to an ideal elastic solid, of which the solidity, or whatever we please to call the property considered, is absolutely perfect.

To shew how implicitly the common theory of elasticity seems to be received by some, I may mention that MM. Lamé and Clapeyron mention Indian rubber among the substances to which it would seem they consider their theory applicable*. I do not know whether the coefficient of elasticity, according to that theory, has been determined experimentally for India rubber, but one would fancy that the cubical compressibility thence deduced, by a method which will be seen in the next article, would turn out comparable with that of a gas.

20. I am not going to enter into the solution of equations (30), but I wish to make a few remarks on the results in some simple cases.

If k be the cubical contraction due to a uniform pressure P, then will

$$k = \frac{3P}{A} .$$

If a wire or rod, of which the boundary is any cylindrical surface, be pulled in the direction of its length by a force of which the value, referred to a unit of surface of a section of the rod, in P, the rod will extend itself uniformly in the direction of its length, and contract uniformly in the perpendicular direction; and if e be the extension in the direction of the length, and c the contraction in any perpendicular direction, both referred to a unit of length, we shall have

$$e = \frac{A + B}{3AB} P, \quad c = \frac{A - 2B}{6AB} P :$$

also, the cubical dilatation $= e - 2c = \dfrac{P}{A} .$

If a cylindrical wire of radius r be twisted by a couple of which

* *Mémoires des savans Étrangers*, Tom. IV. p. 469.

the moment is M, and if θ be the angle of torsion for a length z of the wire, we shall have

$$\theta = \frac{2Mz}{\pi Br^4}.$$

The expressions for k, c, e and θ, and of course all expressions of the same nature, depend on the reciprocals of A and B. Suppose now the value of e, or θ, or any similar quantity not depending on A alone, be given as the result of observation. It will easily be conceived that we might find very nearly the same value for B whether we supposed $A = 5B$ or $A = nB$, where n may be considerably greater than 5, or even infinite. Consequently the observation of two such quantities, giving very nearly the same value of B, might be regarded as confirming the common equations.

If we denote by E the coefficient of elasticity when A is supposed to be equal to $5B$ we have, neglecting the atmospheric pressure*,

$$e = \frac{2P}{5E}, \quad \theta = \frac{2Mz}{\pi Er^4}.$$

If now we denote by E_1 the value of E deduced from observation of the value of e, and by E_2 the value of E obtained by observing the value of θ, or else, which comes to the same, by observing the time of oscillation of a known body oscillating by torsion, we shall have

$$\frac{2}{5E_1} = \tfrac{1}{3}\left(\frac{1}{A} + \frac{1}{B}\right), \quad E_2 = B, \quad \text{whence } \frac{1}{A} = \frac{6}{5E_1} - \frac{1}{E_2}.$$

If A be greater than $5B$, E_1 ought to be a little greater than E_2. This appears to agree with observation. Thus the following numbers are given by M. Lamé† $E_1 = 8000$, $E_2 = 7500$ for iron; $E_1 = 2510$, $E_2 = 2250$ for brass‡. The difference between the values of E_1 and E_2 is attributed by M. Lamé to the errors to which the observation of the small quantity e is liable. If the above numbers may be trusted, we shall have

$$A = 60000, \quad B = 7500, \quad \frac{A}{B} = 8 \text{ for iron;}$$

$$A = 29724, \quad B = 2250, \quad \frac{A}{B} = 13{\cdot}21 \text{ for brass.}$$

---

* Lamé, *Cours de Physique*, Tom. I.
† Lamé, *Cours de Physique*, Tom. I.
‡ These numbers refer to the French units of length and weight.

The cubical contraction k is almost too small to be made the subject of direct observation*, it is therefore usually deduced from the value of e, or from the coefficient of elasticity E found in some other way. On the supposition of a single coefficient E, we have $k/e = \frac{3}{2}$, but retaining the two, A and B, we have

$$\frac{k}{e} = \frac{9B}{A+B} = 9\left(1+\frac{B}{A}\right)^{-1}\frac{B}{A},$$

which will differ greatly from $\frac{3}{2}$ if A/B be much greater than 5. The whole subject therefore requires, I think, a careful examination, before we can set down the values of the coefficients of cubical contraction of different substances in the list of well ascertained physical data. The result, which is generally admitted, that the ratio of the velocity of propagation of normal, to that of tangential vibrations in a solid is equal to $\sqrt{3}$, is another which depends entirely on the supposition that $A = 5B$. The value of m, again, as deduced from observation, will depend upon the ratio of A to B; and it would be highly desirable to have an accurate list of the values of m for different substances, in hopes of thereby discovering in what manner the action of heat on those substances is related to the physical constants belonging to them, such as their densities, atomic weights, &c.

The observations usually made on elastic solids are made on slender pieces, such as wires, rods, and thin plates. In such pieces, all the particles being at no great distance from the surface, it is easy to see that when any small portion is squeezed in one direction it has considerable liberty of expanding itself in a direction perpendicular to this, and consequently the results must depend mainly on the value of B, being not very different from what they would be if A were infinite. This is not so much the case with thick, stout pieces. If therefore such pieces could be put into a state of isochronous vibration, so that the musical notes and nodal lines could be observed, they would probably be better adapted than slender pieces for determining the value of mA. The value of

---

* I find however that direct experiments have been made by Prof. Oersted. According to these experiments the cubical compressibility of solids which would be obtained from Poisson's theory is in some cases as much as 20 or 30 times too great. See the *Report of the British Association* for 1833, p. 353, or *Archives des découvertes*, &c. for 1834, p. 94. [It is to be noted that Oersted's method gives only *differences* of compressibility.]

m might be determined by comparing the value of mA, deduced from the observation of vibrations, with the value of A, deduced from observations made in cases of equilibrium, or, perhaps, of very slow motion.

21. The equations (32) are the same as those which have been obtained by different authors as the equations of motion of the luminiferous ether in vacuum. Assuming for the present that the equations of motion of this medium ought to be determined on the same principles as the equations of motion of an elastic solid, it will be necessary to consider whether the equations (32) are altered by introducing the consideration of a uniform pressure Π existing in the medium when in equilibrium; for we have evidently no right to assume, either that no such pressure exists, or, supposing it to exist, that the medium would expand itself but very slightly if it were removed. It will now no longer be allowable to confound the pressure referred to a unit of surface as it was, in the position of equilibrium of the medium, with the pressure referred to a unit of surface as it actually is. The latter mode of referring the pressure is more natural, and will be more convenient. Let the pressure, referred to a unit of surface at it is, be resolved into a normal pressure $\Pi + p_1$ and a tangential pressure t_1. All the reasoning of Sect. III. will apply to the small forces p_1 and t_1; only it must be remembered that in estimating the whole oblique pressure a normal pressure Π must be compounded with the pressures given by equations (31). In forming the equations of motion, the pressure Π will not appear, because the resultant force due to it acting on the element of the medium which is considered is zero. The equations (32) will therefore be the equations of motion required.

If we had chosen to refer the pressure to a unit of surface in the original state of the surface, and had resolved the whole pressure into a pressure $\Pi + p_1$ normal to the original position of the surface, and a pressure t_1 tangential to that position, the reasoning of Sec. III. would still have applied, and we should have obtained the same expressions as in (31) for the pressures P_1, T_1, &c., but the numerical value of A would have been different. According to this method, the pressure Π would have appeared in the equations of motion. It is when the pressures are measured according to the method which I have adopted that

it is true that the equilibrium of the medium would be unstable
if either A or B were negative. I must here mention that from
some oversight the right-hand sides of Poisson's equations, at
page 68 of the memoir to which I have referred, are wrong. The
first of these equations ought to contain $\dfrac{\Pi}{\rho}\left(\dfrac{d^2u}{dx^2} + \dfrac{d^2u}{dy^2} + \dfrac{d^2u}{dz^2}\right)$,
instead of $\dfrac{\Pi}{\rho}\dfrac{d^2u}{dx^2}$, and similar changes must be made in the other
two equations.

It is sometimes brought as an objection to the equations of
motion of the luminiferous ether, that they are the same as those
employed for the motion of solid bodies, and that it seems un-
natural to employ the same equations for substances which must
be so differently constituted. It was, perhaps, in consequence
of this objection that Poisson proposes, at page 147 of the memoir
which I have cited, to apply to the calculation of the motion of
the lumiferous ether the same principles, with a certain modification, as those which he employed in arriving at his equations (9)
page 152, i.e. the equations (12) of this paper. That modification
consists in supposing that a certain function of the time $\phi(t)$ does
not vary very rapidly compared with the variation of the pressure.
Now the law of the transmission of a motion transversal to the
direction of propagation depending on equations (12) of this paper
is expressed, in the simplest case, by the equation (24); and we
see that this law is the same as that of the transmission of heat,
a law extremely different from that of the transmission of vibratory motions. It seems therefore unlikely that these principles
are applicable to the calculation of the motion of light, unless
the modification which I have mentioned be so great as wholly
to alter the character of the motion, that is, unless we suppose the
pressure to vary extremely fast compared with the function $\phi(t)$,
whereas in ordinary cases of the motion of fluids the function $\phi(t)$
is supposed to vary extremely fast compared with the pressure.

Another view of the subject may be taken which I think
deserves notice. Before explaining this view however it will be
necessary to define what I mean in this paragraph by the word
elasticity. There are two distinct kinds of elasticity; one, that
by which a body which is uniformly compressed tends to
regain its original volume, the other, that by which a body
which is constrained in a manner independent of compres-

sion tends to assume its original form.   The constants A and B of Sec. III. may be taken as measures of these two kinds of elasticity.   In the present paragraph, the word will be used to denote the second kind.   Now many highly elastic substances, as iron, copper, &c., are yet to a very sensible degree plastic.   The plasticity of lead is greater than that of iron or copper, and, as appears from experiment, its elasticity less.   On the whole it is probable that the greater the plasticity of a substance the less its elasticity, and *vice versâ*, although this rule is probably far from being without exception.   When the plasticity of the substance is still further increased, and its elasticity diminished, it passes into a viscous fluid.   There seems no line of demarcation between a solid and a viscous fluid.   In fact, the practical distinction between these two classes of bodies seems to depend on the intensity of the *extraneous* force of gravity, compared with the intensity of the forces by which the parts of the substance are held together.   Thus, what on the Earth is a soft solid might, if carried to the Sun, and retained at the same temperature, be a viscous fluid, the force of gravity at the surface of the Sun being sufficient to make the substance spread out and become level at the top: while what on the Earth is a viscous fluid might on the surface of Pallas be a soft solid.   The gradation of viscous, into what are called perfect fluids seems to present as little abruptness as that of solids into viscous fluids; and some experiments which have been made on the sudden conversion of water and ether into vapour, when enclosed in strong vessels and exposed to high temperatures, go towards breaking down the distinction between liquids and gases.

According to the law of continuity, then, we should expect the property of elasticity to run through the whole series, only, it may become insensible, or else may be masked by some other more conspicuous property.   It must be remembered that the elasticity here spoken of is that which consists in the tangential force called into action by a displacement of continuous sliding : the displacements also which will be spoken of in this paragraph must be understood of such displacements as are independent of condensation or rarefaction.   Now the distinguishing property of fluids is the extreme mobility of their parts.   According to the views explained in this article, this mobility is merely an extremely great plasticity, so that a fluid admits of a *finite*, but exceedingly small amount of constraint before it will be relieved from its state of tension by its molecules assuming new positions of equilibrium. Consequently the same oblique pressures can be called into action in a fluid as in a solid, provided the amount of relative displacement of the parts be exceedingly small. All we know for certain is that the effect of elasticity in fluids, (elasticity of the second kind be it remembered,) is quite insensible in cases of equilibrium, and it is probably insensible in all ordinary cases of fluid motion. Should it be otherwise, equations (8) and (12) will not be true, or only approximately true. But a little consideration will shew that the property of elasticity may be quite insensible in ordinary cases of fluid motion, and may yet be that on which the phenomena of light entirely depend. When we find a vibrating string, the small extent of vibration of which can be actually seen, filling a whole room with sound, and re-member how rapidly the intensity of the vibrations of the air must diminish as the distance from the string increases*, we may easily conceive how small in general must be the amount of the relative motion of adjacent particles of air in the case of sound. Now the extent of the vibration of the ether, in the case of light, may be as small compared with the length of a wave of light as that of the air is compared with the length of a wave of sound : we have no reason to suppose it otherwise. When we remember then that the length of a wave of sound in air varies from some inches to several feet, while the greatest length of a wave of light is about ·00003 of an inch, it is easy to imagine that the *relative* displacement of the particles of ether may be so small as not to reach, nor even come near to the greatest relative dis-placement which could exist without the molecules of the medium assuming new positions of equilibrium, or, to keep clear of the idea of molecules, without the medium assuming a new arrange-ment which might be permanent.

It has been supposed by some that air, like the luminiferous ether, ought to admit of transversal vibrations. According to the views of this article such would, mathematically speaking, be the case ; but the extent of such vibrations would be necessarily so very small as to render them utterly insensible, unless we had

---

* [In all ordinary cases it is to the vibrations of the sounding-board, or of the supporting body acting as a sounding-board, and not to those of the string directly, that the sound is almost wholly due.]

organs with a delicacy equal to that of the retina adapted to receive them.

It has been shewn to be highly probable that the ratio of A to B increases rapidly according as the medium considered is softer and more plastic. For fluids therefore, and among them for the luminiferous ether, we should expect the ratio of A to B to be extremely great. Now if N be the velocity of propagation of normal vibrations in the medium considered in Sect. III., and T that of transversal vibrations, it may be shewn from equations (32) that

$$N^2 = \frac{mA + 4B}{3\rho}, \quad T^2 = \frac{B}{\rho}.$$

This is very easily shewn in the simplest case of plane waves: for if $\beta = \gamma = 0$, $\alpha = f(x)$, the equations (32) give

$$\rho \frac{d^2a}{dt^2} = \tfrac{1}{3}(mA + 4B)\frac{d^2\alpha}{dx^2},$$

whence $\alpha = \phi(Nt - x) + \psi(Nt + x)$; and if $\alpha = \gamma = 0$, $\beta = f(x)$, the same equations give $\rho \dfrac{d^2\beta}{dt^2} = B\dfrac{d^2\beta}{dx^2}$, whence

$$\beta = \zeta(Tt - x) + \xi(Tt + x).$$

Consequently we should expect to find the ratio of N to T extremely great. This agrees with a conclusion of the late Mr Green's*. Since the equilibrium of any medium would be unstable if either A or B were negative, the least possible value of the ratio of N^2 to T^2 is $\tfrac{4}{3}$, a result at which Mr Green also arrived. As however it has been shewn to be highly probable that $A > 5B$ even for the hardest solids, while for the softer ones A/B is much greater than 5, it is probable that N/T is greater than $\sqrt{3}$ for the hardest solids, and much greater for the softer ones.

If we suppose that in the luminiferous ether A/B may be considered infinite, the equations of motion admit of a simplification. For if we put $mA\left(\dfrac{d\alpha}{dx} + \dfrac{d\beta}{dy} + \dfrac{d\gamma}{dz}\right) = -p$ in equations (32), and suppose mA to become infinite while p remains finite, the equations become

---

* *Cambridge Philosophical Transactions*, Vol. VII. Part I. p. 2.

$$\rho \frac{d^2\alpha}{dt^2} = -\frac{dp}{dx} + B\left(\frac{d^2\alpha}{dx^2} + \frac{d^2\alpha}{dy^2} + \frac{d^2\alpha}{dz^2}\right), \ \&c.$$

$$\text{and} \qquad \frac{d\alpha}{dx} + \frac{d\beta}{dy} + \frac{d\gamma}{dz} = 0. \qquad \Bigg\} \ \dots\dots(33).$$

When a vibratory motion is propagated in a medium of which (33) are the equations of motion, it may be shewn that $p = \psi\,(t)$ if the medium be indefinitely extended, or else if there be no motion at its boundaries. In considering therefore the transmission of light in an uninterrupted vacuum the terms involving p will disappear from equations (33); but these terms are, I believe, important in explaining Diffraction, which is the principal phenomenon the laws of which depend only on the equations of motion of the luminiferous ether in vacuum. It will be observed that putting $A = \infty$ comes to the same thing as regarding the ether as incompressible with respect to those motions which constitute Light.

On the Proof of the Proposition that $(Mx + Ny)^{-1}$ is an Integrating Factor of the Homogeneous Differential Equation $M + N\,dy/dx = 0$.

[From the *Cambridge Mathematical Journal*, Vol. IV. p. 241. (*May*, 1845.)]

A FALLACIOUS proof is sometimes given of this proposition, which ought to be examined. The substance of the proof is as follows.

Let us see whether it is possible to find a multiplier V, a homogeneous function of x and y, which shall render $Mdx + Ndy$ an exact differential. Let M and N be of n, and V of p dimensions; let

$$dU = V\,(Mdx + Ndy)\ldots\ldots\ldots\ldots\ldots(1)\,;$$

then, on properly choosing the arbitrary constant in U, U will be a homogeneous function of $n + p + 1$ dimensions, $\left.\right\}\ (A),$ whence, by a known theorem,

$$(n + p + 1)\ U = x\,\frac{dU}{dx} + y\,\frac{dU}{dy} = V\,(Mx + Ny)\ \ldots\ldots(2)\,;$$

therefore, dividing (1) by (2),

$$\frac{dU}{(n + p + 1)\ U} = \frac{Mdx + Ndy}{Mx + Ny}\,;$$

and the first side of this equation being an exact differential, it follows that the second side is so also, and consequently that $(Mx + Ny)^{-1}$ is an integrating factor.

Now the factor so found is of $-n-1$ dimensions; so that the first side of (2) is zero. In fact, we shall see that the statement (A) is not true as applied to the case in question, unless

$$Mx + Ny = 0.$$

The general form of a function of x of n dimensions is Ax^n. The general form of a homogeneous function of x and y of n dimensions is $x^n \psi \left(\frac{y}{x}\right)$. The integral of the first is in general $Ax^{n+1}/(n+1)$, omitting the arbitrary constant; and consequently the dimensions of the function are increased by unity by integration. But in the particular case in which $n = -1$, the integral is $A \log x$, which is not a quantity of 0 dimensions, at least according to the definition just given, *according to which definition only* is the proposition with reference to homogeneous functions assumed in (2) true. Let us now examine in what cases U will be of $n+p+1$ dimensions.

Putting $M = M_0 x^n$, $N = N_0 x^n$, $y = xz$, M_0 and N_0 will be functions of z alone, and we shall have

$$Mdx + Ndy = x^n \{(M_0 + N_0 z) dx + N_0 x \, dz\}.$$

If $M_0 + N_0 z = 0$, i.e. if $Mx + Ny = 0$, we see that x^{-n-1} will be an integrating factor. The integral, being a function of z, will be of 0 dimensions, and both sides of (2) will be zero.

If $Mx + Ny$ is not equal to 0, we may multiply and divide by $(M_0 + N_0 z) x$, and we have

$$Mdx + Ndy = x^{n+1} (M_0 + N_0 z) \left(\frac{dx}{x} + \frac{N_0 dz}{M_0 + N_0 z}\right).$$

Hence we see that $\{x^{n+1} (M_0 + N_0 z)\}^{-1}$ or $(Mx + Ny)^{-1}$ is an integrating factor. For this factor we have

$$U = \log(x) + \phi\left(\frac{y}{x}\right),$$

ϕ denoting the function arising from the integration with respect to z.

In this case we have $x \dfrac{dU}{dx} + y \dfrac{dU}{dy} = 1$, not $= 0$.

It may be of some interest to enquire in what cases an exact differential of any number of independent variables, in which the differential coefficients are homogeneous functions of n dimensions, has an integral which is a homogeneous function of $n+1$ dimensions.

9—2

Let $dU = Mdx + Ndy + Pdz + \dots$ be the exact differential. Let $y = y'x$, $z = z'x \dots$, $M = M_0x^n$, $N = N_0x^n \dots$, so that M_0, $N_0 \dots$ are functions of y', $z' \dots$ only; then

$$dU = x^n \{(M_0 + N_0y' + P_0z' \dots) \, dx + (N_0dy' + P_0dz' \dots) \, x\}.$$

First, suppose the coefficient of dx in this equation to be zero, or $Mx + Ny + Pz \dots = 0$; then the expression for dU cannot be an exact differential unless $n = -1$. In this case U will be a function of y', $z' \dots$, and will therefore be a homogeneous function of $n+1$ or 0 dimensions.

Secondly, suppose the coefficient of dx not to be zero; then

$$dU = x^{n+1} (M_0 + N_0y' \dots) \left(\frac{dx}{x} + \frac{N_0dy' + P_0dz' \dots}{M_0 + N_0y' + P_0z' \dots}\right)$$
$$= (Mx + Ny + Pz \dots) \left(\frac{dx}{x} + \frac{N_0dy' + P_0dz' \dots}{M_0 + N_0y' + P_0z' \dots}\right) \dots \dots (3).$$

Now I say that $\dfrac{N_0dy' + P_0dz' \dots}{M_0 + N_0y' + P_0z' \dots}$ is the exact differential of a function of the independent variables y', $z' \dots$, or, taking y, $z \dots$ for the independent variables instead of y', $z' \dots$, x being supposed constant, and putting for M_0, $N_0, \dots$ their values, that

$$\frac{Ndy + Pdz + \dots}{Mx + Ny + Pz \dots}$$

is an exact differential.

For, putting $Mx + Ny + Pz \dots = D$, in order that the quantity considered should be an exact differential, it is necessary and sufficient that the system of equations of which the type is

$$\frac{d\dfrac{N}{D}}{dz} = \frac{d\dfrac{P}{D}}{dy}$$ should be satisfied. This equation gives

$$D \left(\frac{dN}{dz} - \frac{dP}{dy}\right) + P \frac{dD}{dy} - N \frac{dD}{dz} = 0.$$

Now, since $dN/dz = dP/dy$, by the conditions of $Mdx + Ndy + Pdz \dots$ being an exact differential, the above equation becomes $P\dfrac{dD}{dy} - N\dfrac{dD}{dz} = 0$, or

$$P \left(\frac{dM}{dy} x + \frac{dN}{dy} y + \frac{dP}{dy} z \dots\right) - N \left(\frac{dM}{dz} x + \frac{dN}{dz} y + \frac{dP}{dz} z \dots\right) = 0.$$

Replacing $dM/dy,\ dP/dy\ \ldots$ by $dN/dx,\ dN/dz\ \ldots$ and $dM/dz,$ $dN/dz\ \ldots$ by $dP/dx,\ dP/dy\ \ldots$, this equation becomes

$$P\left(\frac{dN}{dx}\,x+\frac{dN}{dy}\,y+\frac{dN}{dz}\,z\ldots\right)-N\left(\frac{dP}{dx}\,x+\frac{dP}{dy}\,y+\frac{dP}{dz}\,z\ldots\right)=0.$$

Now
$$\frac{dN}{dx}\,x+\frac{dN}{dy}\,y+\ldots=nN,$$

$$\frac{dP}{dx}\,x+\frac{dP}{dy}\,y+\ldots=nP,$$

and therefore the above equation *is* satisfied. Hence

$$\frac{Ndy+Pdz\ldots}{Mx+Ny+Pz\ldots},\ \text{or its equal}\ \frac{N_0dy'+P_0dz'\ldots}{M_0+N_0y'+P_0z'\ldots},$$

is an exact differential $d\psi\ (y',\ z'\ldots)$. Consequently equation (3) becomes

$$dU=(Mx+Ny+Pz\ldots)\,d\{\log x+\psi\,(y',\ z'\ldots)\};$$

which equation being by hypothesis integrable, it follows that

$$Mx+Ny+Pz\ldots=\phi\{\log x+\psi\,(y',\ z'\ldots)\}:$$

and $Mx+Ny\ldots$ being moreover a homogeneous function of $n+1$ dimensions, it is clear that we must have $\phi\,(\alpha)=Ae^{(n+1)\alpha}$. Hence we have

$$dU=Ax^{n+1}\,e^{(n+1)\psi}\,d\,(\log x+\psi).$$

If now $n+1$ is not equal to 0, we have

$$U=\frac{Ax^{n+1}\,e^{(n+1)\psi}}{n+1},$$

omitting the constant; but if $n=-1$, we have

$$U=A\,(\log x+\psi)+C.$$

We see then that if $Mx+Ny+Pz\ldots=0$, which can only happen when $n=-1$, U will be a homogeneous function of $n+1$ or 0 dimensions. If $Mx+Ny+Pz\ldots$ is not equal to 0, then, if $n+1$ is not equal to 0, and the constant in U is properly chosen, U will be a homogeneous function of $n+1$ dimensions, but if $n+1=0$, U will not be a homogeneous function of 0 dimensions, but will contain $\log x$. Of course it might equally have contained the logarithm of y or z, &c.; in fact,

$$\log x+\psi\,(y',\ z'\ldots)=\log y+\log\frac{x}{y}+\psi\,(y',\ z'\ldots)$$

$$=\log y+\chi\,(y',\ z'\ldots).$$

[From the *Philosophical Magazine*, Vol. XXVII. p. 9. (*July*, 1845.)]

ON THE ABERRATION OF LIGHT.

THE general explanation of the phenomenon of aberration is so simple, and the coincidence of the value of the velocity of light thence deduced with that derived from the observations of the eclipses of Jupiter's satellites so remarkable, as to leave no doubt on the mind as to the truth of the explanation. But when we examine the cause of the phenomenon more closely, it is far from being so simple as it appears at first sight. On the theory of emissions, indeed, there is little difficulty; and it would seem that the more particular explanation of the cause of aberration usually given, which depends on the consideration of the motion of a telescope as light passes from its object-glass to its cross wires, has reference especially to this theory; for it does not apply to the theory of undulations, unless we make the rather startling hypothesis that the luminiferous ether passes freely through the sides of the telescope and through the earth itself. The undulatory theory of light, however, explains so simply and so beautifully the most complicated phenomena, that we are naturally led to regard aberration as a phenomenon unexplained by it, but not incompatible with it.

The object of the present communication is to attempt an explanation of the cause of aberration which shall be in accordance with the theory of undulations. I shall suppose that the earth and the planets carry a portion of the ether along with them so that the ether close to their surfaces is at rest relatively to those surfaces, while its velocity alters as we recede from the surface, till, at no great distance, it is at rest in space. According to the undulatory theory, the direction in which a heavenly body is seen

is normal to the fronts of the waves which have emanated from
it, and have reached the neighbourhood of the observer, the ether
near him being supposed to be at rest relatively to him. If
the ether in space were at rest, the front of a wave of light at any
instant being given, its front at any future time could be found
by the method explained in Airy's tracts. If the ether were in
motion, and the velocity of propagation of light were infinitely
small, the wave's front would be displaced as a surface of particles of the ether. Neither of these suppositions is however true,
for the ether moves while light is propagated through it. In the
following investigation I suppose that the displacements of a
wave's front in an elementary portion of time due to the two
causes just considered take place independently.

Let u, v, w be the resolved parts along the rectangular axes of
x, y, z, of the velocity of the particle of ether whose co-ordinates
are x, y, z, and let V be the velocity of light supposing the ether
at rest. In consequence of the distance of the heavenly bodies, it
will be quite unnecessary to consider any waves except those which
are plane, except in so far as they are distorted by the motion of
the ether. Let the axis of z be taken in, or nearly in the direction
of propagation of the wave considered, so that the equation of
a wave's front at any time will be

$$z = C + Vt + \zeta \dots\dots\dots\dots\dots(1),$$

C being a constant, t the time, and ζ a small quantity, a function
of x, y and t. Since u, v, w and ζ are of the order of the aberration, their squares and products may be neglected.

Denoting by α, β, γ the angles which the normal to the wave's
front at the point (x, y, z) makes with the axes, we have, to the
first order of approximation,

$$\cos\alpha = -\frac{d\zeta}{dx}, \quad \cos\beta = -\frac{d\zeta}{dy}, \quad \cos\gamma = 1 \dots\dots\dots\dots(2);$$

and if we take a length Vdt along this normal, the co-ordinates
of its extremity will be

$$x - \frac{d\zeta}{dx}\,Vdt, \quad y - \frac{d\zeta}{dy}\,Vdt, \quad z + Vdt.$$

If the ether were at rest, the locus of these extremities would be
the wave's front at the time $t + dt$, but since it is in motion, the

co-ordinates of those extremities must be further increased by udt, vdt, wdt. Denoting then by x', y', z' the co-ordinates of the point of the wave's front at the time $t + dt$ which corresponds to the point (x, y, z) at the time t, we have

$$x' = x + \left(u - V\frac{d\zeta}{dx}\right)dt, \quad y' = y + \left(v - V\frac{d\zeta}{dy}\right)dt,$$
$$z' = z + (w + V)\,dt\,;$$

and eliminating x, y and z from these equations and (1), and denoting ζ by $f(x, y, t)$, we have for the equation to the wave's front at the time $t + dt$,

$$z' - (w + V)\,dt = C + Vt$$
$$+ f\left\{x' - \left(u + \frac{d\zeta}{dx}\right)dt, \quad y' - \left(v + \frac{d\zeta}{dy}\right)dt, t\right\},$$

or, expanding, neglecting dt^2 and the square of the aberration, and suppressing the accents of x, y and z,

$$z = C + Vt + \zeta + (w + V)\,dt \dots \dots \dots \dots (3).$$

But from the definition of ζ it follows that the equation to the wave's front at the time $t + dt$ will be got from (1) by putting $t + dt$ for t, and we have therefore for this equation

$$z = C + Vt + \zeta + \left(V + \frac{d\zeta}{dt}\right)dt \dots \dots \dots \dots (4).$$

Comparing the identical equations (3) and (4), we have

$$\frac{d\zeta}{dt} = w.$$

This equation gives $\zeta = \int w\,dt$; but in the small term ζ we may replace $\int w\,dt$ by $\int w\,dz \div V$: this comes to taking the approximate value of z given by the equation $z = C + Vt$ instead of t for the parameter of the system of surfaces formed by the wave's front in its successive positions. Hence equation (1) becomes

$$z = C + Vt + \frac{1}{V}\int w\,dz.$$

Combining the value of ζ just found with equations (2), we get, to a first approximation,

$$\alpha - \frac{\pi}{2} = \frac{1}{V}\int \frac{dw}{dx}\,dz, \quad \beta - \frac{\pi}{2} = \frac{1}{V}\int \frac{dw}{dy}\,dz,$$

equations which might very easily be proved directly in a more geometrical manner.

If random values are assigned to u, v and w, the law of aberration resulting from these equations will be a complicated one; but if u, v and w are such that $u\,dx + v\,dy + w\,dz$ is an exact differential, we have,

$$\frac{dw}{dx} = \frac{du}{dz}, \qquad \frac{dw}{dy} = \frac{dv}{dz};$$

whence, denoting by the suffixes 1, 2 the values of the variables belonging to the first and second limits respectively, we obtain

$$\alpha_2 - \alpha_1 = \frac{u_2 - u_1}{V}, \qquad \beta_2 - \beta_1 = \frac{v_2 - v_1}{V} \dots\dots\dots\dots(6).$$

If the motion of the ether be such that $u\,dx + v\,dy + w\,dz$ is an exact differential for one system of rectangular axes, it is easy to prove, by the transformation of co-ordinates, that it is an exact differential for any other system. Hence the formulæ (6) will hold good, not merely for light propagated in the direction first considered, but for light propagated in any direction, the direction of propagation being taken in each case for the axis of z. If we assume that $u\,dx + v\,dy + w\,dz$ is an exact differential for that part of the motion of the ether which is due to the motion of translation of the earth and planets, it does not therefore follow that the same is true for that part which depends on their motions of rotation. Moreover, the diurnal aberration is too small to be detected by observation, or at least to be measured with any accuracy, and I shall therefore neglect it.

It is not difficult to shew that the formulæ (6) lead to the known law of aberration. In applying them to the case of a star, if we begin the integrations in equations (5) at a point situated at such a distance from the earth that the motion of the ether, and consequently the resulting change in the direction of the light, is insensible, we shall have $u_1 = 0$, $v_1 = 0$; and if, moreover, we take the plane xz to pass through the direction of the earth's motion, we shall have

$$v_2 = 0, \quad \beta_2 - \beta_1 = 0,$$

and

$$\alpha_2 - \alpha_1 = \frac{u_2}{V};$$

that is, the star will appear displaced towards the direction in which the earth is moving, through an angle equal to the ratio of the velocity of the earth to that of light, multiplied by the sine of the angle between the direction of the earth's motion and the line joining the earth and the star.

ADDITIONAL NOTE.

[In what precedes *waves* of light are alone considered, and the course of a *ray* is not investigated, the investigation not being required. There follows in the original paper an investigation having for object to shew that in the case of a body like the moon or a planet which is itself in motion, the effect of the distortion of the waves in the neighbourhood of the body in altering the apparent place of the body as determined by observation is insensible. For this, the orthogonal trajectory of the wave in its successive positions from the body to the observer is considered, a trajectory which in its main part will be a straight line, from which it will not differ except in the immediate neighbourhood of the body and of the earth, where the ether is distorted by their respective motions. The perpendicular distance of the further extremity of the trajectory from the prolongation of the straight line which it forms in the intervening quiescent ether is shewn to subtend at the earth an angle which, though not actually 0, is so small that it may be disregarded.

The orthogonal trajectory of a wave in its successive positions does not however represent the course of a ray, as it would do if the ether were at rest. Some remarks made by Professor Challis in the course of discussion suggested to me the examination of the path of a ray, which in the case in which $udx + vdy + wdz$ is an exact differential proved to be a straight line, a result which I had not foreseen when I wrote the above paper, which I may mention was read before the Cambridge Philosophical Society on the 18th of May, 1845 (see *Philosophical Magazine*, vol. XXIX., p. 62). The rectilinearity of the path of a ray in this case, though not expressly mentioned by Professor Challis, is virtually contained in what he wrote. The problem is rather simplified by introducing the consideration of rays, and may be treated from the beginning in the following manner.

The notation in other respects being as before, let α', β' be the small angles by which the direction of the wave-normal at the point (x, y, z) deviates from that of Oz towards Ox, Oy, respectively, so that α', β' are the complements of α, β, and let $\alpha_{,}$, $\beta_{,}$ be the inclinations to Oz of the course of a ray at the same point. By compounding the velocity of propagation through the ether with the velocity of the ether we easily see that

$$\alpha_{,} = \alpha' + \frac{u}{V}, \qquad \beta_{,} = \beta' + \frac{v}{V}.$$

Let us now trace the changes of $\alpha_{,}$, $\beta_{,}$ during the time dt. These depend first on the changes of α', β', and secondly on those of u, v.

As regards the change in the direction of the wave-normal, we notice that the seat of a small element of the wave in its successive positions is in a succession of planes of particles nearly parallel to the plane of x, y. Consequently the direction of the element of the wave will be altered during the time dt by the motion of the ether as much as a plane of particles of the ether parallel to the plane of the wave, or, which is the same to the order of small quantities retained, parallel to the plane xy. Now if we consider a particle of ether at the time t having for coordinates x, y, z, another at a distance dx parallel to the axis of x, and a third at a distance dy parallel to the axis of y, we see that the displacements of these three particles parallel to the axis of z during the time dt will be

$$wdt, \quad \left(w + \frac{dw}{dx}\,dx\right)dt, \quad \left(w + \frac{dw}{dy}\,dy\right)dt\,;$$

and dividing the relative displacements by the relation distances, we have $dw/dx\,.\,dt$, $dw/dy\,.\,dt$ for the small angles by which the normal is displaced, in the planes of xz, yz, from the axes x, y, so that

$$d\alpha' = -\frac{dw}{dx}\,dt, \quad d\beta' = -\frac{dw}{dy}\,dt.$$

We have seen already that the changes of u, v are $du/dz\,.\,Vdt$, $dv/dz\,.\,Vdt$, so that

$$d\alpha_{,} = \left(\frac{du}{dz} - \frac{dw}{dx}\right)dt, \quad d\beta_{,} = \left(\frac{dv}{dz} - \frac{dw}{dy}\right)dt.$$

Hence, provided the motion of the ether be such that

$$udx + vdy + wdz$$

is an exact differential, the change of direction of a ray as it travels along is *nil*, and therefore the course of a ray is a straight line notwithstanding the motion of the ether. The rectilinearity of propagation of a ray of light, which *à priori* would seem very likely to be interfered with by the motion of the ether produced by the earth or heavenly body moving through it, is the tacit assumption made in the explanation of aberration given in treatises of Astronomy, and provided that be accounted for the rest follows as usual*. It follows further that the angle subtended at the earth by the perpendicular distance of the point where a ray leaves a heavenly body from the straight line prolonged which represents its course through the intervening quiescent ether, is not merely too small to be observed, but actually *nil*.]

---

* To make this explanation *quite* complete, we should properly, as Professor Challis remarks, consider the light coming from the wires of the observing telescope, in company with the light from the heavenly body.

[From the *Philosophical Magazine*, Vol. XXVIII. p. 76. (*Feb.* 1846.)]

On Fresnel's Theory of the Aberration of Light.

THE theory of the aberration of light, and of the absence of any influence of the motion of the earth on the laws of refraction, &c., given by Fresnel in the ninth volume of the *Annales de Chimie*, p. 57, is really very remarkable. If we suppose the diminished velocity of propagation of light within refracting media to arise solely from the greater density of the ether within them, the elastic force being the same as without, the density which it is necessary to suppose the ether within a medium of refractive index μ to have is μ^2, the density in vacuum being taken for unity. Fresnel supposes that the earth passes through the ether without disturbing it, the ether penetrating the earth quite freely. He supposes that a refracting medium moving with the earth carries with it a quantity of ether, of density $\mu^2 - 1$, which constitutes the excess of density of the ether within it over the density of the ether in vacuum. He supposes that light is propagated through this ether, of which part is moving with the earth, and part is at rest in space, as it would be if the whole were moving with the velocity of the centre of gravity of any portion of it, that is, with a velocity $(1 - \mu^{-2}) v$, v being the velocity of the earth. It may be observed however that the result would be the same if we supposed the whole of the ether within the earth to move together, the ether entering the earth in front, and being immediately condensed, and issuing from it behind, where it is immediately rarefied, undergoing likewise sudden condensation or rarefaction in passing from one refracting medium to another. On this supposition, the evident condition that a mass v of the ether must pass in a unit of time across a plane of area unity, drawn anywhere within the earth in a direction perpendicular to that of the earth's motion, gives $(1 - \mu^{-2})\, v$ for the velocity of the ether within a refracting medium. As this idea is rather simpler than Fresnel's, I shall adopt it in considering his theory. Also, instead of considering the earth as in motion and the ether outside it as at rest, it will be simpler to conceive a velocity equal and opposite to that of the earth impressed both on the earth and on the ether. On this supposition the earth will be at rest; the ether outside it will be moving with a velocity v, and the ether in a refracting medium with a velocity v/μ^2, in a direction contrary to that of the earth's real motion. On account of the smallness of the coefficient of aberration, we may also neglect the square of the ratio of the earth's velocity to that of light; and if we resolve the earth's velocity in different directions, we may consider the effect of each resolved part separately.

In the ninth volume of the *Comptes Rendus* of the Academy of Sciences, p. 774, there is a short notice of a memoir by M. Babinet, giving an account of an experiment which seemed to present a difficulty in its explanation. M. Babinet found that when two pieces of glass of equal thickness were placed across two streams of light which interfered and exhibited fringes, in such a manner that one piece was traversed by the light in the direction of the earth's motion, and the other in the contrary direction, the fringes were not in the least displaced. This result, as M. Babinet asserts, is contrary to the theory of aberration contained in a memoir read by him before the Academy in 1829, as well as to the other received theories on the subject. I have not been able to meet with this memoir, but it is easy to shew that the result of M. Babinet's experiment is in perfect accordance with Fresnel's theory.

Let T be the thickness of one of the glass plates, V the velocity of propagation of light in vacuum, supposing the ether at rest. Then V/μ would be the velocity with which light would traverse the glass if the ether were at rest; but the ether moving with a velocity v/μ^2, the light traverses the glass with a velocity $\dfrac{V}{\mu} \pm \dfrac{v}{\mu^2}$, and therefore in a time

$$T \div \left(\frac{V}{\mu} \pm \frac{v}{\mu^2} \right) = \frac{\mu T}{V} \left(1 \mp \frac{v}{\mu V} \right).$$

But if the glass were away, the light, travelling with a velocity $V \pm v$, would pass over the space T in the time

$$T \div (V \pm v) = \frac{T}{V}\left(1 \mp \frac{v}{V}\right).$$

Hence the retardation, expressed in time, $= (\mu - 1)\dfrac{T}{V}$, the same as if the earth were at rest. But in this case no effect would be produced on the fringes, and therefore none will be produced in the actual case.

I shall now shew that, according to Fresnel's theory, the laws of reflexion and refraction in singly refracting media are un-influenced by the motion of the earth. The method which I employ will, I hope, be found simpler than Fresnel's; besides it applies easily to the most general case. Fresnel has not given the calculation for reflexion, but has merely stated the result; and with respect to refraction, he has only considered the case in which the course of the light within the refracting medium is in the direction of the earth's motion. This might still leave some doubt on the mind, as to whether the result would be the same in the most general case.

If the ether were at rest, the direction of light would be that of a normal to the surfaces of the waves. When the motion of the ether is considered, it is most convenient to define the direction of light to be that of the line along which the *same portion* of a wave moves relatively to the earth. For this is in all cases the direction which is ultimately observed with a tele-scope furnished with cross wires. Hence, if A is any point in a wave of light, and if we draw AB normal to the wave, and proportional to V or V/μ, according as the light is passing through vacuum or through a refracting medium, and if we draw BC in the direction of the motion of the ether, and proportional to v or v/μ^2, and join AC, this line will give the direction of the ray. Of course, we might equally have drawn AD equal and parallel to BC and in the opposite direction, when DB would have given the direction of the ray.

Let a plane P be drawn perpendicular to the reflecting or refracting surface and to the waves of incident light, which in this investigation may be supposed plane. Let the velocity v of the ether in vacuum be resolved into p perpendicular to the plane P, and q in that plane; then the resolved parts of the velocity v/μ^2 of the ether within a refracting medium will be p/μ^2, q/μ^2. Let us first consider the effect of the velocity p.

It is easy to see that, as far as regards this resolved part of the velocity of the ether, the directions of the refracted and reflected *waves* will be the same as if the ether were at rest. Let BAC (fig. 1) be the intersection of the refracting surface and the plane P; DAE a normal to the refracting surface; AF, AG, AH normals to the incident, reflected and refracted waves. Hence AF, AG, AH will be in the plane P, and

$$\angle \, GAD = FAD, \quad \mu \sin HAE = \sin FAD.$$

Take

$$AG = AF, \quad AH = \frac{1}{\mu} \, AF.$$

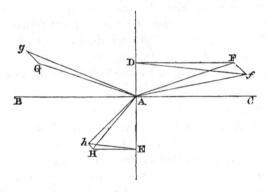

Draw Gg, Hh perpendicular to the plane P, and in the direction of the resolved part p of the velocity of the ether, and Ff in the opposite direction; and take

$$Ff : Hh : FA :: p : \frac{p}{\mu^2} : V, \quad \text{and} \quad Gg = Ff,$$

and join A with f, g and h. Then fA, Ag, Ah will be the directions of the incident, reflected and refracted rays. Draw FD, HE perpendicular to DE, and join fD, hE. Then fDF, hEH will be the inclinations of the planes fAD, hAE to the plane P. Now

$$\tan DFf = \frac{p}{V \sin FAD}, \quad \tan HEh = \frac{\mu^{-2} p}{\mu^{-1} V \sin HAE},$$

and $\sin FAD = \mu \sin HAE$; therefore $\tan FDf = \tan HEh$, and therefore the refracted ray Ah lies in the plane of incidence fAD. It is easy to see that the same is true of the reflected ray Ag. Also $\angle\, gAD = fAD$; and the angles fAD, hAE are sensibly equal to FAD, HAE respectively, and we therefore have without sensible error, $\sin fAD = \mu \sin hAE$. Hence the laws of reflexion and refraction are not sensibly affected by the velocity p.

Let us now consider the effect of the velocity q. As far as depends on this velocity, the incident, reflected and refracted rays will all be in the plane P. Let AH, AK, AL be the intersections of the plane P with the incident, reflected and refracted waves. Let ψ, $\psi_{,}$, ψ' be the inclinations of these waves to the refracting surface; let NA be the direction of the resolved part q of the velocity of the ether, and let the angle $NAC = \alpha$.

The resolved part of q in a direction perpendicular to AH is $q \sin (\psi + \alpha)$. Hence the wave AH travels with the velocity

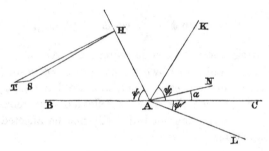

$V + q \sin (\psi + \alpha)$; and consequently the line of its intersection with the refracting surface travels along AB with the velocity $\operatorname{cosec} \psi \{V + q \sin (\psi + \alpha)\}$. Observing that q/μ^2 is the velocity of the ether within the refracting medium, and V/μ the velocity of propagation of light, we shall find in a similar manner that the lines of intersection of the refracting surface with the reflected and refracted waves travel along AB with velocities

$$\operatorname{cosec} \psi_{,}\{V + q \sin (\psi_{,} - \alpha)\}, \quad \operatorname{cosec} \psi' \left\{ \frac{V}{\mu} + \frac{q}{\mu^2} \sin (\psi' + \alpha) \right\}.$$

But since the incident, reflected and refracted waves intersect the refracting surface in the same line, we must have

$$\left. \begin{aligned} \sin \psi_{,}\{V + q \sin (\psi_{,} + \alpha)\} &= \sin \psi \{V + q \sin (\psi_{,} - \alpha)\} \\ \mu \sin \psi' \{V + q \sin (\psi + \alpha)\} &= \sin \psi \left\{ V + \frac{q}{\mu} \sin (\psi' + \alpha) \right\} \end{aligned} \right\} \dots\text{(A)}$$

Draw HS perpendicular to AH, ST parallel to NA, take $ST : HS :: q : V$, and join HT. Then HT is the direction of the incident ray; and denoting the angles of incidence, reflexion and refraction by ϕ, $\phi_{,}$, ϕ', we have

$$\phi - \psi = SHT = \frac{ST \sin S}{SH} = \frac{1}{V} \times \text{resolved part of } q \text{ along } AH$$

$$= \frac{q}{V} \cos (\psi + \alpha).$$

Similarly,

$$\phi_{,} - \psi_{,} = \frac{q}{V} \cos (\psi_{,} - \alpha), \quad \phi' - \psi' = \frac{q}{\mu V} \cos (\psi' + \alpha) ;$$

whence

$$\sin \psi = \sin \phi - \frac{q}{V} \cos \phi \cos (\phi + \alpha),$$

$$\sin \psi_{,} = \sin \phi_{,} - \frac{q}{V} \cos \phi_{,} \cos (\phi_{,} - \alpha),$$

$$\sin \psi' = \sin \phi' - \frac{q}{\mu V} \cos \phi' \cos (\phi' + \alpha).$$

On substituting these values in equations (A), and observing that in the terms multiplied by q we may put $\phi_{,} = \phi$, $\mu \sin \phi' = \sin \phi$, the small terms destroy each other, and we have $\sin \phi_{,} = \sin \phi$, $\mu \sin \phi' = \sin \phi$. Hence the laws of reflexion and refraction at the surface of a refracting medium will not be affected by the motion of the ether.

In the preceding investigation it has been supposed that the refraction is out of vacuum into a refracting medium. But the result is the same in the general case of refraction out of one medium into another, and reflexion at the common surface. For all the preceding reasoning applies to this case if we merely substitute p/μ'^2, q/μ'^2 for p, q, V/μ' for V, and μ/μ' for μ, μ' being the refractive index of the first medium. Of course refraction out of a medium into vacuum is included as a particular case.

It follows from the theory just explained, that the light coming from any star will behave in all cases of reflexion and ordinary refraction precisely as it would if the star were situated in the place which it appears to occupy in consequence of aberration, and the earth were at rest. It is, of course, immaterial whether the star is observed with an ordinary telescope, or with a telescope having its tube filled with fluid. It follows also that terrestrial objects are referred to their true places. All these results would follow immediately from the theory of aberration which I proposed in the July number of this Magazine; nor have I been able to obtain any result, admitting of being compared with experiment, which would be different according to which theory we adopted. This affords a curious instance of two totally different theories running parallel to each other in the explanation of phenomena. I do not suppose that many would be disposed to maintain Fresnel's theory, when it is shewn that it may be dispensed with, inasmuch as we would not be disposed to believe, without good evidence, that the ether moved quite freely through the solid mass of the earth. Still it would have been satisfactory, if it had been possible, to have put the two theories to the test of some decisive experiment.

[From the *Cambridge and Dublin Mathematical Journal*,
Vol. I. p. 183 (May, 1846).]

ON A FORMULA FOR DETERMINING THE OPTICAL CONSTANTS OF DOUBLY REFRACTING CRYSTALS.

IN order to explain the object of this formula, it will be necessary to allude to the common method of determining the optical constants. Two plane faces of the crystal are selected, which are parallel to one of the axes of elasticity; or if such do not present themselves, they are obtained artificially by grinding. A pencil of light is transmitted across these faces in a plane perpendicular to them both, as in the case of an ordinary prism. This pencil is by refraction separated into two, of which one is polarized in the plane of incidence, and follows the ordinary law of refraction, while the other is polarized in a plane perpendicular to the plane of incidence, and follows a different law. It will be convenient to call these pencils respectively the *ordinary* and the *extraordinary*, in the case of biaxal, as well as uniaxal crystals. The minimum deviation of the ordinary pencil is then observed, and one of the optical constants, namely that which relates to the axis of elasticity parallel to the refracting edge, is thus determined by the same formula which applies to ordinary media. This formula will also give one of the other constants, by means of the observation of the minimum deviation of the extraordinary pencil, in the particular case in which one of the principal planes of the crystal bisects the angle between the refracting planes: but if this condition be not fulfilled it will be necessary to employ either two or three prisms, according as the crystal is uniaxal or biaxal, to determine all the constants. The extraordinary pencil, however, need not in any case be rejected, provided only a formula be obtained connecting the minimum deviation observed

with the optical constants. It will thus be possible to determine
all the constants with a smaller number of prisms; the necessity
of using artificial faces may often be obviated; or if two faces
are cut as nearly as may be equally inclined to one of the axes of
elasticity lying in the plane of incidence, or one cut face is used
with a natural face, the errors of cutting may be allowed for.

Let AEB be a section of the prism by the plane of refraction,
(the reader will have no difficulty in drawing a figure,) E being
the refracting edge; let i be the refracting angle; OA, OB, OC
the directions of the axes of elasticity, O being any point within
the prism, the two former of these lines being in, and the latter
perpendicular to, the plane of refraction; a, b, c the optical constants referring to them, that is, according to Fresnel's theory,
the velocities of propagation of waves in which the vibrations
are parallel to the three axes respectively. Everything being
symmetrical with respect to the plane of incidence, we need only
consider what takes place in that plane. This plane will cut
the wave surface in a circle of radius c, and an ellipse whose
semiaxes are a along OB and b along OA. We have only got to
consider the ellipse, since it is it that determines the direction
of the extraordinary ray. The form of the crystal will very often
make known the directions of the axes of elasticity. Supposing
these directions known, let α, β denote the inclinations of OA, OB
to the produced parts of EA, EB respectively; α, β and i being
of course connected by the equation $\alpha + \beta = \frac{1}{2}\pi + i$.

Let ϕ, ψ be the angles of incidence and emergence, the light
being supposed incident on the face EA; ϕ' the inclination of the
refracted wave to EA, ψ' its inclination to EB, D the deviation,
v the velocity of the wave within the crystal, u its velocity in
the outer medium, which may be supposed to be either air, or a
liquid of known refractive power. Then we have

$$D = \phi + \psi - i^* \ \dots\dots\dots\dots\dots(1),$$
$$\phi' + \psi' = i \ \dots\dots\dots\dots\dots\dots(2),$$
$$v \sin \phi = u \sin \phi' \ \dots\dots\dots\dots\dots(3),$$
$$v \sin \psi = u \sin \psi' \ \dots\dots\dots\dots\dots(4),$$
$$v^2 = a^2 \cos^2 (\alpha - \phi') + b^2 \sin^2 (\alpha - \phi') \ \dots\dots\dots\dots(5).$$

* I am indebted to the Rev. P. Frost for the suggestion of employing equations
(1)...(4), rather than making use of the ellipse in which the wave surface is cut by
the plane of incidence.

From (2), (3), (4),

$$u \sin \psi' = v \sin \psi = u \sin (i - \phi') = u \sin i \cos \phi' - v \cos i \sin \phi ;$$

$$\therefore \cos \phi' = \frac{v}{u \sin i} (\sin \psi + \cos i \sin \phi) ;$$

and
$$\sin \phi' = \frac{v}{u \sin i} \sin i \sin \phi :$$

substituting in (5),

$$u^2 \sin^2 i = a^2 \{\cos \alpha (\sin \psi + \cos i \sin \phi) + \sin \alpha \sin i \sin \phi\}^2$$
$$+ b^2 \{\sin \alpha (\sin \psi + \cos i \sin \phi) - \cos \alpha \sin i \sin \phi\}^2,$$

or $u^2 \sin^2 i = a^2 (\cos \alpha \sin \psi + \sin \beta \sin \phi)^2$
$$+ b^2 (\sin \alpha \sin \psi + \cos \beta \sin \phi)^2 \dots\dots\dots\dots(6),$$

the relation between ϕ and ψ. Putting $\psi - \phi = \theta$, and taking account of (1), (6) becomes

$$2u^2 \sin^2 i = \{a^2 \cos^2 \alpha + b^2 \sin^2 \alpha\} \{1 - \cos (D + i + \theta)\}$$
$$+ \{a^2 \sin^2 \beta + b^2 \cos^2 \beta\} \{1 - \cos (D + i - \theta)\}$$
$$+ 2 (a^2 \cos \alpha \sin \beta + b^2 \sin \alpha \cos \beta) \{\cos \theta - \cos (D + i)\},$$

or
$$F \cos \theta + G \sin \theta + H = 0 \dots\dots\dots\dots(7),$$

where

$$F = a^2 \{(\cos^2 \alpha + \sin^2 \beta) \cos (D + i) - 2 \cos \alpha \sin \beta\}$$
$$+ b^2 \{(\sin^2 \alpha + \cos^2 \beta) \cos (D + i) - 2 \sin \alpha \cos \beta\},$$
$$G = (a^2 - b^2)(\sin^2 \beta - \cos^2 \alpha) \sin (D + i),$$
$$H = 2u^2 \sin^2 i - a^2 \{\cos^2 \alpha + \sin^2 \beta - 2 \cos \alpha \sin \beta \cos (D + i)\}$$
$$- b^2 \{\sin^2 \alpha + \cos^2 \beta - 2 \sin \alpha \cos \beta \cos (D + i)\}.$$

Now when D, regarded as a function of θ, is a maximum or minimum $\dfrac{dD}{d\theta} = 0$, whence from (7)

$$- F \sin \theta + G \cos \theta = 0 ;$$

and eliminating θ from this equation and (7), we have

$$F^2 + G^2 = H^2.$$

Putting for F, G and H their values, and reducing, this equation becomes

$$\sin^2 (D + i) a^2 b^2 - \{\cos^2 \alpha + \sin^2 \beta - 2 \cos (D + i) \cos \alpha \sin \beta\} u^2 a^2$$
$$- \{\sin^2 \alpha + \cos^2 \beta - 2 \cos (D + i) \sin \alpha \cos \beta\} u^2 b^2 + \sin^2 i . u^4 = 0 \dots (8).$$

This equation will be rendered more convenient for numerical calculation by replacing products and powers of sines and cosines by sums and differences. Treated in this manner, the equation becomes

$$\text{versin } 2 (D + i)\, a^2 b^2 - (A + B)\, u^2 a^2 - (A - B)\, u^2 b^2$$
$$+ \text{versin } 2i . u^4 = 0 \ ...(9),$$

where $A = \text{versin } D + \text{versin } (D + 2i)$,

$$B = \cos 2\alpha - \cos 2\beta - \cos (D + 2\alpha) + \cos (D + 2\beta).$$

If the principal plane $A\,OC$ of the crystal bisects the angle between the refracting faces, we have

$$\alpha = \frac{i}{2}, \quad \beta = \frac{\pi}{2} + \frac{i}{2},$$

whence from (8), putting $D + i = \Delta$,

$$\left(a^2 \sin^2 \frac{\Delta}{2} - u^2 \sin^2 \frac{i}{2}\right) \left(b^2 \cos^2 \frac{\Delta}{2} - u^2 \cos^2 \frac{i}{2}\right) = 0.$$

The former of these factors is evidently that which corresponds to the problem ; the latter corresponds to refraction through a prism having its faces parallel to those of the actual prism, and having its refracting angle supplemental to i. We have therefore

$$a = u \frac{\sin \dfrac{i}{2}}{\sin \dfrac{\Delta}{2}} \ ;$$

so that the constant a is given by the same formula that applies to ordinary media, as it should.

If the refracting faces are perpendicular to the axes of elasticity which lie in the plane of incidence, the formula (8) or (9) takes a very simple form. In this case we have $\alpha = \beta = i = \frac{1}{2}\pi$, and therefore

$$\cos^2 D . a^2 b^2 - u^2 a^2 - u^2 b^2 + u^4 = 0.$$

Mathematically speaking, one prism would be sufficient for determining the three constants a, b, c. For c would be determined by means of the ordinary pencil; and by observing the extraordinary pencil with the crystal in air, and again with the crystal in some liquid, we should have two equations of the form (8), by combining which we should obtain a^2 and b^2 by the solution of a quadratic equation. But since a is usually nearly equal to b, it is evident that the course of the extraordinary ray within the crystal would be nearly the same in the two observa-

tions, being in each case inclined at nearly equal angles to the
refracting faces, and consequently the errors of observation would
be greatly multiplied in the result. Even if a differed greatly
from b, only one of these constants could be accurately determined
in this manner if the refracting angle were nearly bisected by
a principal plane. But two prisms properly chosen appear amply
sufficient for determining accurately the three constants by the
method of minimum deviations, even should neither prism have
its angle exactly bisected by a principal plane of the crystal.

It is not necessary to observe the deviation when it is a
minimum, as Professor Miller has remarked to me, since the angle
of incidence may be measured very accurately by moving the
telescope employed till the luminous slit, seen directly, appears
on the cross wires, and then turning it till the slit, seen by reflection at the first face of the prism, again appears on the cross
wires, the prism meanwhile remaining fixed*. The angle through
which the telescope has been turned is evidently the supplement
of twice the angle of incidence. If this method of observation be
adopted, ϕ, D, and i will be known by observation, whence ψ
will be got immediately from (1). Thus all the coefficients in
(6) will be known quantities, and this equation furnishes a very
simple relation between a and b. The coefficients may easily be
calculated numerically by treating them like those in equation
(8), or else by employing subsidiary angles.

[* A method of measuring the refractive indices of isotropic media depending on
the measurement of the deviation and angle of incidence is described by Professor
Swan in the *Edinburgh New Philosophical Journal*, Vol. xxxvi. (1844) p. 102.]

[From the *Philosophical Magazine*, Vol. XXIX. p. 6 (July, 1846)].

ON THE CONSTITUTION OF LUMINIFEROUS ETHER, VIEWED WITH
REFERENCE TO THE PHENOMENON OF THE ABERRATION OF
LIGHT.

IN a former communication to this Magazine (July, 1845),[*]
I shewed that the phenomenon of aberration might be explained
on the undulatory theory of light, without making the startling
supposition that the earth in its motion round the sun offers
no resistance to the ether. It appeared that the phenomenon
was fully accounted for, provided we supposed the motion of the
ether such as to make

$$udx + vdy + wdz \quad\quad\quad\quad\quad (a)$$

an exact differential, where u, v, w are the resolved parts, along
three rectangular axes, of the velocity of the particle of ether
whose co-ordinates are x, y, z. It appeared moreover that it
was necessary to make this supposition in order to account in
this way for the phenomenon of aberration. I did not in that
paper enter into any speculations as to the physical causes in
consequence of which (a) might be an exact differential. The
object of the present communication is to consider this question.

The enquiry naturally divides itself into two parts:—*First*,
In what manner does one portion of ether act on another beyond the limits of the earth's atmosphere? *Secondly*, What
takes place in consequence of the mutual action of the air and
the ether?

In order to separate these two questions, let us first conceive
the earth to be destitute of an atmosphere. Before considering
the motion of the earth and the ether, let us take the case of

* *Ante*, p. 134.

a solid moving in an ordinary incompressible fluid, which may be supposed to be infinitely extended in all directions about the solid. If we suppose the solid and fluid to be at first at rest, and the solid to be then moved in any manner, it follows from the three first integrals of the ordinary equations of fluid motion, obtained by M. Cauchy, that the motion of the fluid at any time will be such that (a) is an exact differential. From this it may be easily proved, that if at any instant the solid be reduced to rest, the whole of the fluid will be reduced to rest likewise ; and that the motion of the fluid is the same as it would have been if the solid had received by direct impact the motion which it has at that instant. Practically however the motion of the fluid after some time would differ widely from what would be thus obtained, at least if the motion of the solid be progressive and not oscillatory. This appears to be due to two causes : first, the motion considered would probably be unstable in the part of the fluid behind the solid ; and secondly, a tangential force is called into play by the sliding of one portion of fluid along another ; and this force is altogether neglected in the common equations of hydrodynamics, from which equations the motion considered is deduced. If, instead of supposing the solid to move continuously, we supposed it first to be in motion for a very small interval of time, then to be at rest for another equal interval, then to be in motion for a third interval equal to the former, and so on alternately, theoretically the fluid ought to be at rest at the expiration of the first, third, &c. intervals, but practically a very slight motion would remain at the end of the first interval, would last through the second and third, and would be combined with a slight motion of the same kind, which would have been left at the end of the third interval, even if the fluid immediately before the commencement of it had been at rest ; and the accumulation of these small motions would soon become sensible.

Let us now return to the ether. We know that the transversal vibrations constituting light are propagated with a velocity about 10,000 times as great as the velocity of the earth ; and Mr Green has shewn that the velocity of propagation of normal vibrations is in all probability incomparably greater than that of transversal vibrations (*Cambridge Philosophical Transactions*, vol. VII. p. 2). Consequently, in considering the motion of the ether due to the motion of the earth, we may regard the ether as perfectly incompressible. To explain dynamically the pheno- mena of light, it seems necessary to suppose the motion of the ether subject to the same laws as the motion of an elastic solid. If the views which I have explained at the end of a paper *On the Friction of Fluids*, &c. (*Cambridge Philosophical Transactions*, vol. VIII. part 3)* be correct, it is only for extremely small vi- bratory motions that this is the case, while if the motion be progressive, or not very small, the ether will behave like an ordinary fluid. According to these views, therefore, the earth will set the ether in motion in the same way as a solid would set an ordinary incompressible fluid in motion.

Instead of supposing the earth to move continuously, let us first suppose it to move discontinuously, in the same manner as the solid considered above, the ether being at rest just before the commencement of the first small interval of time. By what precedes, the ether will move during the first interval in the same, or nearly the same, manner as an incompressible fluid would; and when, at the end of this interval, the earth is reduced to rest, the whole of the ether will be reduced to rest, except as regards an extremely small motion, of the same nature as that already considered in the case of an ordinary fluid. But in the present case this small motion will be propagated into space with the velocity of light; so that just before the com- mencement of the third interval the ether may be considered as at rest, and everything will be the same as before. Supposing now the number of intervals of time to be indefinitely increased, and their magnitude indefinitely diminished, we pass to the case in which the earth is supposed to move continuously.

It appears then, from these views of the constitution of the ether, that (a) must be an exact differential, if it be not pre- vented from being so by the action of the air on the ether. We know too little about the mutual action of the ether and material particles to enable us to draw any very probable conclusion respecting this matter; I would merely hazard the following conjecture. Conceive a portion of the ether to be filled with a great number of solid bodies, placed at intervals, and suppose these bodies to move with a velocity which is very small compared

---

* *Ante*, p. 125.

with the velocity of light, then the motion of the ether between the bodies will still be such that (a) is an exact differential. But if these bodies are sufficiently close and numerous, they must impress either the whole, or a considerable portion of their own velocity on the ether between them. Now the molecules of air may act the part of these solid bodies. It may thus come to pass that (a) is an exact differential, and yet the ether close to the surface of the earth is at rest relatively to the earth. The latter of these conditions is however not necessary for the explanation of aberration*.

[* A short demonstration that the path of a ray in the moving ether is a straight line, which here followed, is omitted, as the proposition has already been proved in the additional note printed at p. 138.]

[From the *Report of the British Association for* 1846, Part I. p. 1.]

REPORT ON RECENT RESEARCHES IN HYDRODYNAMICS.

AT the meeting of the British Association held at Cambridge last year, the Committee of the Mathematical Section expressed a wish that a Report on Hydrodynamics should be prepared, in continuation of the reports which Prof. Challis had already presented to the Association on that subject. Prof. Challis having declined the task of preparing this report, in consequence of the pressure of other engagements, the Committee of the Association did me the honour to entrust it to me. In accordance with the wishes of the Committee, the object of the present report will be to notice researches in this subject subsequent to the date of the reports of Prof. Challis. It will sometimes however be convenient, for the sake of giving a connected view of certain branches of the subject, to refer briefly to earlier investigations.

The fundamental hypothesis on which the science of hydrostatics is based may be considered to be, that the mutual action of two adjacent portions of a fluid at rest is normal to the surface which separates them. The equality of pressure in all directions is not an independent hypothesis, but a necessary consequence of the former. This may be easily proved by the method given in the *Exercises* of M. Cauchy*, a method which depends on the consideration of the forces acting on a tetrahedron of the fluid, the dimensions of which are in the end supposed to vanish. This proof applies equally to fluids at rest and fluids in motion ; and thus the hypothesis above-mentioned may be considered as the fundamental hypothesis of the ordinary theory of hydrodynamics, as well as hydrostatics. This hypothesis is fully confirmed by

* Tom. ii. p. 42.

experiment in the case of the equilibrium of fluids; but the comparison of theory and experiment is by no means so easy in the case of their motion, on account of the mathematical difficulty of treating the equations of motion. Still enough has been done to shew that the ordinary equations will suffice for the explanation of a great variety of phænomena; while there are others the laws of which depend on a tangential force, which is neglected in the common theory, and in consequence of which the pressure is different in different directions about the same point. The linear motion of fluids in uniform pipes and canals is a simple instance*. In the following report I shall first consider the common theory of hydrodynamics, and then notice some theories which take account of the inequality of pressure in different directions. It will be convenient to consider the subject under the following heads :—

I. General theorems connected with the ordinary equations of fluid motion.

II. Theory of waves, including tides.

III. The discharge of gases through small orifices.

IV. Theory of sound.

V. Simultaneous oscillations of fluids and solids.

VI. Formation of the equations of motion when the pressure is not supposed equal in all directions.

I. Although the common equations of hydrodynamics have been so long known, their complexity is so great that little has been done with them except in the case in which the expression usually denoted by

$$u dx + v dy + w dz \dots\dots\dots\dots\dots\dots(A)$$

is the exact differential of a function of the independent variables x, y, z†. It becomes then of the utmost importance to inquire in what cases this supposition may be made. Now Lagrange enunciated two theorems, by virtue of which, supposing them true, the supposition may be made in a great number of important cases, in fact, in nearly all those cases which it is most interesting to

---

[* See the footnote at p. 99.]

† In nearly all the investigations of Mr Airy it will be found that (A) is an exact differential, although he does not start with assuming it to be so.

investigate. It must be premised that in these theorems the accelerating forces X, Y, Z are supposed to be such that

$$X dx + Y dy + Z dz$$

is an exact differential, supposing the time constant, and the density of the fluid is supposed to be either constant, or a function of the pressure. The theorems are—

First, that (A) is approximately an exact differential when the motion is so small that squares and products of u, v, w and their differential coefficients may be neglected. By calling (A) approximately an exact differential, it is meant that there exists an expression $u_{,} dx + v_{,} dy + w_{,} dz$, which is accurately an exact differential, and which is such that $u_{,}, v_{,}, w_{,}$ differ from u, v, w respectively by quantities of the second order only.

Secondly, that (A) is accurately an exact differential at all times when it is so at one instant, and in particular when the motion begins from rest.

It has been pointed out by Poisson that the first of these theorems is not true*. In fact, the initial motion, being arbitrary, need not be such as to render (A) an exact differential. Thus those cases coming under the first theorem in which the assertion is true are merged in those which come under the second, at least if we except the case of small motions kept up by disturbing causes, a case in which we have no occasion to consider initial motion at all. This case it is true is very important.

The validity of Lagrange's proof of the second theorem depends on the legitimacy of supposing u, v and w capable of expansion according to positive, integral powers of the time t, for a sufficiently small value of that variable. This proof lies open to objection; for there are functions of t the expansions of which contain fractional powers, and there are others which cannot be expanded according to ascending powers of t, integral or fractional, even though they may vanish when $t = 0$. It has been shewn by Mr Power that Lagrange's proof is still applicable if u, v and w admit of expansion according to ascending powers of t of any kind†. The second objection however still remains: nor does the proof which Poisson has substituted for Lagrange's in his 'Traité de Mécanique' appear at all more satisfactory. Besides, it does not appear

---

* *Mémoires de l'Académie des Sciences*, tom. x. p. 554.

† *Transactions of the Cambridge Philosophical Society*, vol. vii. p. 455.

from these proofs what becomes of the theorem if it is only for a certain portion of the fluid that (A) is at one instant an exact differential.

M. Cauchy has however given a proof of the theorem *, which is totally different from either of the former, and perfectly satisfactory. M. Cauchy first eliminates the pressure by differentiation from the three partial differential equations of motion. He then changes the independent variables in the three resulting equations from x, y, z, t to a, b, c, t, where a, b, c are the initial co-ordinates of the particle whose co-ordinates at the time t are x, y, z. The three transformed equations admit each of being once integrated with respect to t, and the arbitrary functions of a, b, c introduced by integration are determined by the initial motion, which is supposed to be given. The theorem in question is deduced without difficulty from the integrals thus obtained. It is easily proved that if the velocity is suddenly altered by means of impulsive forces applied at the surface of the fluid, the alteration is such as to leave (A) an exact differential if it were such before impact. M. Cauchy's proof shews moreover that if (A) be an exact differential for one portion of the fluid, although not for the whole, it will always remain so for that portion. It should be observed, that although M. Cauchy has proved the theorem for an incompressible fluid only, the same method of proof applies to the more general case in which the density is a function of the pressure.

In a paper read last year before the Cambridge Philosophical Society, I have given a new proof of the same theorem†. This proof is rather simpler than M. Cauchy's, inasmuch as it does not require any integration.

In a paper published in the Philosophical Magazine‡, Prof. Challis has raised an objection to the application of the theorem to the case in which the motion of the fluid begins from rest. According to the views contained in this paper, we are not in general at liberty to suppose (A) to be an exact differential when u, v and w vanish: this supposition can only be made when the limiting value of $t^{-a}(u\,dx + v\,dy + w\,dz)$ is an exact differential, where a is so taken as that one at least of the terms in this expression does not vanish when t vanishes.

* _Mémoires des Savans Étrangers_, tom. i. p. 40.

† _Transactions of the Cambridge Philosophical Society_, vol. viii. p. 307.

‡ Vol. xxiv. New Series, p. 94.

It is maintained by Prof. Challis that the received equations of hydrodynamics are not complete, as regards the analytical principles of the science, and he has given a new fundamental equation, in addition to those received, which he calls the *equation of continuity of the motion**. On this equation Prof. Challis rests a result at which he has arrived, and which all must allow to be most important, supposing it correct, namely that whenever (A) is an exact differential the motion of the fluid is necessarily rectilinear, one peculiar case of circular motion being excepted. As I have the misfortune to differ from Professor Challis on the points mentioned in this and the preceding paragraph, for reasons which cannot be stated here, it may be well to apprise the reader that many of the results which will be mentioned further on as satisfactory lie open to Professor Challis's objections.

By virtue of the equation of continuity of a homogeneous incompressible fluid, the expression $u dy - v dx$ will always be the exact differential of a function of x and y. In the Cambridge Philosophical Transactions † there will be found some applications of this function, and of an analogous function for the case of motion which is symmetrical about an axis, and takes place in planes passing through the axis. The former of these functions had been previously employed by Mr Earnshaw.

II. In the investigations which come under this head, it is to be understood that the motion is supposed to be very small, so that first powers only of small quantities are retained, unless the contrary is stated.

The researches of MM. Poisson and Cauchy were directed to the investigation of the waves produced by disturbing causes acting arbitrarily on a small portion of the fluid, which is then left to itself. The mathematical treatment of such cases is extremely difficult; and after all, motions of this kind are not those which it is most interesting to investigate. Consequently it is the simpler cases of wave motion, and those which are more nearly connected with the phenomena which it is most desirable to explain, that have formed the principal subject of more recent investigations. It is true that there is one memoir by M. Ostrogradsky,

---

* *Transactions of the Cambridge Philosophical Society*, vol. viii. p. 31; and *Philosophical Magazine*, vol. xxvi. New Series, p. 425.

† Vol. vii. p. 439. (*Ante*, p. 1.)

162 REPORT ON RECENT RESEARCHES IN HYDRODYNAMICS.

read before the French Academy in 1826*, to which this character does not apply. In this memoir the author has determined the motion of the fluid contained in a cylindrical basin, supposing the fluid at first at rest, but its surface not horizontal. The interest of the memoir however depends almost exclusively on the mathematical processes employed; for the result is very complicated, and has not been discussed by the author. There is one circumstance mentioned by M. Plana† which increases the importance of the memoir in a mathematical point of view, which is that Poisson met with an apparent impossibility in endeavouring to solve the same problem. I do not know whether Poisson's attempt was ever published.

Theory of Long Waves.—When the length of the waves whose motion is considered is very great compared with the depth of the fluid, we may without sensible error neglect the difference between the horizontal motions of different particles in the same vertical line, or in other words suppose the particles once in a vertical line to remain in a vertical line: we may also neglect the vertical, compared with the horizontal effective force. These considerations extremely simplify the problem; and the theory of long waves is very important from its bearing on the theory of the tides. Lagrange's solution of the problem in the case of a fluid of uniform depth is well known. It is true that Lagrange fell into error in extending his solution to cases to which it does not apply; but there is no question as to the correctness of his result when properly restricted, that is when applied to the case of long waves only. There are however many questions of interest connected with this theory which have not been considered by Lagrange. For instance, what will be the velocity of propagation in a uniform canal whose section is not rectangular? How will the form of the wave be altered if the depth of the fluid, or the dimensions of the canal, gradually alter?

In a paper read before the Cambridge Philosophical Society in May 1837‡, the late Mr Green has considered the motion of long waves in a rectangular canal whose depth and breadth alter very slowly, but in other respects quite arbitrarily. Mr Green arrived at the following results:—If β be the breadth, and γ the depth of

* *Mémoires des Savans Etrangers*, tom. iii. p. 23.
† *Turin Memoirs* for 1835, p. 253.
‡ *Transactions of the Cambridge Philosophical Society*, vol. vi. p. 457.

the canal, then the height of the wave $\propto \beta^{-\frac{1}{2}}\gamma^{-\frac{1}{4}}$, the horizontal velocity of the particles in a given phase of their motion $\propto \beta^{-\frac{1}{2}}\gamma^{-\frac{3}{4}}$, the length of the wave $\propto \gamma^{\frac{1}{2}}$, and the velocity of propagation $= \sqrt{g\gamma}$. With respect to the height of the wave, Mr Russell was led by his experiments to the same law of its variation as regards the breadth of the canal, and with respect to the effect of the depth he observes that the height of the wave increases as the depth of the fluid decreases, but that the variation of the height of the wave is very slow compared with the variation of the depth of the canal.

In another paper read before the Cambridge Philosophical Society in February 1839*, Mr Green has given the theory of the motion of long waves in a triangular canal with one side vertical. Mr Green found the velocity of propagation to be the same as that in a rectangular canal of half the depth.

In a memoir read before the Royal Society of Edinburgh in April 1839†, Prof. Kelland has considered the case of a uniform canal whose section is of any form. He finds that the velocity of propagation is given by the very simple formula $\sqrt{\dfrac{gA}{b}}$, where A is the area of a section of the canal, and b the breadth of the fluid at the surface. This formula agrees with the experiments of Mr Russell, and includes as a particular case the formula of Mr Green for a triangular canal.

Mr Airy, the Astronomer Royal, in his excellent treatise on Tides and Waves, has considered the case of a variable canal with more generality than Mr Green, inasmuch as he has supposed the section to be of any form‡. If A, b denote the same things as in the last paragraph, only that now they are supposed to vary slowly in passing along the canal, the coefficient of horizontal displacement $\propto A^{-\frac{3}{4}}b^{\frac{1}{4}}$, and that of the vertical displacement $\propto A^{-\frac{1}{4}}b^{-\frac{1}{4}}$, while the velocity of propagation at any point of the canal is that given by the formula of the preceding paragraph. Mr Airy has proved the latter formula§ in a more simple manner than Prof. Kelland, and has pointed out the restrictions under which it is

* *Transactions of the Cambridge Philosophical Society*, vol. vii. p. 87.
† *Transactions of the Royal Society of Edinburgh*, vol. xiv. pp. 524, 530.
‡ *Encyclopædia Metropolitana*, article 'Tides and Waves.' Art. 260 of the treatise.
§ Art. 218, &c.

11—2

true. Other results of Mr Airy's will be more conveniently considered in connection with the tides.

Theory of Oscillatory Waves.—When the surface of water is covered with an irregular series of waves of different sizes, the longer waves will be continually overtaking the shorter, and the motion will be very complicated, and will offer no regular laws. In order to obtain such laws we must take a simpler case: we may for instance propose to ourselves to investigate the motion of a series of waves which are propagated with a constant velocity, and without change of form, in a fluid of uniform depth, the motion being in two dimensions and periodical. A series of waves of this sort may be taken as the *type* of oscillatory waves in general, or at least of those for which the motion is in two dimensions: to whatever extent a series of waves propagated in fluid of a uniform depth deviates from this standard form, to the same extent they fail in the characters of uniform propagation and invariable form.

The theory of these waves has long been known. In fact each element of the integrals by which MM. Poisson and Cauchy expressed the disturbance of the fluid denotes what is called by Mr Airy a *standing oscillation*, and a progressive oscillation of the kind under consideration will result from the superposition of two of these standing oscillations properly combined. Or, if we merely replace products of sines and cosines under the integral signs by sums and differences, each element of the new integrals will denote a progressive oscillation of the standard kind. The theory of these waves however well deserves a more detailed investigation. The most important formula connected with them is that which gives the relation between the velocity of propagation, the length of the waves, and the depth of the fluid. If c be the velocity of propagation, λ the length of the waves, measured from crest to crest, h the depth of the fluid, and $m = \dfrac{2\pi}{\lambda}$, then

$$c^2 = \frac{g}{m} \frac{\epsilon^{mh} - \epsilon^{-mh}}{\epsilon^{mh} + \epsilon^{-mh}} \quad\ldots\ldots\ldots\ldots\ldots\ldots\ldots(B).$$

If the surface of the fluid be cut by a vertical plane perpendicular to the ridges of the waves, the section of the surface will be the curve of sines. Each particle of the fluid moves round and round in an ellipse, whose major axis is horizontal. The particle is in its highest position when the crest of the wave is passing over it, and is then moving in the direction of propagation of the wave; it is in its lowest position when the hollow of the wave is passing over it, and is then moving in a direction contrary to the direction of propagation. At the bottom of the fluid the ellipse is reduced to a right line, along which the particle oscillates. When the length of waves is very small compared with the depth of the fluid, the motion at the bottom is insensible, and all the expressions will be sensibly the same as if the depth were infinite. On this supposition the expression for c reduces itself to $\sqrt{\dfrac{g\lambda}{2\pi}}$. The ellipses in which the particles move are replaced by circles, and the motion in each circle is uniform. The motion decreases with extreme rapidity as the point considered is further removed from the surface; in fact, the coefficients of the horizontal and vertical velocity contain as a factor the exponential ϵ^{-my}, where y is the depth of the particle considered below the surface. When the depth of the fluid is finite, the *law* (as to time) of the horizontal and vertical displacements of the particles is the same as when the depth is infinite. When the length of the waves is very great compared with the depth of the fluid, the horizontal motion of different particles in the same vertical line is sensibly the same. The expression for c reduces itself to $\sqrt{gh}$, the same as would have been obtained directly from the theory of long waves. The whole theory is given very fully in the treatise of Mr Airy[*]. The nature of the motion of the individual particles, as deduced from a rigorous theory, was taken notice of, I believe for the first time, by Mr Green[†], who has considered the case in which the depth is infinite.

The oscillatory waves just considered are those which are propagated uniformly in fluid of which the depth is everywhere the same. When this condition is not satisfied, as for instance when the waves are propagated in a canal whose section is not rectangular, it is desirable to know how the velocity of propagation and the form of the waves are modified by this circumstance. There is one such case in which a solution has been obtained. In a paper read before the Royal Society of Edinburgh in January 1841,

---

[*] Tides and Waves, art. 160, &c.

[†] *Transactions of the Cambridge Philosophical Society*, vol. vii. p. 95.

Prof. Kelland has arrived at a solution of the problem in the case of a triangular canal whose sides are inclined at an angle of 45° to the vertical, or of a canal with one side vertical and one side inclined at an angle of 45°, in which the motion will of course be the same as in one half of the complete canal*. The velocity of propagation is given by the formula (B), which applies to a rectangular canal, or to waves propagated without lateral limitation, provided we take for h half the greatest depth in the triangular canal, and for λ, or $2\pi/m$, a quantity less than the length of the waves in the triangular canal in the ratio of 1 to $\sqrt{2}$. As to the form of the waves, a section of the surface made by a vertical plane parallel to the edges of the canal is the curve of sines ; a section made by a vertical plane perpendicular to the former is the common catenary, with its vertex in the plane of the middle of the canal (supposed complete), and its concavity turned upwards or downwards according as the section is taken where the fluid is elevated or where it is depressed. Thus the ridges of the waves do not bend forwards, but are situated in a vertical plane, and they rise higher towards the slanting sides of the canal than in the middle. I shall write down the value of ϕ, the integral of (A), and then any one who is familiar with the subject can easily verify the preceding results. In the following expression x is measured along the bottom line of the canal, y is measured horizontally, and z vertically upwards :—

$$\phi = A \left(\epsilon^{\alpha y} + \epsilon^{-\alpha y}\right) \left(\epsilon^{\alpha z} + \epsilon^{-\alpha z}\right) \sin \sqrt{2}\, \varkappa \, (x - ct) \ldots \ldots (C).$$

I have mentioned these results under the head of oscillatory waves, because it is to that class only that the investigation strictly applies. The length of the waves is however perfectly arbitrary, and when it bears a large ratio to the depth of the fluid, it seems evident that the circumstances of the motion of any one wave cannot be materially affected by the waves which precede and follow it, especially as regards the form of the middle portion, or ridge, of the wave. Now the solitary waves of Mr Russell are long compared with the depth of the fluid ; thus in the case of a rectangular canal he states that the length of the wave is about six times the depth. Accordingly Mr Russell finds that the form of the ridge agrees well with the results of Prof. Kelland.

* *Transactions of the Royal Society of Edinburgh,* vol. xv. p. 121.

It appears from Mr Russell's experiments that there is a certain limit to the slope of the sides of a triangular canal, beyond which it is impossible to propagate a wave in the manner just considered. Prof. Kelland has arrived at the same result from theory, but his mathematical calculation does not appear to be quite satisfactory. Nevertheless there can be little doubt that such a limit does exist, and that if it be passed, the wave will be either continually breaking at the sides of the canal, or its ridge will become bow-shaped, in consequence of the portion of the wave in the middle of the canal being propagated more rapidly than the portions which lie towards the sides. When once a wave has become sufficiently curved it may be propagated without further change, as Mr Airy has shewn*. Thus the case of motion above considered is in nowise opposed to the circumstance that the tide wave assumes a curved form when it is propagated in a broad channel in which the water is deepest towards the centre.

It is worthy of remark, that if in equation (C), we transfer the origin to either of the upper edges of the canal (supposed complete), and then suppose h to become infinite, having previously written $A\epsilon^{-4ah}$ for A, the result will express a series of oscillatory waves propagated in deep water along the edge of a bank having a slope of $45°$, the ridges of the waves being perpendicular to the edge of the fluid. It is remarkable that the disturbance of the fluid decreases with extreme rapidity as the perpendicular distance from the edge increases, and not merely as the distance from the surface increases. Thus the disturbance is sensible only in the immediate neighbourhood of the edge, that is at a distance from it which is a small multiple of λ. The formula may be accommodated to the case of a bank having any inclination by merely altering the coefficients of y and z, without altering the sum of the squares of the coefficients. If i be the inclination of the bank to the vertical, it will be easily found that the velocity of propagation is equal to $\left(\dfrac{g\lambda}{2\pi}\cos i\right)^{\frac{1}{2}}$. When i vanishes these waves pass into those already mentioned as the standard case of oscillatory waves; and when i becomes negative, or the bank overhangs the fluid, a motion of this sort becomes impossible.

I have had occasion to refer to what Mr Airy calls a *standing*

* Tides and Waves, art. 359.

oscillation or *standing wave.* To prevent the possibility of confusion, it may be well to observe that Mr Airy uses the term in a totally different sense from Mr Russell. The standing wave of Mr Airy is the oscillation which would result from the co-existence of two series of progressive waves, which are equal in every respect, but are propagated in opposite directions. With respect to the standing wave of Mr Russell, it cannot be supposed that the elevations observed in mountain streams can well be made the subject of mathematical calculation. Nevertheless in so far as the motion can be calculated, by taking a simple case, the theory does not differ from that of waves of other classes. For if we only suppose a velocity equal and opposite to that of the stream impressed both on the fluid and on the stone ·at the bottom which produces the disturbance, we pass to the case of a forced wave produced in still water by a solid dragged through it. There is indeed one respect in which the theory of these standing waves offers a peculiarity, which is, that the velocity of a current is different at different depths. But the theory of such motions is one of great complexity and very little interest.

Theory of Solitary Waves.—It has been already remarked that the length of the solitary wave of Mr Russell is considerable compared with the depth of the fluid. Consequently we might expect that the theory of long waves would explain the main phenomena of solitary waves. Accordingly it is found by experiment that the velocity of propagation of a solitary wave in a rectangular canal is that given by the formula of Lagrange, the height of the wave being very small, or that given by Prof. Kelland's formula when the canal is not rectangular. Moreover, the laws of the motion of a solitary wave, deduced by Mr Green from the theory of long waves, agree with the observations of Mr Russell. Thus Mr Green found, supposing the canal rectangular, that the particles in a vertical plane perpendicular to the length of the canal remain in a vertical plane; that the particles begin to move when the wave reaches them, remain in motion while the wave is passing over them, and are finally deposited in new positions; that they move in the direction of propagation of the wave, or in the contrary direction, according as the wave consists of an elevation or a depression[*]. But when we attempt to introduce into our calculations

---

[*] *Transactions of the Cambridge Philosophical Society,* vol. vii. p. 87.

the finite length of the wave, the problem becomes of great difficulty. Attempts have indeed been made to solve it by the introduction of discontinuous functions. But whenever such functions are introduced, there are certain conditions of continuity to be satisfied at the common surface of two portions of fluid to which different analytical expressions apply; and should these conditions be violated, the solution will be as much in fault as it would be if the fluid were made to penetrate the bottom of the canal. No doubt, the theory is contained, to a first approximation, in the formulæ of MM. Poisson and Cauchy; but as it happens the obtaining of these formulæ is comparatively easy, their discussion forms the principal difficulty. When the height of the wave is not very small, so that it is necessary to proceed to a second approximation, the theory of long waves no longer gives a velocity of propagation agreeing with experiment. It follows, in fact, from the investigations of Mr Airy, that the velocity of propagation of a long wave is, to a second approximation, $\sqrt{g(h+3k)}$, where h is the depth of the fluid when it is in equilibrium, and $h+k$ the height of the crest of the wave above the bottom of the canal[*].

The theory of the two great solitary waves of Mr Russell forms the subject of a paper read by Mr Earnshaw before the Cambridge Philosophical Society in December last[†]. Mr Russell found by experiment that the horizontal motion of the fluid particles was sensibly the same throughout the whole of a vertical plane perpendicular to the length of the canal. He attributed the observed degradation of the wave, and consequent diminution of the velocity of propagation, entirely to the imperfect fluidity of the fluid, and its adhesion to the sides and bottom of the canal. Mr Earnshaw accordingly investigates the motion of the fluid on the hypotheses, —first, that the particles once in a vertical plane, perpendicular to the length of the canal, remain in a vertical plane; secondly, that the wave is propagated with a constant velocity and without

---

[*] Tides and Waves, art. 208. In applying this formula to a *solitary* wave, it is necessary to take for h the depth of the undisturbed portion of the fluid. In the treatise of Mr Airy the formula is obtained for a particular law of disturbance, but the same formula would have been arrived at, by the same reasoning, had the law not been restricted. This formula is given as expressing the velocity of propagation of the phase of high water, which it is true is not quite the same as the velocity of propagation of the crest of the wave; but the two velocities are the same to the second order of approximation.

[†] *Transactions of the Cambridge Philosophical Society*, vol. viii. p. 326.

change of form. It is important to observe that these hypotheses are used not as a *foundation for calculation*, but as a *means of selecting* a particular kind of motion for consideration. The equations of fluid motion admit of integration in this case in finite terms, without any approximation, and it turns out that the motion is possible, *so far as the wave itself is concerned*, and everything is determined in the result except two constants, which remain arbitrary. However, in order that the motion in question should actually take place, it is necessary that there should be an instantaneous generation or destruction of a finite velocity, and likewise an abrupt change of pressure, at the junction of the portion of fluid which constitutes the wave with the portions before and behind which are at rest, both which are evidently impossible. It follows of course that one at least of the two hypotheses must be in fault. Experiment shewing that the first hypothesis is very nearly true, while the second (from whatever cause) is sensibly erroneous, the conclusion is that in all probability the degradation of the wave is not to be attributed wholly to friction, but that it is an essential characteristic of the motion. Nevertheless the formula for the velocity of propagation of the positive wave, at which Mr Earnshaw has arrived, agrees very well with the experiments of Mr Russell; the formula for the negative wave also agrees, but not closely. These two formulæ can be derived from each other only by introducing imaginary quantities.

It is the opinion of Mr Russell that the solitary wave is a phenomenon *sui generis*, in nowise deriving its character from the circumstances of the generation of the wave. His experiments seem to render this conclusion probable. Should it be correct, the analytical character of the solitary wave remains to be discovered. A complete theory of this wave should give, not only its velocity of propagation, but also the law of its degradation, at least of that part of the degradation which is independent of friction, which is probably by far the greater part. With respect to the importance of this peculiar wave however, it must be remarked that the term *solitary wave*, as so defined, must not be extended to the tide wave, which is nothing more (as far as regards the laws of its propagation) than a very long wave, of which the form may be arbitrary. It is hardly necessary to remark that the mechanical theories of the solitary wave and of the aërial sound wave are altogether different.

Theory of River and Ocean Tides.—The treatise of Mr Airy already referred to is so extensive, and so full of original matter, that it will be impossible within the limits of a report like the present to do more than endeavour to give an idea of the nature of the calculations and methods of explanation employed, and to mention some of the principal results.

On account of the great length of the tide wave, the horizontal motion of the water will be sensibly the same from top to bottom. This circumstance most materially simplifies the calculation. The partial differential equation for the motion of long waves, when the motion is very small, is in the simplest case the same as that which occurs in the theory of the rectilinear propagation of sound; and in Mr Airy's investigations the arbitrary functions which occur in its integral are determined by the conditions to be satisfied at the ends of the canal in which the waves are propagated, in a manner similar to that in which the arbitrary functions are determined in the case of a tube in which sound is propagated. When the motion is not very small, the partial differential equation of wave motion may be integrated by successive approximations, the arbitrary functions being determined at each order of approximation as before.

To proceed to some of the results. The simplest conceivable case of a tidal river is that in which the river is regarded as a uniform, indefinite canal, without any current. The height of the water at the mouth of the canal will be expressed, as in the open sea, by a periodic function of the time, of the form $a \sin (nt + a)$. The result of a first approximation of course is that the disturbance at the mouth of the canal will be propagated uniformly up it, with the velocity due to half the depth of the water. But on proceeding to a second approximation*, Mr Airy finds that the form of the wave will alter as it proceeds up the river. Its front will become shorter and steeper, and its rear longer and more gently sloping. When the wave has advanced sufficiently far up the river, its surface will become horizontal at one point in the rear, and further on the wave will divide into two. At the mouth of the river the greatest velocities of the ebb and flow of the tide are equal, and occur at low and high water respectively; the time during which the water is rising is also equal to the time during

* Art. 198, &c.

which it is falling. But at a station up the river the velocity of
the ebb-stream is greater than that of the flow-stream, and the
rise of the water occupies less time than its fall. If the station
considered is sufficiently distant from the mouth of the river, and
the tide sufficiently large, the water after it has fallen some way
will begin to rise again : there will in fact be a double rise and
fall of the water at each tide. This explains the double tides
observed in some tidal rivers. The velocity with which the phase of
high water travels up the river is found to be $\sqrt{gk(1+3b)}$, k being
the depth of the water when in equilibrium, and bk the greatest
elevation of the water at the mouth of the river above its mean
level. The same formula will apply to the case of low water if we
change the sign of b. This result is very important, since it shews
that the interval between the time of the moon's passage over the
meridian of the river station and the time of high water will be
affected by the height of the tide. Mr Airy also investigates the
effect of the current in a tidal river. He finds that the difference
between the times of the water's rising and falling is increased by
the current.

When the canal is stopped by a barrier the circumstances are
altered. When the motion is supposed small, and the disturbing
force of the sun and moon is neglected, it is found in this case
that the tide-wave is a stationary wave*, so that there is high or
low water at the same instant at every point of the canal; but
if the length of the canal exceeds a certain quantity, it is high
water in certain parts of the canal at the instant when it is low
water in the remainder, and *vice versâ*. The height of high water
is different in different parts of the canal : it increases from the
mouth of the canal to its extremity, provided the canal's length
does not exceed a certain quantity. If four times the length of
the canal be any odd multiple of the length of a free wave whose
period is equal to that of the tide, the denominator of the expression for the tidal elevation vanishes. Of course friction would
prevent the elevation from increasing beyond a certain amount,
but still the tidal oscillation would in such cases be very large.

When the channel up which the tide is propagated decreases
in breadth or depth, or in both, the height of the tide increases in
ascending the channel. This accounts for the great height of the

* Art. 307.

tides observed at the head of the Bristol Channel, and in such places. In some of these cases however the great height may be partly due to the cause mentioned at the end of the last paragraph.

When the tide-wave is propagated up a broad channel, which becomes shallow towards the sides, the motion of the water in the centre will be of the same nature as the motion in a free canal, so that the water will be flowing up the channel with its greatest velocity at the time of high water. Towards the coasts however there will be a considerable flow of water to and from the shore; and as far as regards this motion, the shore will have nearly the same effect as a barrier in a canal, and the oscillation will be of the nature of a stationary wave, so that the water will be at rest when it is at its greatest height. If, now, we consider a point at some distance from the shore, but still not near the middle of the channel, the velocity of the water up and down the channel will be connected with its height in the same way as in the case of a progressive wave, while the velocity to and from the shore will be connected with the height of the water in the same way as in a stationary wave. Combining these considerations, Mr Airy is enabled to explain the apparent rotation of the water in such localities, which arises from an actual rotation in the direction of its motion*.

When the motion of the water is in two dimensions the mathematical calculation of the tidal oscillations is tolerably simple, at least when the depth of the water is uniform. But in the case of nature the motion is in three dimensions, for the water is distributed over the surface of the earth in broad sheets, the boundaries of which are altogether irregular. On this account a complete theory of the tides appears hopeless, even in the case in which the depth is supposed uniform. Laplace's theory, in which the whole earth is supposed to be covered with water, the depth of which follows a very peculiar law, gives us no idea of the effect of the limitation of the ocean by continents. Mr Airy consequently investigates the motion of the water on the supposition of its being confined to narrow canals of uniform depth, which in the calculation are supposed circular. The case in which the canal forms a great circle is especially considered. This method enables us in

* Art. 360, &c.

some degree to estimate the effect of the boundaries of the sea; and it has the great advantage of leading to calculations which can be worked out. There can be no doubt, too, that the conclusions arrived at will apply, *as to their general nature*, to the actual case of the earth.

With a view to this application of the theory, Mr Airy calculates the motion of the water in a canal when it is under the action of a disturbing force, which is a periodic function of the time. The disturbing force at a point whose abscissa, measured along the canal from a fixed point, is x, is supposed to be expressed by a function of the form $A \sin (nt - mx + a)$. This supposition is sufficiently general for the case of the tides, provided the canal on the earth be supposed circular. In all cases the disturbing force will give rise to an oscillation in the water having the same period as the force itself. This oscillation is called by Mr Airy a *forced wave*. It will be sufficient here to mention some of the results of this theory as applied to the case of the earth.

In all cases the expression for the tidal elevation contains as a denominator the difference of the squares of two velocities, one the velocity of propagation of a free wave along the canal, the other the velocity with which a particular phase of the disturbing force travels along the canal, or, which is the same, the velocity of propagation of the forced wave. Hence the height of the tides will not depend simply on the *magnitude* of the disturbing force, but also on its *period*. Thus the mass of the moon cannot be inferred *directly* from the comparison of spring and neap tides, since the heights of the solar and lunar tides are affected by the different motions of the sun and moon in right ascension, and consequently in hour-angle. When the canal under consideration is equatorial the diurnal tide vanishes. The height of high water is the same at all points of the canal, and there is either high or low water at the point of the canal nearest to the attracting body, according as the depth of the water is greater or less than that for which a free wave would be propagated with the same velocity as the forced wave. In the general case there is both a diurnal and a semidiurnal tide, and the height of high water, as well as the interval between the transit of the attracting body over the meridian of the place considered and the time of high water, is different at different points of the canal. When the canal is a great circle passing through the poles, the tide-wave is a stationary wave.

When the coefficient of the disturbing force is supposed to vary slowly, in consequence of the change in declination, &c. of the disturbing body, it is found that the greatest tide occurs on the day on which the disturbing force is the greatest.

The preceding results have been obtained on the supposition of the absence of all friction; but Mr Airy also takes friction into consideration. He supposes it to be represented by a horizontal force, acting uniformly from top to bottom of the water, and varying as the first power of the horizontal velocity. Of course this supposition is not exact: still there can be no doubt that it represents generally the effect of friction. When friction is taken into account, the denominator of the expressions for the tidal elevation is essentially positive, so that the motion can never become infinite. In the case of a uniform tidal river stopped by a barrier, the high water is no longer simultaneous at all points, but the phase of high water always travels up the river. But of all the results obtained by considering friction, the most important appears to be, that when the slow variation of the disturbing force is taken into account, the greatest tide, instead of happening on the day when the disturbing force is greatest, will happen later by a certain time ρ_1. Moreover, in calculating the tides, we must use, not the relative positions of the sun and moon for the instant for which the tide is calculated, but their relative positions for a time earlier by the same interval ρ_1 as in the preceding case. The expression for ρ_1 depends both on the depth of the canal and on the period of the tide, and therefore its value for the diurnal tide cannot be inferred from its value for the semidiurnal. It appears also that the phase of the tide is accelerated by friction.

The mechanical theory of the tides of course belongs to hydrodynamics; but I do not conceive that the consideration of the reduction and discussion of tidal observations falls within the province of this report.

Before leaving the investigations of Mr Airy, I would call attention to a method which he sometimes employs very happily in giving a general explanation of phenomena depending on motions which are too complicated to admit of accurate calculation. It is evident that any arbitrary motion may be assigned to a fluid, (with certain restrictions as to the absence of abruptness,) provided we suppose certain forces to act so as to produce them. The values of these forces are given by the equations of motion. In some cases the forces thus obtained will closely resemble some known forces; while in others it will be possible to form a clear conception of the kind of motion which must take place in the absence of such forces. For example, supposing that there is propagated a series of oscillatory waves of the standard kind, except that the height of the waves increases proportionably to their distance from a fixed line, remaining constant at the same point as the time varies, Mr Airy finds for the force requisite to maintain such a motion an expression which may be assimilated to the force which wind exerts on water. This affords a general explanation of the increase in the height of the waves in passing from a windward to a lee shore*. Again, by supposing a series of waves, as near the standard kind as circumstances will admit, to be propagated along a canal whose depth decreases slowly, and examining the force requisite to maintain this motion, he finds that a force must be applied to hold back the heads of the waves. In the absence, then, of such a force the heads of the waves will have a tendency to shoot forwards. This explains the tendency of waves to break over a sunken shoal or along a sloping beach†. The word *tendency* is here used, because when a wave comes at all near breaking, but little reliance can be placed in any investigation which depends upon the supposition of the motion being small. To take one more example of the application of this method, by supposing a wave to travel, unchanged in form, along a canal, with a velocity different from that of a free wave, and examining the force requisite to maintain such a motion, Mr Airy is enabled to give a general explanation of some very curious circumstances connected with the motion of canal boats‡, which have been observed by Mr Russell.

III.   In the 16th volume of the 'Journal de l'École Polytechnique'§, will be found a memoir by MM. Barré de Saint-Venant and Wantzel, containing the results of some experiments on the discharge of air through small orifices, produced by considerable differences of pressure. The formula for the velocity of efflux derived from the theory of steady motion, and the supposition that the mean pressure at the orifice is equal to the pressure at a distance from the orifice in the space into which the discharge

---

* Art. 265, &c.   † Art. 238, &c.
‡ Art. 405, &c.   § Cahier xxvii. p. 85.

takes place, leads to some strange results of such a nature as to make us doubt its correctness. If we call the space from which the discharge takes place the *first* space, and that into which it takes place the *second* space, and understand by the term *reduced velocity* the velocity of efflux diminished in the ratio of the density in the second space to the density in the first, so that the reduced velocity measures the rate of discharge, provided the density in the first space remain constant, it follows from the common formula that the reduced velocity vanishes when the density in the second space vanishes, so that a gas cannot be discharged into a vacuum. Moreover, if the density of the first space is given, the reduced velocity is a maximum when the density in the second space is rather more than half that in the first. The results remain the same if we take account of the contraction of the vein, and they are not materially altered if we take into account the cooling of the air by its rapid dilatation. The experiments above alluded to were made by allowing the air to enter an exhausted receiver through a small orifice, and observing simultaneously the pressure and temperature of the air in the receiver, and the time elapsed since the opening of the orifice. It was found that when the exhaustion was complete the reduced velocity had a certain value, depending on the orifice employed, and that the velocity did not sensibly change till the pressure of the air in the receiver became equal to about ⅔ths of the atmospheric pressure. The reduced velocity then began to decrease, and finally vanished when the pressure of the air in the receiver became equal to the atmospheric pressure.

These experiments shew that when the difference of pressure in the first and second spaces is considerable, we can by no means suppose that the mean pressure at the orifice is equal to the pressure at a distance in the second space, nor even that there exists a contracted vein, at which we may suppose the pressure to be the same as at a distance. The authors have given an empirical formula, which represents very nearly the reduced velocity, whatever be the pressure of the air in the space into which the discharge takes place.

The orifices used in these experiments were generally about one millimetre in diameter. It was found that widening the mouth of the orifice, so as to make it funnel-shaped, produced a much greater proportionate increase of velocity when the velocity

178 REPORT ON RECENT RESEARCHES IN HYDRODYNAMICS.

of efflux was small than when it was large. The authors have
since repeated their experiments with air coming from a vessel in
which the pressure was four atmospheres: they have also tried
the effect of using larger orifices of four or five millimetres
diameter. The general results were found to be the same as
before*.

IV. In the 6th volume of the *Transactions of the Cambridge
Philosophical Society*, p. 403, will be found a memoir by Mr Green
on the reflection and refraction of sound, which is well worthy of
attention. This problem had been previously considered by Poisson in an elaborate memoir†. Poisson treats the subject with
extreme generality, and his analysis is consequently very complicated. Mr Green, on the contrary, restricts himself to the case of
plane waves, a case evidently comprising nearly all the phenomena
connected with this subject which are of interest in a physical
point of view, and thus is enabled to obtain his results by a very
simple analysis. Indeed Mr Green's memoirs are very remarkable,
both for the elegance and rigour of the analysis, and for the ease
with which he arrives at most important results. This arises in a
great measure from his divesting the problems he considers of all
unnecessary generality: where generality is really of importance
he does not shrink from it. In the present instance there is one
important respect in which Mr Green's investigation is more general
than Poisson's, which is, that Mr Green has taken the case of any
two fluids, whereas Poisson considered the case of two elastic fluids,
in which equal condensations produce equal increments of pressure.
It is curious, that Poisson, forgetting this restriction, applied his
formulæ to the case of air and water. Of course his numerical
result is altogether erroneous. Mr Green easily arrives at the
ordinary laws of reflection and refraction. He obtains also a very
simple expression for the intensity of the reflected sound. If A is
the ratio of the density of the second medium to that of the first,
and B the ratio of the cotangent of the angle of refraction to the
cotangent of the angle of incidence, then the intensity of the
reflected sound is to the intensity of the incident as $A - B$ to
$A + B$. In this statement the intensity is supposed to be measured by the first power of the maximum displacement. When

* *Comptes Rendus*, tom. xvii. p. 1140.
† *Mémoires de l'Académie des Sciences*, tom. x. p. 317.

the velocity of propagation in the first medium is less than in the second, and the angle of incidence exceeds what may be called the critical angle, Mr Green restricts himself to the case of vibrations following the cycloidal law. He finds that the sound suffers total internal reflection. The expression for the disturbance in the second medium involves an exponential with a negative index, and consequently the disturbance becomes quite insensible at a distance from the surface equal to a small multiple of the length of a wave. The phase of vibration of the reflected sound is also accelerated by a quantity depending on the angle of incidence. It is remarkable, that when the fluids considered are ordinary elastic fluids, or rather when they are such that equal condensations produce equal increments of pressure, the expressions for the intensity of the reflected sound, and for the acceleration of phase when the angle of incidence exceeds the critical angle, are the same as those given by Fresnel for light polarized in a plane perpendicular to the plane of incidence.

V. Not long after the publication of Poisson's memoir on the simultaneous motions of a pendulum and of the surrounding air*, a paper by Mr Green was read before the Royal Society of Edinburgh, which is entitled 'Researches on the Vibration of Pendulums in Fluid Media†.' Mr Green does not appear to have been at that time acquainted with Poisson's memoir. The problem which he has considered is one of the same class as that treated by Poisson. Mr Green has supposed the fluid to be incompressible, a supposition, however, which will apply without sensible error to air, in considering motions of this sort. Poisson regarded the fluid as elastic, but in the end, in adapting his formula to use, he has neglected as insensible the terms by which the effect of an elastic differs from that of an inelastic fluid. The problem considered by Mr Green is, however, in one respect much more general than that solved by Poisson, since Mr Green has supposed the oscillating body to be an ellipsoid, whereas Poisson considered only a sphere. Mr Green has obtained a complete solution of the problem in the case in which the ellipsoid has a motion of translation only, or in which the small motion of the fluid due to its motion

---

* *Mémoires de l'Académie des Sciences*, tom. xi. p. 521.

† This paper was read in December, 1833, and is printed in the 13th volume of the Society's *Transactions*, p. 54, &c.

of rotation is neglected. The result is that the resistance of the fluid will be allowed for if we suppose the mass of the ellipsoid increased by a mass bearing a certain ratio to that of the fluid displaced. In the general case this ratio depends on three transcendental quantities, given by definite integrals. If, however, the ellipsoid oscillates in the direction of one of its principal axes, the ratio depends on one only of these transcendents. When the ellipsoid passes into a spheroid, the transcendents above mentioned can be expressed by means of circular or logarithmic functions. When the spheroid becomes a sphere, Mr Green's result agrees with Poisson's. It is worthy of remark, that Mr Green's formula will enable us to calculate the motion of an ellipse or circle oscillating in a fluid, in a direction perpendicular to its plane, since a material ellipse or circle may be considered as a limiting form of an ellipsoid. In this case, however, the motion would probably have to be extremely small, in order that the formula should apply with accuracy.

In a paper 'On the Motion of a small Sphere acted on by the Vibrations of an Elastic Medium,' read before the Cambridge Philosophical Society in April 1841*, Prof. Challis has considered the motion of a ball pendulum, retaining in his solution small quantities to the second order. The principles adopted by Prof. Challis in the solution of this problem are at variance with those of Poisson, and have given rise to a controversy between him and Mr Airy, which will be found in the 17th, 18th, and 19 volumes of the *Philosophical Magazine* (New Series). In the paper just referred to, Prof. Challis finds that when the fluid is incompressible there is no decrement in the arc of oscillation, except what arises from friction and capillary attraction. In the case of air there is a slight theoretical decrement; but it is so small that Prof. Challis considers the observed decrement to be mainly owing to friction. This result follows also from Poisson's solution. Prof. Challis also finds that a small sphere moving with a uniform velocity experiences no resistance, and that when the velocity is partly uniform and partly variable, the resistance depends on the variable part only. The problem, however, referred to in the title of this paper, is that of calculating the motion of a small sphere situated in an elastic fluid, and acted on by no forces except the pressure of the

---

* *Transactions of the Cambridge Philosophical Society*, vol. vii. p. 333.

fluid, in which an indefinite series of plane condensing and rarefying waves is supposed to be propagated. This problem is solved by the author on principles similar to those which he has adopted in the problem of an oscillating sphere. The views of Prof. Challis with respect to this problem, which he considers a very important one, are briefly stated at the end of a paper published in the *Philosophical Magazine**.

In a paper 'On some Cases of Fluid Motion,' published in the *Transactions of the Cambridge Philosophical Society*†, I have considered some modifications of the problem of the ball pendulum, adopting in the main the principles of Poisson, of the correctness of which I feel fully satisfied, but supposing the fluid incompressible from the first. In this paper the effect of a distant rigid plane interrupting the fluid in which the sphere is oscillating is given to the lowest order of approximation with which the effect is sensible. It is shewn also that when the ball oscillates in a concentric spherical envelope, the effect of the resistance of the fluid is to add to the mass of the sphere a mass equal to

$$\frac{b^3 + 2a^3}{b^3 - a^3} \frac{m}{2},$$

where a is the radius of the ball, b that of the envelope, and m the mass of the fluid displaced. Poisson, having reasoned on the very complicated case of an elastic fluid, had come to the conclusion that the envelope would have no effect.

One other instance of fluid motion contained in this paper will here be mentioned, because it seems to afford an accurate means of comparing theory and experiment in a class of motions in which they have not hitherto been compared, so far as I am aware. When a box of the form of a rectangular parallelepiped, filled with fluid and closed on all sides, is made to perform small oscillations, it appears that the motion of the box will be the same as if the fluid were replaced by a solid having the same mass, centre of gravity, and principal axes as the solidified fluid but different principal moments of inertia. These moments are given by infinite series, which converge with extreme rapidity, so that the numerical calculation is very easy. The oscillations most convenient to employ would probably be either oscillations by torsion, or bifilar oscillations.

VI. M. Navier was, I believe, the first to give equations for the motion of fluids without supposing the pressure equal in all directions. His theory is contained in a memoir read before the French Academy in 1822*. He considers the case of a homogeneous incompressible fluid. He supposes such a fluid to be made up of ultimate molecules, acting on each other by forces which, when the molecules are at rest, are functions simply of the distance, but which, when the molecules recede from, or approach to each other, are modified by this circumstance, so that two molecules repel each other less strongly when they are receding, and more strongly when they are approaching, than they do when they are at rest†. The alteration of attraction or repulsion is supposed to be, for a given distance, proportional to the velocity with which the molecules recede from, or approach to each other; so that the mutual repulsion of two molecules will be represented by $f(r) - VF'(r)$, where r is the distance of the molecules, V the velocity with which they recede from each other, and $f(r)$, $F(r)$ two unknown functions of r depending on the molecular force, and as such becoming insensible when r has become sensible. This expression does not suppose the molecules to be necessarily receding from each other, nor their mutual action to be necessarily repulsive, since V and $F(r)$ may be positive or negative. It is not absolutely necessary that $f(r)$ and $F(r)$ should always have the same sign. In forming the equations of motion M. Navier adopts the hypothesis of a symmetrical arrangement of the particles, or at least, which leads to the same result, neglects the irregular part of the mutual action of neighbouring molecules. The equations at which he arrives are those which would be obtained from the common equations by writing $\dfrac{dp}{dx} - A\left(\dfrac{d^2u}{dx^2} + \dfrac{d^2u}{dy^2} + \dfrac{d^2u}{dz^2}\right)$ in place of $\dfrac{dp}{dx}$ in the first, and making similar changes in the second and third. A is here an unknown constant depending on the nature of the fluid.

The same subject has been treated on by Poisson‡, who has adopted hypotheses which are very different from those of M.

* *Mémoires de l'Académie des Sciences*, tom. vi. p. 389.

† This idea appears to have been borrowed from Dubuat. See his *Principes d'Hydraulique*, tom. ii. p. 60.

‡ *Journal de l'École Polytechnique*, tom. xiii. cah. 20, p. 139.

Navier. Poisson's theory is of this nature. He supposes the
time t to be divided into n equal parts, each equal to τ. In
the first of these he supposes the fluid to be displaced in the same
manner as an elastic solid, so that the pressures in different
directions are given by the equations which he had previously
obtained for elastic solids. If the causes producing the displacement were now to cease to act, the molecules would very
rapidly assume a new arrangement, which would render the
pressure equal in all directions, and while this re-arrangement
was going on, the pressure would alter in an unknown manner
from that belonging to a displaced elastic solid to the pressure
belonging to the fluid in its new state. The causes of displacement are however going on during the second interval τ;
but since these different small motions will take place independently, the new displacement which will take place in the
second interval τ will be the same as if the molecules were not
undergoing a re-arrangement. Supposing now n to become infinite, we pass to the case in which the fluid is continually beginning to be displaced like an elastic solid, and continually
re-arranging itself so as to make the pressure equal in all directions. The equations at which Poisson arrived are, in the cases
of a homogeneous incompressible fluid, and of an elastic fluid
in which the change of density is small, those which would be
derived from the common equations by replacing dp/dx in the
first by

$$\frac{dp}{dx} - A\left(\frac{d^2u}{dx^2} + \frac{d^2u}{dy^2} + \frac{d^2u}{dz^2}\right) - B\frac{d}{dx}\left(\frac{du}{dx} + \frac{dv}{dy} + \frac{dw}{dz}\right),$$

and making similar changes in the second and third. In these
equations A and B are two unknown constants. It will be
observed that Poisson's equations reduce themselves to Navier's
in the case of an incompressible fluid.

The same subject has been considered in a quite different
point of view by M. Barré de Saint-Venant, in a communication
to the French Academy in 1843, an abstract of which is contained
in the *Comptes Rendus* *. The principal difficulty is to connect
the oblique pressures in different directions about the same point
with the differential coefficients du/dx, du/dy, &c., which express
the relative motion of the fluid particles in the immediate neigh-

* Tom. xvii. p. 1240.

bourhood of that point. This the author accomplishes by assuming that the tangential force on any plane passing through
the point in question is in the direction of the principal sliding
(*glissement*) along that plane. The sliding along the plane xy
is measured by $\dfrac{dw}{dx} + \dfrac{du}{dz}$ in the direction of x, and $\dfrac{dw}{dy} + \dfrac{dv}{dz}$ in the
direction of y. These two slidings may be compounded into one,
which will form the principal sliding along the plane xy. It
is then shewn, by means of M. Cauchy's theorems connecting
the pressures in different directions in any medium, that the
tangential force on any plane passing through the point considered,
resolved in any direction in that plane, is proportional to the
sliding along that plane resolved in the same direction, so that
if T represents the tangential force, referred to a unit of surface,
and S the sliding, $T = \epsilon S$. The pressure on a plane in any direction is then found. This pressure is compounded of a normal
pressure, alike in all directions, and a variable oblique pressure,
the expression for which contains the one unknown quantity ϵ.
If the fluid be supposed incompressible, and ϵ constant, the
equations which would be obtained by the method of M. Barré
de Saint-Venant agree with those of M. Navier. It will be
observed that this method does not require the consideration of
ultimate molecules at all.

When the motion of the fluid is very small, Poisson's equations
agree with those given by M. Cauchy for the motion of a solid
entirely destitute of elasticity*, except that the latter do not
contain the pressure p. These equations have been obtained
by M. Cauchy without the consideration of molecules. His
method would apply, with very little change, to the case of
fluids.

In a paper read last year before the Cambridge Philosophical
Society†, I have arrived at the equations of motion in a different
manner. The method employed in this paper does not necessarily require the consideration of ultimate molecules. Its principal feature consists in eliminating from the relative motion
of the fluid about any particular point the relative motion which
corresponds to a certain motion of rotation, and examining the
nature of the relative motion which remains. The equations

---

* *Exercices de Mathématiques*, tom. iii. p. 187.
† *Transactions of the Cambridge Philosophical Society*, vol. viii. p. 287.

finally adopted in the cases of a homogeneous incompressible
fluid, and of an elastic fluid in which the change of density is
small, agree with those of Poisson, provided we suppose in the
latter $A = 3B$. It is shewn that this relation between A and B
may be obtained on Poisson's own principles.

The equations hitherto considered are those which must be
satisfied at any point in the interior of the fluid mass; but there
is hardly any instance of the practical application of the equations,
in which we do not want to know also the particular conditions
which must be satisfied at the surface of the fluid. With respect
to a free surface there can be little doubt: the condition is simply
that there shall be no tangential force on a plane parallel to the
surface, taken immediately within the fluid. As to the case
of a fluid in contact with a solid, the condition at which Navier
arrived comes to this: that if we conceive a small plane drawn
within the fluid parallel to the surface of the solid, the tangential
force on this plane, referred to a unit of surface, shall be in the
same direction with, and proportional to the velocity with which
the fluid flows past the surface of the solid. The condition obtained by Poisson is essentially the same.

Dubuat stated, as a result of his experiments, that when the
velocity of water flowing through a pipe is less than a certain
quantity, the water adjacent to the surface of the pipe is at rest*.
This result agrees very well with an experiment of Coulomb's.
Coulomb found that when a metallic disc was made to oscillate
very slowly in water about an axis passing through its centre
and perpendicular to its plane, the resistance was not altered
when the disc was smeared with grease; and even when the
grease was covered with powdered sandstone the resistance was
hardly increased†. This is just what one would expect on the
supposition that the water close to the disc is carried along with
it, since in that case the resistance must depend on the internal
friction of the fluid; but the result appears very extraordinary on
the supposition that the fluid in contact with the disc flows
past it with a finite velocity. It should be observed, however,
that this result is compatible with the supposition that a thin
film of fluid remains adhering to the disc, in consequence of
capillary attraction, and becomes as it were solid, and that the

* See the Table given in tom. i. of his *Principes d'Hydraulique*, p. 93.
† *Mémoires de l'Institut*, 1801, tom. iii. p. 286.

fluid in contact with this film flows past it with a finite velocity. If we consider Dubuat's supposition to be correct, the condition to be assumed in the case of a fluid in contact with a solid is that the fluid does not move relatively to the solid. This con-·dition will be included in M. Navier's, if we suppose the coefficient of the velocity when M. Navier's condition is expressed analytically, which he denotes by E, to become infinite. It seems probable from the experiments of M. Girard, that the condition to be satisfied at the surface of fluid in contact with a solid is different according as the fluid does or does not moisten the surface of the solid.

M. Navier has applied his theory to the results of some experiments of M. Girard's on the discharge of fluids through capillary tubes. His theory shews that if we suppose E to be finite, the discharge through extremely small tubes will depend only on E, and not on A. The law of discharge at which he arrives agrees with the experiments of M. Girard, at least when the tubes are extremely small. M. Navier explained the difference observed by M. Girard in the discharge of water through tubes of glass and tubes of copper of the same size by supposing the value of E different in the two cases. This difference was explained by M. Girard himself by supposing that a thin film of fluid remains adherent to the pipe, in consequence of molecular action, and that the thickness of this film differs with the substance of which the tube is composed, as well as with the liquid employed*. If we adopt Navier's explanation, we may reconcile it with the experiments of Coulomb by supposing that E is very large, so that unless the fluid is confined in a very narrow pipe, the results will depend mainly on A, being sensibly the same as they would be if E were infinite.

There is one circumstance connected with the motion of a ball-pendulum oscillating in air, which has not yet been accounted for, the explanation of which seems to depend on this theory. It is found by experiment that the correction for the inertia of the air is greater for small than for large spheres, that is to say, the mass which we must suppose added to that of the sphere bears a greater ratio to the mass of the fluid displaced in the former, than in the latter case. According to the common theory of fluid motion, in which everything is supposed

---

* *Mémoires de l'Académie des Sciences*, tom i. pp. 203 and 234.

to be perfectly smooth, the ratio ought to be independent of the magnitude of the sphere. In the imperfect theory of friction in which the friction of the fluid on the sphere is taken into account, while the equal and opposite friction of the sphere on the fluid is neglected, it is shewn that the arc of oscillation is diminished, while the time of oscillation is sensibly the same as before. But when the tangential action of the sphere on the fluid, and the internal friction of the fluid itself are considered, it is clear that one consequence will be, to speak in a general way, that a portion of the fluid will be dragged along with the sphere. Thus the correction for the inertia of the fluid will be increased, since the same moving force has now to overcome the inertia of the fluid dragged along with the sphere, and not only, as in the former case, the inertia of the sphere itself, and of the fluid pushed away from before it, and drawn in behind it. Moreover the additional correction for inertia must depend, speaking approximately, on the *surface* of the sphere, whereas the first correction depended on its *volume*, and thus the effect of friction in altering the time of oscillation will be more conspicuous in the case of small, than in the case of large spheres, other circumstances being the same. The correction for inertia, when friction is taken into account, will not, however, depend solely on the magnitude of the sphere, but also on the time of oscillation. With a given sphere it will be greater for long, than for short oscillations.

[From the *Transactions of the Cambridge Philosophical Society*, Vol. VIII. p. 409.]

SUPPLEMENT TO A MEMOIR ON SOME CASES OF FLUID MOTION.

Read *Nov.* 3, 1846.

IN a memoir which the Society did me the honour to publish in their *Transactions**, I shewed that when a box whose interior is of the form of a rectangular parallelepiped is filled with fluid and made to perform small oscillations the motion of the box will be the same as if the fluid were replaced by a solid having the same mass, centre of gravity, and principal axes as the solidified fluid, but different moments of inertia about those axes. The box is supposed to be closed on all sides, and it is also supposed that the box itself and the fluid within it were both at rest at the beginning of the motion. The investigation was founded upon the ordinary equations of Hydrodynamics, which depend upon the hypothesis of the absence of any tangential force exerted between two adjacent portions of a fluid in motion, an hypothesis which entails as a necessary consequence the equality of pressure in all directions. The particular case of motion under consideration appears to be of some importance, because it affords an accurate means of comparing with experiment the common theory of fluid motion, which depends upon the hypothesis just mentioned. In my former paper, I gave a series by means of which the numerical values of the principal moments of the solid which may be substituted for the fluid might be calculated with facility. The present supplement contains a different series for the same purpose, which is more easy of numerical calculation than the former. The comparison of the

* Vol. VIII. Part I. p. 105. (*Ante*, p. 17.)

two series may also be of some interest in an analytical point
of view, since they appear under very different forms. I have
taken the present opportunity of mentioning the results of some
experiments which I have performed on the oscillations of a box,
such as that under consideration. The experiments were not
performed with sufficient accuracy to entitle them to be described
in detail.

The calculation of the motion of fluid in a rectangular box
is given in the 13th article of my former paper. I shall not
however in the first instance restrict myself to a rectangular
parallelepiped, since the simplification which I am about to give
applies more generally. Suppose then the problem to be solved
to be the following. A vessel whose interior surface is composed
of any cylindrical surface and of two planes perpendicular to the
generating lines of the cylinder is filled with a homogeneous,
incompressible fluid; the vessel and the fluid within it having
been at first at rest, the former is then moved in any manner;
required to determine the motion of the fluid at any instant,
supposing that at that instant the vessel has no motion of rotation
about an axis parallel to the generating lines of the cylinder.

I shall adopt the notation of my former paper. u, v, w are
the resolved parts of the velocity at any point along the rectangular axes of x, y, z. Since the motion begins from rest we
shall have $udx + vdy + wdz$ an exact differential $d\phi$. Let the
rectangular axes to which the fluid is referred be fixed relatively
to the vessel, and let the axis of x be parallel to the generating
lines of the cylindrical surface. The instantaneous motion of
the vessel may be decomposed into a motion of translation, and
two motions of rotation about the axes of y and z respectively;
for by hypothesis there is no motion of rotation about the axis
of x. According to the principles of my former paper, the instantaneous motion of the fluid will be the same as if it had
been produced directly by impact, the impact being such as
to give the vessel the velocity which it has at the instant considered. We may also consider separately the motion of translation of the vessel, and each of the motions of rotation; the
actual motion of the fluid will be compounded of those which
correspond to each of the separate motions of the vessel. For
my present purpose it will be sufficient to consider one of the

motions of rotation, that which takes place round the axis of z for instance. Let ω be the angular velocity about the axis of z, ω being considered positive when the vessel turns from the axis of x to that of y. It is easy to see that the instantaneous motion of the cylindrical surface is such as not to alter the volume of the interior of the vessel, supposing the plane ends fixed, and that the same is true of the instantaneous motion of the ends. Consequently we may consider separately the motion of the fluid due to the motion of the cylindrical surface, and to that of the ends. Let ϕ_c be the part of ϕ due to the motion of the cylindrical surface, ϕ_e the part due to the motion of the ends. Then we shall have

$$\phi = \phi_c + \phi_e \quad\dotsb\dotsb(1).$$

Consider now the motion corresponding to a value of ϕ, ωxy. It will be observed that ωxy satisfies the equation, {(36) of my former paper,} which ϕ is to satisfy. Corresponding to this value of ϕ we have

$$u = \omega y, \quad v = \omega x, \quad w = 0.$$

Hence the velocity, corresponding to this motion, of a particle of fluid in contact with the cylindrical surface of the vessel, resolved in a direction perpendicular to the surface, is the same as the velocity of the surface itself resolved in the same direction, and therefore the fluid does not penetrate into, nor separate from the cylindrical surface. The velocity of a particle in contact with either of the plane ends, resolved in a direction perpendicular to the surface, is equal and opposite to the velocity of the surface itself resolved in the same direction. Hence we shall get the complete value of ϕ by adding the part already found, namely ωxy, to *twice* the part due to the motion of the plane ends. We have therefore,

$$\phi = \omega xy + 2\phi_e = 2\phi_c - \omega xy, \text{ by (1)} \quad\dotsb\dotsb(2),$$

and
$$\phi_c - \phi_e = \omega xy \quad\dotsb\dotsb(3).$$

Hence whenever either ϕ_c or ϕ_e can be found, the complete solution of the problem will be given by (2). And even when both these functions can be obtained independently, (2) will enable us to dispense with the use of one of them, and (3) will give a relation between them. In this case (3) will express a theorem in pure analysis, a theorem which will sometimes be very curious, since the analytical expressions for ϕ_c and ϕ_e will generally be totally different in form. The problem admits of solution in the case of a circular cylinder terminated by planes perpendicular to its axis, and in the case of a rectangular parallelepiped. In the former case, the numerical calculation of the moments of inertia of the solid by which the fluid may be replaced would probably be troublesome, in the latter it is extremely easy. I proceed to consider this case in particular.

Let the rectangular axes to which the fluid is referred coincide with three adjacent edges of the parallelepiped, and let a, b, c be the lengths of the edges. The motion which it is proposed to calculate is that which arises from a motion of rotation of the box about an axis parallel to that of z and passing through the centre of the parallelepiped. Consequently in applying (2) we must for a moment conceive the axis of z to pass through the centre of the parallelepiped, and then transfer the origin to the corner, and we must therefore write $\omega (x - \frac{1}{2}a)(y - \frac{1}{2}b)$ for ωxy. In the present case the cylindrical surface consists of the four faces which are parallel to the axis of x, and the remaining faces form the plane ends. The motion of the face xy and the opposite face has evidently no effect on the fluid, so that ϕ_c will be the part of ϕ due to the motion of the face xz and the opposite face. The value of this quantity is given near the middle of page 62 in my former paper. We have then by the second of the formulæ (2)

$$\phi = \frac{8\omega a^2}{\pi^3} \Sigma_0 \frac{1}{n^3} \frac{(e^{n\pi b/a} - 1)e^{-n\pi y/a} + (e^{-n\pi b/a} - 1)e^{n\pi y/a}}{e^{n\pi b/a} - e^{-n\pi b/a}} \cos n\pi x/a$$

$$- \omega (x - \tfrac{1}{2}a)(y - \tfrac{1}{2}b)\ldots\ldots\ldots\ldots(4),$$

the sign Σ_0 denoting the sum corresponding to all odd integral values of n from 1 to ∞. This value of ϕ expresses completely the motion of the fluid due to a motion of rotation of the box about an axis parallel to that of z, and passing through the centre of its interior.

Suppose now the motion to be very small, so that the square of the velocity may be neglected. Then, p denoting the part of the pressure due to the motion, we shall have $p = - \rho \, d\phi/dt$. Also in finding $d\phi/dt$ we may suppose the axes to be fixed in space, since by taking account of their motion we should only introduce terms depending on the square of the velocity. In fact, if for the sake of distinction we denote the co-ordinates of a fluid particle referred to the moveable axes by x', y', while x, y denote its co-ordinates referred to axes fixed in space, which after differentiation with respect to t we may suppose to coincide with the moveable axes at the instant considered, and if we denote the differential coefficient of ϕ with respect to t by $(d\phi/dt)$ when x, y, t are the independent variables, and by $d\phi/dt$ when x', y', t are the independent variables, we shall have

$$\left(\frac{d\phi}{dt}\right) = \frac{d\phi}{dt} + \frac{d\phi}{dx'}\frac{dx'}{dt} + \frac{d\phi}{dy'}\frac{dy'}{dt} = \frac{d\phi}{dt} + u\frac{dx'}{dt} + v\frac{dy'}{dt} \; ;$$

for $d\phi/dx'$, $d\phi/dy'$ mean absolutely the same as $d\phi/dx$, $d\phi/dy$, and are therefore equal to u, v respectively. Now dx'/dt, dy'/dt, depending on the motion of the axes, are small quantities of the order ω; their values are in fact ωy, $-\omega x$; so that, omitting small quantities of the order ω^2, we have

$$\left(\frac{d\phi}{dt}\right) = \frac{d\phi}{dt} \; .$$

We shall therefore find the value of p from that of ϕ by merely writing $-\rho\, d\omega/dt$ for ω. In order to determine the motion of the box it will be necessary to find the resultant of the fluid pressures on its several faces. As shewn in my former paper, these pressures will have no resultant force, but only a resultant couple, of which the axis will evidently be parallel to that of z. In calculating this couple, it is immaterial whether we take the moments about the axis of z, or about a line parallel to it passing through the centre of the parallelepiped: suppose that we adopt the latter plan. If we reckon the couple positive when it tends to turn the box from the axis of x to that of y we shall evidently have $-\int_0^a\int_0^c p_{y=0}\left(x - \frac{a}{2}\right)dx\,dz$ for the part arising from the

---

* It may be very easily proved by means of this equation, combined with the general equation which determines p, that whether the velocity be great or small the fluid will have the same effect on the motion of the box as the solid of which the moment of inertia is determined in this paper on the supposition that the motion is small.

pressure on the face xz, and $\int_0^b \int_0^c p_{x=0} \left(y - \dfrac{b}{2} \right) dy\,dz$ for the part arising from the pressure on the face yz. It is easily seen from (4) that $p_{x=a} = -p_{x=0}$, and $p_{y=b} = -p_{y=0}$, so that the couples due to the pressures on the faces xz, yz are equal to the couples due to the pressures on the opposite faces respectively. In order, therefore, to find the whole couple we have only got to double the part already found. As the integrations do not present the slightest difficulty, it will be sufficient to write down the result. It will be found that the whole couple is equal to $-C\,d\omega/dt$, where

$$C = \frac{\rho abc}{12}(b^2 - 3a^2) + \frac{64\rho a^4 c}{\pi^5} \Sigma_0 \frac{1}{n^5} \frac{1 - e^{-n\pi b/a}}{1 + e^{-n\pi b/a}} \quad\text{........(5).}$$

This expression has been simplified after integration by putting for $\Sigma_0\, 1/n^4$ its value $\pi^4/96$.

It appears then that the effect of the inertia of the fluid is to increase the moment of inertia of the box about an axis passing through its centre and parallel to the edge c by the quantity C. In equation (40) of my former paper, there is given an expression for C which is apparently very different from that given by (5), but the numerical values of the two expressions are necessarily the same. If we denote the moment of inertia of the fluid supposed to be solidified by $C_{,}$, we shall have $C_{,} = \rho abc\,(a^2 + b^2)/12$; and if we put

$$\frac{a}{b} = r, \qquad \frac{C}{C_{,}} = f(r),$$

and treat (5) as equation (40) of my former paper was treated, we shall find

$$f(r) = (1 + r^2)^{-1} \{ 1 - 3r^2 + 2r^3\, (1.260497 - 1.254821\, \Sigma_0 \frac{1}{n^5} \text{versin}\, 2\theta_n) \}$$
$$\text{.................(6),}$$

where $\qquad$ tab. log tan $\theta_n = 10 - .6821882 \dfrac{n}{r}$.

The equation (6) is true, (except as regards the decimals omitted,) whatever be the value of r; but for convenience of calculation it will be proper to take r less than 1, that is, to choose for a the smaller of the two a, b. The value of $f(r)$ given by (6) is apparently very different from that given at the bottom of page 64 of my former paper, but any one may easily satisfy himself as to equivalence of the two expressions by assigning to r a value at random, and calculating the value of $f(r)$ from the two expressions separately. The expression (6) is however preferable to the other, especially when we have to calculate the value of $f(r)$ for small values of r. The infinite series contained in (6) converges with such rapidity that in the most unfavourable case, that is, when $r = 1$ nearly, the omission of all terms after the first would only introduce an error of about .000003 in the value of $f(r)$.

For the sake of shewing the manner in which $f(r)$ alters with r, I have calculated the following values of the function. The expression (6) shews that $f'(r) = 0$, when $r = 0$; and $f'(r)$ is also $= 0$ when $r = 1$, since $f\left(\dfrac{1}{r}\right) = f(r)$.

r	$f(r)$	r	$f(r)$
0.0	1	0.6	0.3374
0.1	0.9629	0.7	0.2521
0.2	0.8655	0.8	0.1958
0.3	0.7322	0.9	0.1655
0.4	0.5873	1	0.1565
0.5	0.4512		

The experiments to which I have alluded were made with a wooden box measuring inside 8 inches by 4 square. The box weighed not quite 1 lb., and contained about $4\frac{1}{2}$ lbs. of water, so that the inertia of the water which had to be overcome was by no means small compared with that of the box. The box was suspended by two parallel threads 3 inches apart and between 4 and 5 feet long: it was twisted a little, and then left to itself, so that it oscillated about a vertical axis midway between the threads. The points of attachment of the threads were in a line drawn through the centre of the upper face parallel to one of its sides, and were equidistant from the centre. The weight of the box when empty, the length and distance of the threads, the time of oscillation, and the known length of the seconds' pendulum are data sufficient for determining the moment of inertia of the box about a vertical axis passing through its centre. When the box is filled with water the same quantities determine the moment of inertia of the box and the water it contains, whence the moment of inertia of the water alone is obtained by subtraction. It is supposed here that the centre of gravity of the box coincides with the centre of gravity of its interior volume. In the following experiments a different face of the box was uppermost each time. In Nos. 1 and 2 the long edges of the box were vertical, in Nos. 3 and 4 they were horizontal. In all cases the inertia determined by experiment was a little greater than that resulting from theory : the difference will be given in fractional parts of the latter. The difference was 1/21 in No. 1, 1/13 in No. 2, 1/17 in No. 3, and 1/21 in No. 4. On referring to the table at the end of the last paragraph, it will be seen that the ratio of the moment of inertia of the fluid to what it would be if the fluid were solid is about three times as great in the last two experiments as in the first two.

I had expected beforehand to find the inertia determined by experiment a little greater than that given by theory, for this reason. In the theory, it is supposed that both the fluid itself and the surface of the box are perfectly smooth. This however is not strictly true. The box by its roughness exerts a tangential force on the fluid immediately in contact with it, and this force produces an effect on the fluid at a small distance from the surface of the box, in consequence of the internal friction of the fluid itself. We may conceive the effect of this force on the time of oscillation in a general way by supposing a thin film of fluid close to the surface of the box to be dragged along with it. Consequently, the moment of inertia determined by experiment will be a little greater than it would have been had the fluid and the surface of the box been perfectly smooth.

These experiments are sufficient to shew that in the case of a vessel of about the size and shape of the one I used, filled with water, and performing small oscillations of the duration of about one second (as was the case in my experiments), the time of oscillation is not much increased by friction ; at least, if we suppose, as there is reason for supposing, that the effect of friction

13—2

does not depend on the nature of the surface of the box. They are not however sufficiently exact to allow us to place any reliance on the accuracy of the small differences between the results of experiment, and of the common theory of fluid motion, and consequently they are useless as tests of any theory of friction.

[From the *Transactions of the Cambridge Philosophical Society*, Vol. VIII. p. 441.]

ON THE THEORY OF OSCILLATORY WAVES.

[Read March 1, 1847.]

In the Report of the Fourteenth Meeting of the British Association for the Advancement of Science it is stated by Mr Russell, as a result of his experiments, that the velocity of propagation of a series of oscillatory waves does not depend on the height of the waves*. A series of oscillatory waves, such as that observed by Mr Russell, does not exactly agree with what it is most convenient, as regards theory, to take as the type of oscillatory waves. The extreme waves of such a series partake in some measure of the character of solitary waves, and their height decreases as they proceed. In fact it will presently appear that it is only an indefinite series of waves which possesses the property of being propagated with a uniform velocity, and without change of form: at least this is the case when the waves are such as can be propagated along the surface of a fluid which was previously at rest. The middle waves, however, of a series such as that observed by Mr Russell agree very nearly with oscillatory waves of the standard form. Consequently, the velocity of propagation determined by the observation of a number of waves, according to Mr Russell's method, must be very nearly the same as the velocity of propagation of a series of oscillatory waves of the standard form, and whose length is equal to the mean length of the waves observed, which are supposed to differ from each other but slightly in length.

* Page 369 (note), and page 370.

On this account I was induced to investigate the motion of
oscillatory waves of the above form to a second approximation,
that is, supposing the height of the waves finite, though small.
I find that the expression for the velocity of propagation is independent of the height of the waves to a second approximation.
With respect to the form of the waves, the elevations are no
longer similar to the depressions, as is the case to a first approximation, but the elevations are narrower than the hollows,
and the height of the former exceeds the depth of the latter.
This is in accordance with Mr Russell's remarks at page 448 of
his first Report*. I have proceeded to a third approximation
in the particular case in which the depth of the fluid is very
great, so as to find in this case the most important term, depending on the height of the waves, in the expression for the
velocity of propagation. This term gives an increase in the
velocity of propagation depending on the square of the ratio of
the height of the waves to their length.

There is one result of a second approximation which may
possibly be of practical importance. It appears that the forward
motion of the particles is not altogether compensated by their
backward motion ; so that, in addition to their motion of oscillation, the particles have a progressive motion in the direction
of propagation of the waves. In the case in which the depth of
the fluid is very great, this progressive motion decreases rapidly
as the depth of the particle considered increases. Now when a
ship at sea is overtaken by a storm, and the sky remains overcast,
so as to prevent astronomical observations, there is nothing to
trust to for finding the ship's place but the dead reckoning. But
the estimated velocity and direction of motion of the ship are
her velocity and direction of motion relatively to the water. If
then the whole of the water near the surface be moving in the
direction of the waves, it is evident that the ship's estimated
place will be erroneous. If, however, the velocity of the water
can be expressed in terms of the length and height of the waves,
both which can be observed approximately from the ship, the
motion of the water can be allowed for in the dead reckoning.

As connected with this subject, I have also considered the
motion of oscillatory waves propagated along the common surface
of two liquids, of which one rests on the other, or along the upper

* *Reports of the British Association*, Vol. VI.

surface of the upper liquid. In this investigation there is no object in going beyond a first approximation. When the specific gravities of the two fluids are nearly equal, the waves at their common surface are propagated so slowly that there is time to observe the motions of the individual particles. The second case affords a means of comparing with theory the velocity of propagation of oscillatory waves in extremely shallow water. For by pouring a little water on the top of the mercury in a trough we can easily procure a sheet of water of a small, and strictly uniform depth, a depth, too, which can be measured with great accuracy by means of the area of the surface and the quantity of water poured in. Of course, the common formula for the velocity of propagation will not apply to this case, since the motion of the mercury must be taken into account.

---

1. In the investigations which immediately follow, the fluid is supposed to be homogeneous and incompressible, and its depth uniform. The inertia of the air, and the pressure due to a column of air whose height is comparable with that of the waves are also neglected, so that the pressure at the upper surface of the fluid may be supposed to be zero, provided we afterwards add the atmospheric-pressure to the pressure so determined. The waves which it is proposed to investigate are those for which the motion is in two dimensions, and which are propagated with a constant velocity, and without change of form. It will also be supposed that the waves are such as admit of being excited, independently of friction, in a fluid which was previously at rest. It is by these characters of the waves that the problem will be rendered determinate, and not by the initial disturbance of the fluid, supposed to be given. The common theory of fluid motion, in which the pressure is supposed equal in all directions, will also be employed.

Let the fluid be referred to the rectangular axes of x, y, z, the plane xz being horizontal, and coinciding with the surface of the fluid when in equilibrium, the axis of y being directed downwards, and that of x taken in the direction of propagation of the waves, so that the expressions for the pressure, &c. do not contain z. Let p be the pressure, ρ the density, t the time, u, v the resolved parts of the velocity in the directions of the axes of x, y; g the force of gravity, h the depth of the fluid when in equilibrium. From the character of the waves which was mentioned last, it follows by a known theorem that $u\,dx + v\,dy$ is an exact differential $d\phi$. The equations by which the motion is to be determined are well known. They are

$$p = g\rho y - \rho\frac{d\phi}{dt} - \frac{\rho}{2}\left\{\left(\frac{d\phi}{dx}\right)^2 + \left(\frac{d\phi}{dy}\right)^2\right\} \quad \ldots\ldots\ldots(1);$$

$$\frac{d^2\phi}{dx^2} + \frac{d^2\phi}{dy^2} = 0 \ldots\ldots\ldots\ldots\ldots(2);$$

$$\frac{d\phi}{dy} = 0, \text{ when } y = h \ldots\ldots\ldots\ldots(3);$$

$$\frac{dp}{dt} + \frac{d\phi}{dx}\frac{dp}{dx} + \frac{d\phi}{dy}\frac{dp}{dy} = 0, \text{ when } p = 0 \ldots\ldots(4);$$

where (3) expresses the condition that the particles in contact with the rigid plane on which the fluid rests remain in contact with it, and (4) expresses the condition that the same surface of particles continues to be the free surface throughout the motion, or, in other words, that there is no generation or destruction of fluid at the free surface.

If c be the velocity of propagation, u, v and p will be by hypothesis functions of $x - ct$ and y. It follows then from the equations $u = d\phi/dx$, $v = d\phi/dy$ and (1), that the differential coefficients of ϕ with respect to x, y and t will be functions of $x - ct$ and y; and therefore ϕ itself must be of the form

$$f(x - ct,\ y) + Ct.$$

The last term will introduce a constant into (1); and if this constant be expressed, we may suppose ϕ to be a function of $x - ct$ and y. Denoting $x - ct$ by x', we have

$$\frac{dp}{dx} = \frac{dp}{dx'},\quad \frac{dp}{dt} = -c\frac{dp}{dx'},$$

and similar equations hold good for ϕ. On making these substitutions in (1) and (4), omitting the accent of x, and writing $-gk$ for C, we have

$$p = g\rho\,(y + k) + c\rho\frac{d\phi}{dx} - \frac{\rho}{2}\left\{\left(\frac{d\phi}{dx}\right)^2 + \left(\frac{d\phi}{dy}\right)^2\right\} \quad \ldots\ldots(5),$$

$$\left(\frac{d\phi}{dx} - c\right)\frac{dp}{dx} + \frac{d\phi}{dy}\frac{dp}{dy} = 0, \text{ when } p = 0 \ldots\ldots\ldots(6).$$

Substituting in (6) the value of p given by (5), we have

$$g\frac{d\phi}{dy} - c^2\frac{d^2\phi}{dx^2} + 2c\left(\frac{d\phi}{dx}\frac{d^2\phi}{dx^2} + \frac{d\phi}{dy}\frac{d^2\phi}{dxdy}\right)$$
$$- \left(\frac{d\phi}{dx}\right)^2\frac{d^2\phi}{dx^2} - 2\frac{d\phi}{dx}\frac{d\phi}{dy}\frac{d^2\phi}{dxdy} - \left(\frac{d\phi}{dy}\right)^2\frac{d^2\phi}{dy^2} = 0 \ldots(7),$$

when
$$g(y+k) + c\frac{d\phi}{dx} - \tfrac{1}{2}\left\{\left(\frac{d\phi}{dx}\right)^2 + \left(\frac{d\phi}{dy}\right)^2\right\} = 0 \ldots\ldots\ldots(8).$$

The equations (7) and (8) are exact; but if we suppose the motion small, and proceed to the second order only of approximation, we may neglect the last three terms in (7), and we may easily eliminate y between (7) and (8). For putting ϕ', $\phi_,$, &c. for the values of $d\phi/dx$, $d\phi/dy$, &c. when $y = 0$, the number of accents above marking the order of the differential coefficient with respect to x, and the number below its order with respect to y, and observing that k is a small quantity of the first order at least, we have from (8)

$$g(y+k) + c(\phi' + \phi_,'y) - \tfrac{1}{2}(\phi'^2 + \phi_,^2) = 0,$$

whence $\quad y = -k - \dfrac{c}{g}\phi' + \dfrac{c}{g}\phi_,'\left(k + \dfrac{c}{g}\phi'\right) + \dfrac{1}{2g}(\phi'^2 + \phi_,^2)^* \ldots.(9).$

Substituting the first approximate value of y in the first two terms of (7), putting $y = 0$ in the next two, and reducing, we have

$$g\phi_, - c^2\phi'' - (g\phi_{,,} - c^2\phi_,'')\left(k + \frac{c}{g}\phi'\right) + 2c(\phi'\phi'' + \phi_,\phi_,') = 0 \ldots(10).$$

ϕ will now have to be determined from the general equation (2) with the particular conditions (3) and (10). When ϕ is known, y, the ordinate of the surface, will be got from (9), and k will then be determined by the condition that the mean value of y shall be zero. The value of p, if required, may then be obtained from (5).

2. In proceeding to a first approximation we have the equations (2), (3) and the equation obtained by omitting the small terms in (10), namely,

$$g\frac{d\phi}{dy} - c^2\frac{d^2\phi}{dx^2} = 0, \text{ when } y = 0 \ldots\ldots\ldots(11).$$

---

* The reader will observe that the y in this equation is the ordinate of the surface, whereas the y in (1) and (2) is the ordinate of any point in the fluid. The context will always shew in which sense y is employed.

The general integral of (2) is

$$\phi = \Sigma A \epsilon^{mx+ny},$$

the sign Σ extending to all values of A, m and n, real or imaginary, for which $m^2 + n^2 = 0$: the particular values of ϕ, $Cx + C'$, $Dy + D'$, corresponding respectively to $n = 0$, $m = 0$, must also be included, but the constants C', D' may be omitted. In the present case, the expression for ϕ must not contain real exponentials in x, since a term containing such an exponential would become infinite either for $x = -\infty$, or for $x = +\infty$, as well as its differential coefficients which would appear in the expressions for u and v; so that m must be wholly imaginary. Replacing then the exponentials in x by circular functions, we shall have for the part of ϕ corresponding to any one value of m,

$$(A\epsilon^{my} + A'\epsilon^{-my}) \sin mx + (B\epsilon^{my} + B'\epsilon^{-my}) \cos mx,$$

and the complete value of ϕ will be found by taking the sum of all possible particular values of the above form and of the particular value $Cx + Dy$. When the value so formed is substituted in (3), which has to hold good for all values of x, the coefficients of the several sines and cosines, and the constant term must be separately equated to zero. We have therefore

$$D = 0, \quad A' = \epsilon^{2mh}A, \quad B' = \epsilon^{2mh}B ;$$

so that if we change the constants we shall have

$$\phi = Cx + \Sigma \left(\epsilon^{m(h-y)} + \epsilon^{-m(h-y)}\right)(A \sin mx + B \cos mx)...(12),$$

the sign Σ extending to all real values of m, A and B, of which m may be supposed positive.

3. To the term Cx in (12) corresponds a uniform velocity parallel to x, which may be supposed to be impressed on the fluid in addition to its other motions. If the velocity of propagation be defined merely as the velocity with which the wave form is propagated, it is evident that the velocity of propagation is perfectly arbitrary. For, for a given state of relative motion of the parts of the fluid, the velocity of propagation, as so defined, can be altered by altering the value of C. And in proceeding to the higher orders of approximation it becomes a question what we shall define the velocity of propagation to be. Thus, we might define it to be the velocity with which the wave form is propa-

gated when the mean horizontal velocity of a particle in the upper surface is zero, or the velocity of propagation of the wave form when the mean horizontal velocity of a particle at the bottom is zero, or in various other ways. The following two definitions appear chiefly to deserve attention.

First, we may define the velocity of propagation to be the velocity with which the wave form is propagated in space, when the mean horizontal velocity *at each point of space occupied by the fluid* is zero. The term "mean" here refers to the variation of the time. This is the definition which it will be most convenient to employ in the investigation. I shall accordingly suppose $C = 0$ in (12), and c will represent the velocity of propagation according to the above definition.

Secondly, we may define the velocity of propagation to be the velocity of propagation of the wave form in space, when the mean horizontal velocity of the mass of fluid comprised between two very distant planes perpendicular to the axis of x is zero. The mean horizontal velocity of the mass means here the same thing as the horizontal velocity of its centre of gravity. This appears to be the most natural definition of the velocity of propagation, since in the case considered there is no current in the mass of fluid, taken as a whole. I shall denote the velocity of propagation according to this definition by c'. In the most important case to consider, namely, that in which the depth is infinite, it is easy to see that $c' = c$, whatever be the order of approximation. For when the depth becomes infinite, the velocity of the centre of gravity of the mass comprised between any two planes parallel to the plane yz vanishes, provided the expression for u contain no constant term.

4. We must now substitute in (11) the value of ϕ.

$$\phi = \Sigma \left(\epsilon^{m(h-y)} + \epsilon^{-m(h-y)} \right) (A \sin mx + B \cos mx) \dots (13);$$

but since (11) has to hold good for all values of x, the coefficients of the several sines and cosines must be separately equal to zero : at least this must be true, provided the series contained in (11) are convergent. The coefficients will vanish for any one value of m, provided

$$c^2 = \frac{g}{m} \frac{\epsilon^{mh} - \epsilon^{-mh}}{\epsilon^{mh} + \epsilon^{-mh}} \dots (14).$$

Putting for shortness $2mh = \mu$, we have

$$\frac{d \log c^2}{d\mu} = -\frac{1}{\mu} + \frac{2}{\epsilon^\mu - \epsilon^{-\mu}},$$

which is positive or negative, μ being supposed positive, according as

$$2\mu > < \epsilon^\mu - \epsilon^{-\mu} > < 2 \left(\mu + \frac{\mu^3}{1 \cdot 2 \cdot 3} + \ldots \right),$$

and is therefore necessarily negative. Hence the value of c given by (14) decreases as μ or m increases, and therefore (11) cannot be satisfied, for a given value of c, by more than one positive value of m. Hence the expression for ϕ must contain only one value of m. Either of the terms $A \cos mx$, $B \sin mx$ may be got rid of by altering the origin of x. We may therefore take, for the most general value of ϕ,

$$\phi = A \left(\epsilon^{m(h-y)} + \epsilon^{-m(h-y)}\right) \sin mx \ldots\ldots\ldots(15).$$

Substituting in (8), we have for the ordinate of the surface

$$y = -\frac{mAc}{g} \left(\epsilon^{mh} + \epsilon^{-mh}\right) \cos mx \ldots\ldots\ldots(16),$$

k being $= 0$, since the mean value of y must be zero. Thus everything is known in the result except A and m, which are arbitrary.

5. It appears from the above, that of all waves for which the motion is in two dimensions, which are propagated in a fluid of uniform depth, and which are such as could be propagated into fluid previously at rest, so that $udx + vdy$ is an exact differential, there is only one particular kind, namely, that just considered, which possesses the property of being propagated with a constant velocity, and without change of form ; so that a solitary wave cannot be propagated in this manner. Thus the degradation in the height of such waves, which Mr Russell observed, is not to be attributed wholly, (nor I believe chiefly,) to the imperfect fluidity of the fluid, and its adhesion to the sides and bottom of the canal, but is an essential characteristic of a solitary wave. It is true that this conclusion depends on an investigation which applies strictly to indefinitely small motions only : but if it were true in general that a solitary wave could be propagated uniformly, without degradation, it would be true in

the limiting case of indefinitely small motions; and to disprove a general proposition it is sufficient to disprove a particular case.

6. In proceeding to a second approximation we must substitute the first approximate value of ϕ, given by (15), in the small terms of (10). Observing that $k = 0$ to a first approximation, and eliminating g from the small terms by means of (14), we find

$$g\phi_{,} - c^2\phi'' - 6A^2m^3c \sin 2mx = 0 \ldots\ldots\ldots\ldots(17).$$

The general value of ϕ given by (13), which is derived from (2) and (3), must now be restricted to satisfy (17). It is evident that no new terms in ϕ involving $\sin mx$ or $\cos mx$ need be introduced, since such terms may be included in the first approximate value, and the only other term which can enter is one of the form

$$B \left(\epsilon^{2m(h-y)} + \epsilon^{-2m(h-y)}\right) \sin 2mx.$$

Substituting this term in (17), and simplifying by means of (14), we find

$$B = \frac{3mA^2}{c \left(\epsilon^{mh} - \epsilon^{-mh}\right)^2}.$$

Moreover since the term in ϕ containing $\sin mx$ must disappear from (17), the equation (14) will give c to a second approximation.

If we denote the coefficient of $\cos mx$ in the first approximate value of y, the ordinate of the surface, by a, we shall have

$$A = - \frac{ga}{mc \left(\epsilon^{mh} + \epsilon^{-mh}\right)} = - \frac{ca}{\left(\epsilon^{mh} - \epsilon^{-mh}\right)};$$

and substituting this value of A in that of ϕ, we have

$$\phi = - ac \frac{\epsilon^{m(h-y)} + \epsilon^{-m(h-y)}}{\epsilon^{mh} - \epsilon^{-mh}} \sin mx + 3ma^2c \frac{\epsilon^{2m(h-y)} + \epsilon^{-2m(h-y)}}{\left(\epsilon^{mh} - \epsilon^{-mh}\right)^4} \sin 2mx$$

$$\ldots\ldots (18).$$

The ordinate of the surface is given to a second approximation by (9). It will be found that

$$y = a \cos mx - ma^2 \frac{\left(\epsilon^{mh} + \epsilon^{-mh}\right)\left(\epsilon^{2mh} + \epsilon^{-2mh} + 4\right)}{2 \left(\epsilon^{mh} - \epsilon^{-mh}\right)^3} \cos 2mx \ldots(19),$$

$$k = \frac{ma^2}{\epsilon^{2mh} - \epsilon^{-2mh}}.$$

7. The equation to the surface is of the form

$$y = a \cos mx - Ka^2 \cos 2mx \ldots\ldots\ldots\ldots(20),$$

where K is necessarily positive, and a may be supposed to be positive, since the case in which it is negative may be reduced to that in which it is positive by altering the origin of x by the quantity π/m or $\lambda/2$, λ being the length of the waves. On referring to (20) we see that the waves are symmetrical with respect to vertical planes drawn through their ridges, and also with respect to vertical planes drawn through their lowest lines. The greatest depression of the fluid occurs when $x = 0$ or $= \pm \lambda$, &c., and is equal to $a - a^2 K$: the greatest elevation occurs when $x = \pm \lambda/2$ or $= \pm 3\lambda/2$, &c., and is equal to $a + a^2 K$. Thus the greatest elevation exceeds the greatest depression by $2a^2 K$. When the surface cuts the plane of mean level, $\cos mx - aK \cos 2mx = 0$. Putting in the small term in this equation the approximate value $mx = \pi/2$, we have $\cos mx = - aK = \cos(\pi/2 + aK)$, whence

$$x = \pm (\lambda/4 + aK\lambda/2\pi), = \pm (5\lambda/4 + aK\lambda/2\pi), \ \&c.$$

We see then that the breadth of each hollow, measured at the height of the plane of mean level, is $\lambda/2 + aK\lambda/\pi$, while the breadth of each elevated portion of the fluid is $\lambda/2 - aK\lambda/\pi$.

It is easy to prove from the expression for K, which is given in (19), that for a given value of λ or of m, K increases as h decreases. Hence the difference in form of the elevated and depressed portions of the fluid is more conspicuous in the case in which the fluid is moderately shallow than in the case in which its depth is very great compared with the length of the waves.

8.   When the depth of the fluid is very great compared with the length of a wave, we may without sensible error suppose h to be infinite. This supposition greatly simplifies the expressions already obtained. We have in this case

$$\phi = - ac\epsilon^{-my} \sin mx \dots\dots\dots\dots\dots\dots\dots\dots(21),$$

$$y = a \cos mx - \tfrac{1}{2} ma^2 \cos 2mx \dots\dots\dots\dots\dots(22),$$

$$k = 0, \quad K = \frac{m}{2} = \frac{\pi}{\lambda}, \quad c^2 = \frac{g\lambda}{2\pi},$$

the y in (22) being the ordinate of the surface.

It is hardly necessary to remark that the state of the fluid at any time will be expressed by merely writing $x - ct$ in place of x in all the preceding expressions.

9. To find the nature of the motion of the individual particles, let $x + \xi$ be written for x, $y + \eta$ for y, and suppose x and y to be independent of t, so that they alter only in passing from one particle to another, while ξ and η are small quantities depending on the motion. Then taking the case in which the depth is infinite, we have

$$\frac{d\xi}{dt} = u = - mac\epsilon^{-m(y+\eta)} \cos m(x + \xi - ct) = - mac\epsilon^{-my} \cos m(x - ct)$$

$$+ m^2 ac\epsilon^{-my} \sin m(x - ct) . \xi + m^2 ac\epsilon^{-my} \cos m(x - ct) . \eta, \text{ nearly,}$$

$$\frac{d\eta}{dt} = v = mac\epsilon^{-m(y+\eta)} \sin m(x + \xi - ct) = mac\epsilon^{-my} \sin m(x - ct)$$

$$+ m^2 ac\epsilon^{-my} \cos m(x - ct) . \xi - m^2 ac\epsilon^{-my} \sin m(x - ct) . \eta, \text{ nearly.}$$

To a first approximation

$$\xi = a\epsilon^{-my} \sin m(x - ct), \qquad \eta = a\epsilon^{-my} \cos m(x - ct),$$

the arbitrary constants being omitted. Substituting these values in the small terms of the preceding equations, and integrating again, we have

$$\xi = a\epsilon^{-my} \sin m(x - ct) + m^2 a^2 ct\epsilon^{-2my},$$

$$\eta = a\epsilon^{-my} \cos m(x - ct).$$

Hence the motion of the particles is the same as to a first approximation, with one important difference, which is that in addition to the motion of oscillation the particles are transferred forwards, that is, in the direction of propagation, with a constant velocity depending on the depth, and decreasing rapidly as the depth increases. If U be this velocity for a particle whose depth below the surface in equilibrium is y, we have

$$U = m^2 a^2 c\epsilon^{-2my} = a^2 \left(\frac{2\pi}{\lambda}\right)^{\frac{3}{2}} g^{\frac{1}{2}} \epsilon^{-\frac{4\pi y}{\lambda}} \dots\dots\dots\dots(23).$$

The motion of the individual particles may be determined in a similar manner when the depth is finite from (18). In this case the values of ξ and η contain terms of the second order, involving respectively $\sin 2m(x - ct)$ and $\cos 2m(x - ct)$, besides the term in ξ which is multiplied by t. The most important thing to consider is the value of U, which is

$$U = m^2 a^2 c \frac{\epsilon^{2m(y-h)} + \epsilon^{-2m(y-h)}}{\left(\epsilon^{mh} - \epsilon^{-mh}\right)^2} \dots\dots\dots\dots\dots(24).$$

Since U is a small quantity of the order a^2, and in proceeding to a second approximation the velocity of propagation is given to the order a only, it is immaterial which of the definitions of velocity of propagation mentioned in Art. 3 we please to adopt.

10. The waves produced by the action of the wind on the surface of the sea do not probably differ very widely from those which have just been considered, and which may be regarded as the typical form of oscillatory waves. On this supposition the particles, in addition to their motion of oscillation, will have a progressive motion in the direction of propagation of the waves, and consequently in the direction of the wind, supposing it not to have recently shifted, and this progressive motion will decrease rapidly as the depth of the particle considered increases. If the pressure of the air on the posterior parts of the waves is greater than on the anterior parts, in consequence of the wind, as unquestionably it must be, it is easy to see that some such progressive motion must be produced. If then the waves are not breaking, it is probable that equation (23), which is applicable to deep water, may give approximately the mean horizontal velocity of the particles; but it is difficult to say how far the result may be modified by friction. If then we regard the ship as a mere particle, in the first instance, for the sake of simplicity, and put U_0 for the value of U when $y = 0$, it is easy to see that after sailing for a time t, the ship must be a distance $U_0 t$ to the lee of her estimated place. It will not however be sufficient to regard the ship as a mere particle, on account of the variation of the factor ϵ^{-2my}, as y varies from 0 to the greatest depth of the ship below the surface of the water. Let δ be this depth, or rather a depth something less, in order to allow for the narrowing of the ship towards the keel, and suppose the effect of the progressive motion of the water on the motion of the ship to be the same as if the water were moving with a velocity the same as all depths, and equal to the mean value of the velocity U from $y = 0$ to $y = \delta$. If U_1 be this mean velocity,

$$U_1 = \frac{1}{\delta} \int_0^\delta U dy = \frac{ma^2 c}{2\delta} (1 - \epsilon^{-2m\delta}).$$

On this supposition, if a ship be steered so as to sail in a direction making an angle θ with the direction of the wind, supposing the water to have no current, and if V be the velocity with which the ship moves through the water, her actual velocity will be the resultant of a velocity V in the direction just mentioned, which, for shortness, I shall call the direction of steering, and of a velocity U_1 in the direction of the wind. But the ship's velocity as estimated by the log-line is her velocity relatively to the water at the surface, and is therefore the resultant of a velocity V in the direction of steering, and a velocity $U_0 - U_1$ in a direction opposite to that in which the wind is blowing. If then E be the estimated velocity, and if we neglect U^2,

$$E = V - (U_0 - U_1) \cos \theta.$$

But the ship's velocity is really the resultant of a velocity $V + U_1 \cos \theta$ in the direction of steering, and a velocity $U_1 \sin \theta$ in the perpendicular direction, while her estimated velocity is E in the direction of steering. Hence, after a time t, the ship will be a distance $U_0 t \cos \theta$ ahead of her estimated place, and a distance $U_1 t \sin \theta$ aside of it, the latter distance being measured in a direction perpendicular to the direction of steering, and on the side towards which the wind is blowing.

I do not suppose that the preceding formula can be employed in practice; but I think it may not be altogether useless to call attention to the importance of having regard to the magnitude and direction of propagation of the waves, as well as to the wind, in making the allowance for lee-way.

11. The formulæ of Art. 6 are perfectly general as regards the ratio of the length of the waves to the depth of the fluid, the only restriction being that the height of the waves must be sufficiently small to allow the series to be rapidly convergent. Consequently, they must apply to the limiting case, in which the waves are supposed to be extremely long. Hence long waves, of the kind considered, are propagated without change of form, and the velocity of propagation is independent of the height of the waves to a second approximation. These conclusions might seem, at first sight, at variance with the results obtained by Mr Airy for the case of long waves*. On proceeding to a second approximation, Mr Airy finds that the form of long waves alters as they proceed, and that the expression for the velocity of propagation contains a

---

* *Encyclopædia Metropolitana, Tides and Waves,* Articles 198, &c.

term depending on the height of the waves. But a little attention will remove this apparent discrepancy. If we suppose mh very small in (19), and expand, retaining only the most important terms, we shall find for the equation to the surface

$$y = a \cos mx - \frac{3a^2}{4m^2h^3} \cos 2mx.$$

Now, in order that the method of approximation adopted may be legitimate, it is necessary that the coefficient of $\cos 2mx$ in this equation be small compared with a. Hence a/m^2h^3, and therefore $\lambda^2 a/h^3$, must be small, and therefore a/h must be small compared with $(h/\lambda)^2$. But the investigation of Mr Airy is applicable to the case in which λ/h is very large; so that in that investigation a/h is large compared with $(h/\lambda)^2$. Thus the difference in the results obtained corresponds to a difference in the physical circumstances of the motion.

12. There is no difficulty in proceeding to the higher orders of approximation, except what arises from the length of the formulæ. In the particular case in which the depth is considered infinite, the formulæ are very much simpler than in the general case. I shall proceed to the third order in the case of an infinite depth, so as to find in that case the most important term, depending on the height of the waves, in the expression for the velocity of propagation.

For this purpose it will be necessary to retain the terms of the third order in the expansion of (7). Expanding this equation according to powers of y, and neglecting terms of the fourth, &c. orders, we have

$$g\phi_, - c^2\phi'' + (g\phi_{,,} - c^2\phi_{,}'')y + (g\phi_{,,,} - c^2\phi_{,,}'')\frac{y^2}{2} + 2c(\phi'\phi'' + \phi_,\phi_,')$$

$$+ 2c(\phi_,'\phi'' + \phi'\phi_,'' + \phi_{,,}\phi_,' + \phi_,\phi_{,,}')y - \phi'^2\phi'' - 2\phi'\phi_,\phi_,' - \phi_,^2\phi_{,,} = 0$$
$$\dots\dots\dots\dots(25).$$

In the small terms of this equation we must put for ϕ and y their values given by (21) and (22) respectively. Now since the value of ϕ to a second approximation is the same as its value to a first approximation, the equation $g\phi_, - c^2\phi'' = 0$ is satisfied to terms of the second order. But the coefficients of y and $y^2/2$, in the first line of (25), are derived from the left-hand member of the preceding equation by inserting the factor ϵ^{-my}, differentiating either once or twice with respect to y, and then putting $y = 0$. Consequently these coefficients contain no terms of the second order, and therefore the terms involving y in the first line of (25) are to be neglected. The next two terms are together equal to $cd(\phi'^2 + \phi_,^2)/dx$. But

$$\phi'^2 + \phi_,^2 = m^2 a^2 c^2,$$

which does not contain x, so that these two terms disappear. The coefficient of y in the second line of (25) may be derived from the two terms last considered in the manner already indicated, and therefore the terms containing y will disappear from (25). The only small terms remaining are the last three, and it will easily be found that their sum is equal to $m^4 a^3 c^3 \sin mx$, so that (25) becomes

$$g\phi_, - c^2 \phi'' + m^4 a^3 c^3 \sin mx = 0 \dots\dots\dots\dots\dots(26).$$

The value of ϕ will evidently be of the form $A\epsilon^{-my} \sin mx$. Substituting this value in (26), we have

$$(m^2 c^2 - mg)A + m^4 a^3 c^3 = 0.$$

Dividing by mA, and putting for A and c^2 their approximate values $- ac$, g/m respectively in the small term, we have

$$mc^2 = g + m^2 a^2 g,$$

whence $\qquad c = \left(\dfrac{g}{m}\right)^{\frac{1}{2}} (1 + \tfrac{1}{2} m^2 a^2) = \left(\dfrac{g\lambda}{2\pi}\right)^{\frac{1}{2}} \left(1 + \dfrac{2\pi^2 a^2}{\lambda^2}\right).$

The equation to the surface may be found without difficulty. It is

$$y = a \cos mx - \tfrac{1}{2} m a^2 \cos 2mx + \tfrac{3}{8} m^2 a^3 \cos 3mx* \dots\dots(27):$$

we have also

$$k = 0, \quad \phi = - ac(1 - \tfrac{5}{8} m^2 a^2) \epsilon^{-my} \sin mx.$$

---

* It is remarkable that this equation coincides with that of the prolate cycloid, if the latter equation be expanded according to ascending powers of the distance of the tracing point from the centre of the rolling circle, and the terms of the fourth order be omitted. The prolate cycloid is the form assigned by Mr Russell to waves of the kind here considered. *Reports of the British Association*, Vol. VI. p. 448. When the depth of the fluid is not great compared with the length of a wave, the form of the surface does not agree with the prolate cycloid even to a second approximation.

The following figure represents a vertical section of the waves propagated along the surface of deep water. The figure is drawn

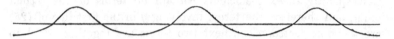

for the case in which $a = \dfrac{7\lambda}{80}$. The term of the third order in (27) is retained, but it is almost insensible. The straight line represents a section of the plane of mean level.

13. If we consider the manner in which the terms introduced by each successive approximation enter into equations (7) and (8), we shall see that, whatever be the order of approximation, the series expressing the ordinate of the surface will contain only cosines of mx and its multiples, while the expression for ϕ will contain only sines. The manner in which y enters into the coefficient of $\cos rmx$ in the expression for ϕ is determined in the case of a finite depth by equations (2) and (3). Moreover, the principal part of the coefficient of $\cos\ rmx$ or $\sin\ rmx$ will be of the order a^r at least. We may therefore assume

$$\phi = \Sigma_1^{\infty} a^r A_r \left(\epsilon^{rm(h-y)} + \epsilon^{-rm(h-y)} \right) \sin rmx,$$
$$y = a \cos mx + \Sigma_2^{\infty} a^r B_r \cos rmx,$$

and determine the arbitrary coefficients by means of equations (7) and (8), having previously expanded these equations according to ascending powers of y. The value of c^2 will be determined by equating to zero the coefficient of $\sin mx$ in (7).

Since changing the sign of a comes to the same thing as altering the origin of x by $\frac{1}{2}\lambda$, it is plain that the expressions for A_r, B_r and c^2 will contain only even powers of a. Thus the values of each of these quantities will be of the form

$$C_0 + C_1 a^2 + C_2 a^4 + \dots$$

It appears also that, whatever be the order of approximation, the waves will be symmetrical with respect to vertical planes passing through their ridges, as also with respect to vertical planes passing through their lowest lines.

14. Let us consider now the case of waves propagated at the common surface of two liquids, of which one rests on the

other. Suppose as before that the motion is in two dimensions, that the fluids extend indefinitely in all horizontal directions, or else that they are bounded by two vertical planes parallel to the direction of propagation of the waves, that the waves are propagated with a constant velocity, and without change of form, and that they are such as can be propagated into, or excited in, the fluids supposed to have been previously at rest. Suppose first that the fluids are bounded by two horizontal rigid planes. Then taking the common surface of the fluids when at rest for the plane xz, and employing the same notation as before, we have for the under fluid

$$\frac{d^2\phi}{d^2x} + \frac{d^2\phi}{dy^2} = 0 \dots\dots\dots\dots(28),$$

$$\frac{d\phi}{dy} = 0 \text{ when } y = h \dots\dots\dots\dots(29),$$

$$p = C + g\rho y + c\rho \frac{d\phi}{dx},$$

neglecting the squares of small quantities. Let $h_{,}$ be the depth of the upper fluid when in equilibrium, and let $p_{,}$, $\rho_{,}$, $\phi_{,}$, $C_{,}$ be the quantities referring to the upper fluid which correspond to p, ρ, ϕ, C referring to the under: then we have for the upper fluid

$$\frac{d^2\phi_{,}}{dx^2} + \frac{d^2\phi_{,}}{dy^2} = 0 \dots\dots\dots\dots(30),$$

$$\frac{d\phi_{,}}{dy} = 0 \text{ when } y = -h_{,} \dots\dots\dots\dots(31),$$

$$p_{,} = C_{,} + g\rho_{,}y + c\rho_{,}\frac{d\phi_{,}}{dx}.$$

We have also, for the condition that the two fluids shall not penetrate into, nor separate from each other,

$$\frac{d\phi}{dy} = \frac{d\phi_{,}}{dy}, \text{ when } y = 0 \dots\dots\dots\dots(32).$$

Lastly, the condition answering to (11) is

$$g\left(\rho \frac{d\phi}{dy} - \rho_{,}\frac{d\phi_{,}}{dy}\right) - c^2\left(\rho \frac{d^2\phi}{dx^2} - \rho_{,}\frac{d^2\phi_{,}}{dx^2}\right) = 0 \dots\dots\dots(33),$$

when $\qquad C - C_{,} + g\left(\rho - \rho_{,}\right)y + c\left(\rho \frac{d\phi}{dx} - \rho_{,}\frac{d\phi_{,}}{dx}\right) = 0 \dots\dots(34).$

Since $C - C'$ is evidently a small quantity of the first order at least, the condition is that (33) shall be satisfied when $y = 0$. Equation (34) will then give the ordinate of the common surface of the two liquids when y is put $= 0$ in the last two terms.

The general value of ϕ suitable to the present case, which is derived from (28) subject to the condition (29), is given by (13) if we suppose that the fluid is free from a uniform horizontal motion compounded with the oscillatory motion expressed by (18). Since the equations of the present investigation are linear, in consequence of the omission of the squares of small quantities, it will be sufficient to consider one of the terms in (13). Let then

$$\phi = A \left(\epsilon^{m(h-y)} + \epsilon^{-m(h-y)} \right) \sin mx \ldots\ldots(35).$$

The general value of $\phi_{,}$ will be derived from (13) by merely writing $-h_{,}$ for h. But in order that (32) may be satisfied, the value of $\phi_{,}$ must reduce itself to a single term of the same form as the second side of (35). We may take then for the value of $\phi_{,}$

$$\phi_{,} = A_{,} \left(\epsilon^{m(h_{,}+y)} + \epsilon^{-m(h_{,}+y)} \right) \sin mx \ldots\ldots\ldots(36).$$

Putting for shortness

$$\epsilon^{mh} + \epsilon^{-mh} = S, \quad \epsilon^{mh} - \epsilon^{-mh} = D,$$

and taking $S_{,}, D_{,}$ to denote the quantities derived from S, D by writing $h_{,}$ for h, we have from (32)

$$DA + D_{,}A_{,} = 0 \ldots\ldots\ldots\ldots(37),$$

and from (33)

$$\rho \left(gD - mc^2 S \right) A + \rho_{,} \left(gD_{,} + mc^2 S_{,} \right) A_{,} = 0 \ldots\ldots(38).$$

Eliminating A and $A_{,}$ from (37) and (38), we have

$$c^2 = \frac{g}{m} \frac{(\rho - \rho_{,}) \, DD_{,}}{\rho SD_{,} + \rho_{,} S_{,} D} \ldots\ldots\ldots\ldots(39).$$

The equation to the common surface of the liquids will be obtained from (34). Since the mean value of y is zero, we have in the first place

$$C_{,} = C \ldots\ldots\ldots\ldots\ldots\ldots(40).$$

We have then, for the value of y,

$$y = a \cos mx \ldots\ldots\ldots\ldots\ldots(41),$$

where

$$a = \frac{mc}{g}\frac{\rho_, A_, S_, - \rho AS}{\rho - \rho_,} = \frac{DD_,}{c}\frac{\rho_, A_, S_, - \rho AS}{\rho S D_, + \rho_, S_, D}\ldots\ldots(42).$$

Substituting in (35) and (36) the values of A and $A_,$ derived from (37) and (42), we have

$$\phi = -\frac{ac}{D}\left(\epsilon^{m(h-y)} + \epsilon^{-m(h-y)}\right)\sin mx\ldots\ldots\ldots(43),$$

$$\phi_, = \frac{ac}{D_,}\left(\epsilon^{m(h_,+y)} + \epsilon^{-m(h_,+y)}\right)\sin mx\ldots\ldots\ldots(44).$$

Equations (39), (40), (41), (43) and (44) contain the solution of the problem. It is evident that C remains arbitrary. The values of p and $p_,$ may be easily found if required.

If we differentiate the logarithm of c^2 with respect to m, and multiply the result by the product of the denominators, which are necessarily positive, we shall find a quantity of the form $P\rho + P_,\rho_,$, where P and $P_,$ do not contain ρ or $\rho_,$. It may be proved in nearly the same manner as in Art. 4, that each of the quantities P, $P_,$ is necessarily negative. Consequently c will decrease as m increases, or will increase with λ. It follows from this that the value of ϕ cannot contain more than two terms, one of the form (35), and the other derived from (35) by replacing $\sin mx$ by $\cos mx$, and changing the constant A: but the latter term may be got rid of by altering the origin of x.

The simplest case to consider is that in which both h and h' are regarded as infinite compared with λ. In this case we have

$$\phi = -ac\epsilon^{-my}\sin mx, \quad \phi_, = ac\epsilon^{my}\sin mx,$$

$$c^2 = \frac{\rho - \rho_,}{\rho + \rho_,}\frac{g}{m}, \quad y = a\cos mx,$$

the latter being the equation to the surface.

15. The preceding investigation applies to two incompressible fluids, but the results are applicable to the case of the waves propagated along the surface of a liquid exposed to the air, provided that in considering the effect of the air we neglect terms which, in comparison with those retained, are of the order of the ratio of the length of the waves considered to the length of

a wave of sound of the same period in air. Taking then ρ for the density of the liquid, $\rho_{,}$ for that of the air at the time, and supposing $h_{,} = \infty$, we have

$$c^2 = \frac{g}{m} \frac{(\rho - \rho_{,}) D}{\rho S + \rho_{,} D} = \frac{gD}{mS}\left\{1 - \left(1 + \frac{D}{S}\right)\frac{\rho_{,}}{\rho}\right\}, \quad \text{nearly.}$$

If we had considered the buoyancy only of the air, we should have had to replace g in the formula (14) by $\frac{\rho - \rho_{,}}{\rho} g$. We should have obtained in this manner

$$c^2 = \frac{g}{m} \frac{(\rho - \rho_{,}) D}{\rho S} = \frac{gD}{mS}\left(1 - \frac{\rho_{,}}{\rho}\right).$$

Hence, in order to allow for the inertia of the air, the correction for buoyancy must be increased in the ratio of 1 to $1 + D/S$. The whole correction therefore increases as the ratio of the length of a wave to the depth of the fluid decreases. For very long waves the correction is that due to buoyancy alone, while in the case of very short waves the correction for buoyancy is doubled. Even in this case the velocity of propagation is altered by only the fractional part $\rho_{,}/\rho$ of the whole; and as this quantity is much less than the unavoidable errors of observation, the effect of the air in altering the velocity of propagation may be neglected.

16. There is a discontinuity in the density of the fluid mass considered in Art. 14, in passing from one fluid into the other; and it is easy to shew that there is a corresponding discontinuity in the velocity. If we consider two fluid particles in contact with each other, and situated on opposite sides of the surface of junction of the two fluids, we see that the velocities of these particles resolved in a direction normal to that surface are the same; but their velocities resolved in a direction tangential to the surface are different. These velocities are, to the order of approximation employed in the investigation, the values of $d\phi/dx$ and $d\phi_{,}/dx$ when $y = 0$. We have then from (43) and (44), for the velocity with which the upper fluid slides along the under,

$$mac\left(\frac{S_{,}}{D_{,}} + \frac{S}{D}\right)\cos mx.$$

17. When the upper surface of the upper fluid is free, the equations by which the problem is to be solved are the same as those of Art. 14, except that the condition (31) is replaced by

$$g \frac{d\phi_i}{dy} - c^2 \frac{d^2\phi_i}{dx^2} = 0, \text{ when } y = -h_i \ldots\ldots\ldots(45);$$

and to determine the ordinate of the upper surface, we have

$$C_i + g\rho_i y + c\rho_i \frac{d\phi_i}{dx} = 0,$$

where y is to be replaced by $-h_i$ in the last term. Let us consider the motion corresponding to the value of ϕ given by (35). We must evidently have

$$\phi_i = (A_i \epsilon^{my} + B_i \epsilon^{-my}) \sin mx,$$

where A_i and B_i have to be determined. The conditions (32), (33) and (45) give

$$DA + A_i - B_i = 0,$$

$$\rho(gD - mc^2 S) A + \rho_i (g + mc^2) A_i - \rho_i (g - mc^2) B_i = 0,$$

$$(g + mc^2) \epsilon^{-mh_i} A_i - (g - mc^2) \epsilon^{mh_i} B_i = 0.$$

Eliminating A, A_i and B_i from these equations, and putting

$$c^2 = \frac{g\zeta}{m},$$

we find

$$(\rho SS_i + \rho_i DD_i) \zeta^2 - \rho(SD_i + S_i D) \zeta + (\rho - \rho_i) DD_i = 0 \ldots (46).$$

The equilibrium of the fluid being supposed to be stable, we must have $\rho_i < \rho$. This being the case, it is easy to prove that the two roots of (46) are real and positive. These two roots correspond to two systems of waves of the same length, which are propagated with the same velocity.

In the limiting case in which $\rho/\rho_i = \infty$, (46) becomes

$$SS_i \zeta^2 - (SD_i + S_i D) \zeta + DD_i = 0,$$

the roots of which are D/S and D_i/S_i, as they evidently ought to be, since in this case the motion of the under fluid will not be affected by that of the upper, and the upper fluid can be in motion by itself.

When $\rho_i = \rho$ one root of (46) vanishes, and the other becomes $\frac{SD_i + S_i D}{SS_i + DD_i}$ or $\frac{\epsilon^{m(h+h_i)} - \epsilon^{-m(h+h_i)}}{\epsilon^{m(h+h_i)} + \epsilon^{-m(h+h_i)}}$. The former of these roots corresponds to the waves propagated at the common surface of the fluids, while the latter gives the velocity of propagation belonging to a single fluid having a depth equal to the sum of the depths of the two considered.

When the depth of the upper fluid is considered infinite, we must put $D_{/}/S_{/} = 1$ in (46). The two roots of the equation so transformed are 1 and $\dfrac{(\rho - \rho')\,D}{\rho S + \rho_{/} D}$, the former corresponding to waves propagated at the upper surface of the upper fluid, and the latter agreeing with Art. 15.

When the depth of the under fluid is considered infinite, and that of the upper finite, we must put $D/S = 1$ in (46). The two roots will then become 1 and $\dfrac{(\rho - \rho_{/})\,D_{/}}{\rho S_{/} + \rho_{/} D_{/}}$. The value of the former root shews that whatever be the depth of the upper fluid, one of the two systems of waves will always be propagated with the same velocity as waves of the same length at the surface of a single fluid of infinite depth. This result is true even when the motion is in three dimensions, and the form of the waves changes with the time, the waves being still supposed to be such as could be excited in the fluids, supposed to have been previously at rest, by means of forces applied at the upper surface. For the most general small motion of the fluids in this case may be regarded as the resultant of an infinite number of systems of waves of the kind considered in this paper. It is remarkable that when the depth of the upper fluid is very great, the root $\zeta = 1$ is that which corresponds to the waves for which the upper fluid is disturbed, while the under is sensibly at rest; whereas, when the depth of the upper fluid is very small, it is the other root which corresponds to those waves which are analogous to the waves which would be propagated in the upper fluid if it rested on a rigid plane.

When the depth of the upper fluid is very small compared with the length of a wave, one of the roots of (46) will be very small; and if we neglect squares and products of $mh_{/}$ and ζ, the equation becomes $2\rho D\zeta - 2(\rho - \rho_{/})\,mh_{/} D = 0$, whence

$$\zeta = \frac{\rho - \rho_{/}}{\rho}\,mh_{/}, \quad c^2 = \frac{\rho - \rho_{/}}{\rho} \quad \dots\dots\dots\dots\dots\dots\dots(47).$$

These formulæ will not hold good if mh be very small as well as $mh_{/}$, and comparable with it, since in that case all the terms of

(46) will be small quantities of the second order, $mh_{,}$ being regarded as a small quantity of the first order. In this case, if we neglect small quantities of the third order in (46), it becomes

$$4\rho\zeta^2 - 4m\rho\,(h + h_{,})\,\zeta + 4(\rho - \rho_{,})\,m^2hh_{,} = 0,$$

whence

$$c^2 = \frac{g}{2}\left\{h + h_{,} \pm \sqrt{(h - h_{,})^2 + \frac{4\rho_{,}}{\rho}\,hh_{,}}\right\}\dots\dots\dots(48).$$

Of these values of c^2, that in which the radical has the negative sign belongs to that system of waves to which the formulæ (47) apply when $h_{,}$ is very small compared with h.

If the two fluids are water and mercury, $\rho/\rho_{,}$ is equal to about 13·57. If the depth of the water be very small compared both with the length of the waves and with the depth of the mercury, it appears from (47) that the velocity of propagation will be less than it would have been, if the water had rested on a rigid plane, in the ratio of ·9624 to 1, or 26 to 27 nearly.

APPENDIX.

[A. *On the relation of the preceding investigation to a case of wave motion of the oscillatory kind in which the disturbance can be expressed in finite terms.*

In the *Philosophical Transactions* for 1863, p. 127, is a paper by the late Professor Rankine in which he has shewn that it is possible to express in finite terms, without any approximation, the motion of a particular class of waves of the oscillatory kind. It is remarkable that the results for waves of this kind were given as long ago as in 1802, by Gerstner*, whose investigation however seems to have been but little noticed for a long time. This case of motion has latterly attracted a good deal of attention, partly no doubt from the facility of dealing with it, but partly, it would seem, from misconceptions as to its intrinsic importance.

---

* See Weber's *Wellenlehre auf Experimente gegründet*, p. 338.

The investigation may be presented in very short compass in the following manner.

Let us confine our attention to the case of a mass of liquid, regarded as a perfect fluid of a depth practically infinite, in which an indefinite series of regular periodic waves is propagated along the surface, the motion being in two dimensions, and vanishing at an infinite depth. Taking the plane of motion for the plane of xy, y being measured vertically downwards, let us seek to express the actual co-ordinates x, y of any particle in terms of two parameters h, k particularising that particle, and of the time t. Let us assume for trial

$$x = h + K \sin m (h - ct), \qquad y = k + K \cos m (h - ct)\ldots\ldots(49),$$

where m, c are two constants, and K a function of k only. It will be easily seen that these equations, regarded merely as expressing the geometrical motion of points, and apart from the physical possibility of the motion, represent a wave disturbance of periodic character travelling in the direction of OX with a velocity of propagation c.

As the disturbance is in two dimensions, we may speak of areas as representing volumes. Let us consider first the condition of constancy of the mass. The four loci corresponding to constant values h, $h + dh$, k, $k + dk$, of the two parameters respectively enclose a quadrangular figure which is ultimately a parallelogram, the area of which must be independent of the time. Now the area is $S\,dh\,dk$ where

$$S = \frac{dx}{dh}\frac{dy}{dk} - \frac{dx}{dk}\frac{dy}{dh} \ldots\ldots\ldots\ldots\ldots\ldots(50).$$

On performing the differentiations we find

$$S = 1 + (mk + K')\cos m (h - ct) + mKK' \ldots\ldots\ldots(51),$$

where K' stands for dK/dk. In order that this may be independent of the time it is necessary and sufficient that

$$mK + K' = 0 \ldots\ldots\ldots\ldots\ldots\ldots(52),$$

whence

$$K = a\epsilon^{-mk} \ldots\ldots\ldots\ldots\ldots\ldots(52'),$$

and

$$S = 1 - m^2 K^2 = 1 - m^2 a^2 \epsilon^{-2mk}\ldots\ldots\ldots\ldots(53).$$

The dynamical equations give

$$\frac{dp}{\rho} = g\,dy - \left(\frac{d^2x}{dt^2}\,dx + \frac{d^2y}{dt^2}\,dy\right)$$

$$= g\,dy + m^2c^2K\{\sin m(h - ct)\,dx + \cos m(h - ct)\,dy\}$$

$$= g\,dy + m^2c^2\{(x - h)\,d(x - h) + (y - k)\,d(y - k)\}$$

$$+ m^2c^2\{(x - h)\,dh + (y - k)\,dk\}.$$

The last line becomes by (49) and (52),

$$mc^2\{mK\sin m(h - ct)\,dh - K'\cos m(h - ct)\,dk\},$$

or $\qquad\qquad - mc^2d\,.\,K\cos m(h - ct).$

The dynamical equations are therefore satisfied, the expression for dp being a perfect differential, and we have

$$\frac{p}{\rho} = gy + \tfrac{1}{2}m^2c^2\{(x - h)^2 + (y - k)^2\} - mc^2K\cos m(h - ct) + C$$

$$= gk + \tfrac{1}{2}m^2c^2K^2 + (g - mc^2)K\cos m(h - ct) + C.$$

It remains to consider the equations of condition at the boundaries of the fluid. The expression for K satisfies the condition of giving a disturbance which decreases indefinitely as the depth increases, and we have only to see if it be possible to satisfy the condition at the free surface. Now the particles at the free surface differ only by the value of the parameter h, as follows from the fundamental conception of wave motion, and therefore for some one value of k we must have $p = 0$ independently of the time. This requires that

$$c^2 = \frac{g}{m} = \frac{g\lambda}{2\pi},$$

and if we please to take $k = 0$ at the surface, and determine C accordingly, we have

$$\frac{p}{\rho} = gk - \tfrac{1}{2}ga^2m(1 - \epsilon^{-2mk}) \quad\ldots\ldots\ldots(54).$$

Since p is independent of the time, not merely for $k = 0$, but for *any* constant value of k, it follows that when the wave motion is converted into steady motion by superposing a velocity equal and opposite to that of propagation, it is not merely the line of motion or stream-line which forms the surface but *all* the streamlines that are lines of constant pressure. This is undoubtedly no necessary property of wave-motion converted into steady motion, which only requires that the particular stream-line at the surface

shall be one for which the pressure is constant, though Gerstner
has expressed himself as if he supposed it necessarily true; it is
merely a character of the special case investigated by Gerstner
and Rankine. Nevertheless in the case of *deep* water it must be
very approximately true. For in the first place it is strictly true
at the surface, and in the second place, it must be sensibly true
at a very moderate depth and for all greater depths, since the
disturbance very rapidly diminishes on passing from the surface
downwards; so that unless the amount of disturbance be excessive
the supposition that all the stream-lines are lines of constant
pressure will not be much in error.

In the case investigated by the mathematicians just mentioned,
each particle returns periodically to the position it had at a given
instant; there is no progressive motion combined with a periodic
disturbance, such as was found in the case investigated in the present paper : and for deep water the absence of progressive motion
is doubtless peculiar to the former case, as will presently more
clearly appear.

If we suppose a regular periodic wave motion to be going on,
and then suppose small suitable pressures applied to the surface in
such a manner as to check the motion, we may evidently produce
a secular subsidence of the wave disturbance while still leaving it
at any moment regular and periodic, save as to secular change,
provided the opposing pressures are suitably chosen. The wave-
length will be left unchanged, but not so, in general, the periodic
time. If the amount of disturbance in one wave period be insensible, the particles which at one time have a common mean depth
must at any future time have a common mean depth, and must
ultimately lie in a horizontal plane when the wave motion has
wholly subsided. In this condition therefore there can be no
motion except a horizontal flow with a velocity which is some
function of the depth. By a converse process we may imagine a
regular periodic wave motion of given wave-length excited in a
fluid in which there previously was none; and according to the
nature of the arbitrary flow with which we start, we shall obtain
as the result a wave motion of such or such a kind*.

In any given case of wave motion, the flow which remains

* To prevent possible misconception I may observe that I am not here con-
templating the actual mode of excitement of waves by wind, which in some respects
is essentially different.

when the waves have been caused to subside in the manner above
explained is easily determined, since we know that in the motion
of a liquid in two dimensions the angular velocity is not affected
by forces applied to the surface.   If ω be the angular velocity

$$2\omega = \frac{dv}{dx} - \frac{du}{dy} = \frac{1}{S}\left\{\frac{dy}{dk}\frac{dv}{dh} - \frac{dy}{dh}\frac{dv}{dk} + \frac{dx}{dk}\frac{du}{dh} - \frac{dx}{dh}\frac{du}{dk}\right\}$$

S being defined by (50).   In Gerstner and Rankine's solution

$$u = -mac\epsilon^{-mk}\cos m(h - ct), \quad v = mac\epsilon^{-mk}\sin m(h - ct),$$

and on effecting the differentiations and substituting for S from
(53) we find

$$\omega = -\frac{m^3a^2c\epsilon^{-2mk}}{1 - m^2a^2\epsilon^{-2mk}} \quad\dots\dots\dots\dots\dots(55).$$

Let y' be the depth and u' the horizontal velocity, after the
wave-motion has been destroyed as above explained, of the line of
particles which had k for a parameter; then we must have

$$\omega = -\tfrac{1}{2}\frac{du'}{dy'} \quad\dots\dots\dots\dots\dots\dots(56).$$

Since in a horizontal length which may be deemed infinite compared with λ the area between the ordinates y', $y' + dy'$ must
be the same as between the lines of particles which have k, $k + dk$
for their k-parameter

$$dy' = Sdk,$$

S being defined by (50).   Putting for S its value given by (53)
we have

$$dy' = (1 - m^2a^2\epsilon^{-2mk})\,dk\dots\dots\dots\dots\dots(57),$$

$$y' = k - \tfrac{1}{2}ma^2(1 - \epsilon^{-2mk}) \quad\dots\dots\dots\dots\dots(58).$$

We have then from (56) by (55) and (57),

$$u' = 2m^3a^2c\int\epsilon^{-2mk}dk = -m^2a^2c\epsilon^{-2mk}\dots\dots\dots\dots(59),$$

since u' vanishes when $k = \infty$.

It appears then that in order that it should be possible to
excite these waves in deep water previously free from wave disturbance, by means of pressures applied to the surface, a preparation must be laid in the shape of a horizontal velocity decreasing
from the surface downwards according to the value of ϵ^{-2mk}, where
k is a function of the depth y' determined by the transcendental
equation in k (58), and moreover a velocity decreasing downwards
according to this law will serve for waves of the present kind of

only one particular height depending on the coefficient of the exponential in the expression for the flow. Under these conditions the horizontal velocity depending (when we adopt approximations) on the square and higher powers of the elevation, which belongs to the wave-motion, is exactly neutralized by the pre-existing horizontal velocity in a contrary direction, pre-existing, that is, when we think of the waves as having been excited in a fluid previously destitute of wave-motion, not as having gone on as they are from a time indefinitely remote. The absence of any forward horizontal motion of the individual particles in waves of this kind, though attractive at first sight, is not of any real physical import, because we are not concerned with the *biographies* so to speak of the individual particles.

The oscillatory waves which most naturally present themselves to our attention are those which are excited in the ocean or on a lake by the action of the wind, or those which having been so excited are propagated into (practically, though not in a rigorous mathematical sense) still water. Of the latter kind are the surf which breaks upon our western coasts as a result of storms out in the Atlantic, or the grand rollers which are occasionally observed at St Helena and Ascension Island. The motion in these cases having been produced from rest, by forces applied to the surface, there is no molecular rotation, and therefore the investigation of the present paper strictly applies. Moreover, if we conceive the waves gradually produced by suitable forces applied to the surface, in the manner explained at p. 222, the investigation applies to the waves (secular change apart) at any period of their growth, and not merely when they have attained one particular height.

There can be no question, it seems to me, that this is the class of oscillatory waves which on merely physical grounds we should naturally select for investigation. The interest of the solution first given by Gerstner, and it *is* of great interest, arises not from any physical pre-eminence of the class of waves to which it relates, but from the imperfection of our analysis, which renders it important to discuss a case in which all the circumstances of the motion can be simply expressed in mathematical terms without any approximation. And though this motion is not exactly that which on purely physical grounds we should prefer to investigate, namely, that in which the molecular rotation is *nil*, yet unless the height of the waves be extravagant, it agrees so nearly with it that for many purposes the simpler expressions of Rankine may be used without material error, even when we are investigating wave motion of the irrotational kind.

B. *Considerations relative to the greatest height of oscillatory irrotational waves which can be propagated without change of form.*

In a paper published in the *Philosophical Magazine,* Vol. XXIX. (1865), p. 25, Rankine gave an investigation which led him to the conclusion that in the steepest possible oscillatory waves of the irrotational kind, the crests become at the vertex infinitely curved in such a manner that a section of the crest by the plane of motion presents two branches of a curve which meet at a right angle*.

In this investigation it is assumed in the first place that the steepness may be pushed to the limit of an infinite curvature at a particular point, and in the second place that the variations

* It is not quite clear whether Rankine supposed his proposition, that "all waves in which molecular rotation is null, begin to break when the two slopes of the crest meet at right angles," to apply only to free waves, or to forced waves as well. One would have supposed the former, were it not that a figure is referred to representing forced waves of one particular kind. It is readily shewn that the contour of a forced wave is arbitrary, even though the motion be restricted to be irrotational. Let $U = C$ (p. 4) be the general equation of the stream lines when the wave motion is converted into steady motion. Then in the general case of a finite depth, which includes as a limiting and therefore particular case that of an infinite depth, the parameter C has one constant value at the upper surface, and another at the bottom, and it satisfies the partial differential equation (5) of p. 4. Hence the problem of finding U is the same as that of determining the permanent temperature, varying in two dimensions only, of a homogeneous isotropic solid the section of which is bounded below by a horizontal line at a finite or infinite depth, and above by a given arbitrary contour, the bounding surfaces being at two given constant temperatures. The latter problem is evidently determinate, and therefore also the former, so that *forced* waves may present in their contour sharp angles, not merely of 90°, but of any value we please to take.

S. 15

of the components of the velocity, in passing from the crest to
a point infinitely close to it, may be obtained by differentiation,
or in other words from the second terms of the expansion by
Taylor's Theorem applied to infinitely small increments of the
variables.

The first assumption might perhaps be called in question,
but it would appear likely to give at any rate a superior limit
to the steepest form possible, if not the steepest form itself.
But as regards the second it would seem à priori very likely
that the crest might just be one of those singular points where
Taylor's Theorem fails; and that such must actually be the case
may be shewn by simple considerations.

Let us suppose that a fluid of either finite or infinite depth
is disturbed by a wave motion which is propagated uniformly
without change, the motion of the fluid being either rotational
or not, and let us suppose further that the crests are perfectly
sharp, so that a crest is formed by two branches of a curve which
either meet at a finite angle (their prolongations belonging to the
region of space where the fluid is not), or else touch, forming
a cusp.

Reduce the wave motion to steady motion by superposing
a velocity equal and opposite to that of propagation. Then
a particle at the surface may be thought of as gliding along a
fixed smooth curve: this follows directly from physical considerations, or from the ordinary equation of steady motion. On
arriving at a crest the particle must be momentarily at rest, and
on passing it must be ultimately in the condition of a particle
starting from rest down an inclined or vertical plane. Hence the
velocity must vary ultimately as the square root of the distance
from the crest.

Hitherto the motion has been rotational or not, let us now
confine ourselves to the case of irrotational motion. Place the
origin at the crest, refer the function ϕ to polar co-ordinates r, θ;
θ being measured from the vertical, and consider the value of ϕ
very near the origin, where ϕ may be supposed to vanish, as the
arbitrary constant may be omitted. In general ϕ will be of the
form $\Sigma A_n r^n \sin n\theta + \Sigma B_n \cos n\theta$. In the present case ϕ must contain sines only on account of the symmetry of the motion, as

already shewn (p. 212), so that retaining only the most important term we may take $\phi = Ar^n \sin n\theta$. Now for a point in the section of the profile we must have $d\phi/d\theta = 0$, and $d\phi/dr$ varying ultimately as $r^{\frac{1}{2}}$. This requires that $n = \frac{3}{2}$, and for the profile that $\frac{3}{2}\theta = \frac{1}{2}\pi$, so that the two branches are inclined at angles of $\pm 60°$ to the vertical, and at an angle of $120°$ to each other, not of $90°$ as supposed by Rankine.

This however leaves untouched the question whether the disturbance can actually be pushed to the extent of yielding crests with sharp edges, or whether on the other hand there exists a limit, for which the outline is still a smooth curve, beyond which no waves of the oscillatory irrotational kind can be propagated without change of form.

After careful consideration I feel satisfied that there is no such earlier limit, but that we may actually approach as near as we please to the form in which the curvature at the vertex becomes infinite, and the vertex becomes a multiple point where the two branches with which alone we are concerned enclose an angle of $120°$. But whether in the limiting form the inclination of the wave to the horizon continually increases from the trough to the summit, and is consequently limited to $30°$, or whether on the other hand the points of inflexion which the profile presents in the general case remain at a finite distance from the summit when the limiting form is reached, so that on passing from the trough to the summit the inclination attains a maximum from which it begins to decrease before the summit is reached, is a question which I cannot certainly decide, though I feel little doubt that the former alternative represents the truth.

In Rankine's case of wave motion the limiting form presents crests which are cusped. For the maximum wave $ma = 1$ or $a = \lambda/2\pi$. We see from (55) that in this case the angular velocity becomes infinite at the surface, where k vanishes; and if we suppose such waves excited in the manner already explained in a fluid initially destitute of wave motion, the horizontal velocity u' which must exist in preparation for the waves must be such that du'/dy' becomes infinite at the surface. It appears to be this circumstance which renders it possible for even rotational waves to attain in the limit to an infinite thinness of crest without losing the property of uniform propagation.

When swells are propagated towards a smooth, very gently shelving shore, the height increases when the finiteness of depth begins to take effect. Presently the limiting height for uniformly propagated irrotational waves is passed, and then the form of the wave changes independently of the mere secular change due to diminishing depth. The tendency is now for the high parts to overtake the less high in front of them, and thereby to become higher still, until at last the crest topples over and the wave finally breaks. The breaking is no doubt influenced by friction against the bottom (denoting by "friction" the effect of the eddies produced), but I do not believe that it is wholly or even mainly due to this cause. Before the wave breaks altogether the top gets very thin, but the maximum height for uniform propagation is probably already passed by a good deal, so that we must guard against being misled by this observation as to the character of the limiting form.

In watching many years ago a grand surf which came rolling in on a sandy beach near the Giant's Causeway, without any storm at the place itself, I recollect being struck with the blunt wedge-like form of the waves where they first lost their flowing outline, and began to show a little broken water at the very summit. It is only I imagine on an oceanic coast, and even there on somewhat rare occasions, that the form of waves of this kind, of nearly the maximum height, can be studied to full advantage. The observer must be stationed nearly in a line with the ridges of the waves where they begin to break.

C. *Remark on the method of Art. I.*

There appears to be a slight advantage in employing the function U or ψ ($= \int (u\,dy - v\,dx)$) instead of ϕ, the wave motion having been reduced to steady motion as is virtually done in Art. 1. The general equation for ψ is the same as for ϕ, (2), and the general expression for ψ answering to that given for ϕ on p. 212 is

$$\psi = -cy + \Sigma_1^\infty C_r \left(\epsilon^{rm(h-y)} - \epsilon^{-rm(h-y)} \right) \cos rmx.$$

The expression for p in terms of ψ is almost identical with that in terms of ϕ. So far there is nothing to choose between the two. But for the two equations which have to be satisfied simultaneously at the surface, instead of $p = 0$ and the somewhat complicated equation (7), we have $p = 0$ and $\psi = \mathrm{const.}$, which constant we may take $= 0$ if we leave open the origin of y. The substitution of this equation of simpler form for (7) is a gain in proceeding to higher orders of approximation. I remember however thinking as I was working at the paper that as far as the approximation there went the gain was not such as to render it worth while to make the change.

But while these sheets were going through the press I devised a totally different method of conducting the approximation, which I find possesses very substantial advantages in proceeding to higher orders of approximation. The reader will find this new method after the paper "on the critical values of the sums of periodic series."]

[From the *Report of the British Association* for 1847, Part II. p. 6.]

On the Resistance of a Fluid to Two Oscillating Spheres.

THE object of this communication was to shew the application of Professor Thomson's method of images to the solution of certain problems in hydrodynamics. Suppose that there exists in an infinite mass of incompressible fluid a point from which, or to which the fluid is flowing with a velocity alike in all directions. Conceive now two such points, of intensities equal in magnitude and opposite in sign, to coexist in the fluid; and then suppose these points to approach, and ultimately coalesce, their intensities varying inversely as the distance between them. Let the resulting point be called a *singular point of the second order*. The motion of a fluid about a solid, oscillating sphere is the same as if the solid sphere were replaced by fluid, in the centre of which existed such a point. It is easy to shew that the motion of the fluid due to a point of this kind, when the fluid is interrupted by a sphere having its centre in the axis of the singular point, is the same as if the sphere's place were occupied by fluid containing one singular point of the second order. By the application of this principle may be found the resistance experienced by a sphere oscillating in presence of a fixed sphere or plane, or within a spherical envelope, the oscillation taking place in the line joining the centres, or perpendicular to the plane. In a similar manner may be found the resistance to two spheres which touch, or are connected by a rod, or to the solid made up of two spheres which cut, provided the exterior angle of the surfaces be a submultiple of two right angles, the oscillation in these cases also taking place in the line joining the centres. The numerical calculation is very simple, and may be carried to any degree of accuracy.

The investigation mentioned in the preceding paper arose out of the communication to me by Sir William Thomson of his beautiful method of electrical images before he had published it. Having myself paid more attention to the motion of fluids than to electricity, I endeavoured to find if it would in any manner apply to the solution of problems in the motion of fluids. I found that what is called above a singular point of the second order had a perfect image in a sphere when its axis was in the direction of a radius, which led to a complete solution of the problem mentioned in the paper when one sphere lay wholly outside or inside the other. I shewed this to Professor Thomson, who pointed out to me that a solution was also attainable, and that in finite terms, when the spheres intersected, provided the angle of intersection was a submultiple of two right angles. He saw that the property of a singular point of the second order of giving a perfect image in the case mentioned, admitted of an application to the theory of magnetism, which he has published in a short paper in the second volume of the *Cambridge and Dublin Mathematical Journal*, (1847) p. 240.

Although the mathematical result is contained in the paper just mentioned, I subjoin the process by which I found it out.

The expression (see p. 41) for the function ϕ around a sphere which moves in a perfect fluid previously at rest may be thought of as applying to the whole of an infinite mass of fluid, provided we conceive what has here been called a singular point of the second order to exist at the origin. Let us conceive a spherical surface S with its centre at O and having a radius a to exist in the fluid; let P be the singular point, lying either within or without the sphere S, and having its axis in the line OP. Let r', θ' be polar co-ordinates originating at P, θ' being measured from OP produced, and let r, θ be polar co-ordinates originating at O; let m be a constant, and $OP = c$, then ϕ being the function due to the singular point we have

$$\phi = -\frac{m \cos \theta'}{r'^2} = -\frac{m \cdot r' \cos \theta'}{r'^3} = -m \frac{r \cos \theta - c}{(r^2 - 2cr \cos \theta + c^2)^{\frac{3}{2}}}$$

$$= -m \frac{d}{dc} (r^2 - 2cr \cos \theta + c^2)^{-\frac{1}{2}}.$$

Now if e be less than 1,

$$(1 - 2e \cos \theta + e^2)^{-\frac{1}{2}} = P_0 + e P_1 + e^2 P_2 + \dots,$$

where P_0, P_1, P_2... are Laplace's, or in this case more properly Legendre's, coefficients*. Hence by expanding and differentiating with respect to c, we have

$$\phi = - m \left(\frac{1P_1}{r^2} + \frac{2cP_2}{r^3} + \frac{3c^2P_3}{r^4} + \ldots \right), \text{ if } r > c \ldots \ldots (1),$$

$$\phi = \ \ m \left(\frac{1P_0}{c^2} + \frac{2rP_1}{c^3} + \frac{3r^2P_2}{c^4} + \ldots \right), \text{ if } r < c \ldots \ldots (2).$$

We are not of course concerned with the constant term in the latter of these two expressions. For the normal velocity (ν) at the surface of the sphere we get by differentiating with respect to r, and then putting $r = a$

$$\nu = m \left(\frac{1 \cdot 2P_1}{a^3} + \frac{2 \cdot 3cP_2}{a^4} + \frac{3 \cdot 4c^2P_3}{a^5} + \ldots \right), \text{ if } a > c \ldots (3),$$

$$\nu = m \left(\frac{1 \cdot 2P_1}{c^3} + \frac{2 \cdot 3aP_2}{c^4} + \frac{3 \cdot 4a^2P_3}{c^5} + \ldots \right), \text{ if } a < c \ldots (4).$$

First suppose the point P outside the sphere, let the sphere be thought of as a solid sphere, and consider the motion "reflected" (p. 28) from it. The reflected motion being symmetrical about the axis, we must have for it

$$\phi = \frac{Q_0}{r} + \frac{Q_1}{r^2} + \frac{Q_2}{r^3} + \ldots \ldots \ldots (5),$$

where Q_0, Q_1, Q_2... are Laplace's functions involving θ only. This gives for the normal velocity (ν') in the reflected motion at the surface of the sphere

$$\nu' = - \frac{1 Q_0}{a^2} - \frac{2 Q_1}{a^3} - \frac{3 Q_2}{a^4} - \ldots \ldots \ldots (6);$$

and since we must have $\nu' = - \nu$ we get from (4) and (6)

$$Q_0 = 0, \quad Q_1 = m \frac{1a^3P_1}{c^3}, \quad Q_2 = m \frac{2a^5P_2}{c^4}, \quad Q_3 = m \frac{3a^7P_3}{c^5} \ldots$$

which reduces (5) to

$$\phi = m \frac{a^3}{c^3} \left(\frac{1P_1}{r^2} + \frac{2a^2P_2}{cr^3} + \frac{3a^4P_3}{c^2r^4} + \ldots \right) \ldots \ldots \ldots (7).$$

---

* The functions which in Art. 9 of the paper "On some Cases of Fluid Motion" (p. 38) I called "Laplace's coefficients," following the nomenclature of Pratt's *Mechanical Philosophy*, are more properly called "Laplace's functions;" the term "Laplace's coefficients" being used to mean the coefficients in the expansion of

$$[1 - 2e \{ \cos\theta \cos\theta' + \sin\theta \sin\theta' \cos(\omega - \omega') \} + e^2]^{-\frac{1}{2}},$$

to be understood according to the usual notation and not as in the text.

This is identical with what (1) becomes on writing m', c' for m, c provided that

$$m' = - m \frac{a^3}{c^3}, \qquad c' = \frac{a^2}{c}.$$

Hence the reflected motion is perfectly represented by supposing the sphere's place occupied by fluid within which, at the point P' in the line OP determined by $OP' = c'$, there exists a singular point of the same character as P, but of opposite sign, and of intensity less in the ratio of a^3 to c^3.

The case of a spherical mass of fluid within a rigid enclosure and containing a singular point of the second order with its axis in a radial direction might be treated in a manner precisely similar, by supposing the space exterior to the sphere filled with fluid, taking to represent the reflected motion in this case, instead of (5), the corresponding expression according to ascending powers of r, and comparing the resulting normal velocity at the surface of the sphere with (3) instead of (4). This is however unnecessary, since we see that the relation between the two singular points P, P' is reciprocal, so that either may be regarded as the image of the other.

Suppose now that we have two solid spheres, S, S', exterior to each other, immersed in a fluid. Suppose that S' is at rest, and that S moves in the direction of the line joining the centres, the fluid being at rest except as depends on the motion of S. The motion of the fluid may be determined by the method of successive reflections (p. 28), which in this case becomes greatly simplified in consequence of the existence of a perfect image representing each reflected motion, so that the process is identical with that of Thomson's method of images, except that the decrease of intensity of the successive images takes place according to the cubes of the ratios of the successive quantities such as a, c, instead of the first powers.

If a sphere move inside a spherical envelope, in the direction of the line joining the centres, the space between being filled with fluid which is otherwise at rest, the motion may be determined in a precisely similar manner.

If two spheres outside each other, or just touching, be connected by an infinitely thin rod, and move in a fluid in the direction of the line joining their centres, we have only to find the motion due to the motion of each sphere supposing the other at rest, and to superpose the results.

I should probably not have thought of applying the method to the solid bounded by the outer portions of two intersecting spheres, had not Professor Thomson shewn me that it was not limited to the cases in which each sphere is complete; and that although it fails from non-convergence when the spheres intersect, yet when the exterior angle of intersection is a submultiple of two right angles the places of the successive images recur in a cycle, and a solution of the problem may be obtained in finite terms by placing singular points of the second order at the places of the images in a complete cycle.

The simplest case is that in which the spheres are generated by the revolution round their common axis of two circles which intersect at right angles. In this case if S, S' are the spheres, O, O' their centres, O_1 the middle point of the common chord of the circles, the image of O in S' will be at O_1, and the image of O_1 in S will be at O'.

Let a, b be the radii of the spheres; c the distance $\sqrt{(a^2 + b^2)}$ of their centres; e, f the distances a^2/c, b^2/c of O_1 from O, O'; C the velocity of the spheres; r, θ the polar co-ordinates of any point measured from O; r_1, θ_1 the co-ordinates measured from O_1; r', θ' the co-ordinates measured from O'; θ, θ_1, θ' being all measured from the line OO'. If S' were away, we should have for the fluid exterior to S

$$\phi = -\, Ca^3 \frac{\cos\theta}{2r^2}.$$

For the image of this in S' we have a singular point at O_1 for which

$$\phi = \frac{Ca^3 b^3}{c^3} \cdot \frac{\cos\theta_1}{2r_1^2},$$

and for the image of this again in S we have a singular point at O' for which

$$\phi = -\, Cb^3 \frac{\cos\theta'}{2r'^2},$$

which is precisely what is required to give the right normal velocity at the surface of S'. Moreover all the singular points lie inside the space bounded by the exterior portions of the intersecting spheres. Hence the three motions together satisfy all the conditions of the problem, so that for the complete solution we have

$$\phi = -\tfrac{1}{2} C \left\{ \frac{a^3 \cos \theta}{r^2} - \frac{a^3 b^3 \cos \theta_1}{c^3 r_1^2} + \frac{b^3 \cos \theta'}{r'^2} \right\}.$$

Just as in the case of a sphere, if a force act on the solid in the direction of its axis, causing a change in the velocity C, the only part of the expression for the resistance of the fluid which will have a resultant will be that depending upon dC/dt. This follows at once, as at pp. 50, 51, from the consideration that when there is no change of C the *vis viva* is constant, and therefore the resultant pressure is *nil*. If we denote by $M' dC/dt$ the resultant pressure acting backwards, we get for the part of M' due to the pressure of the fluid on the exposed portion of the surface of S',

$$\pi \rho b^2 \iint \left\{ \frac{a^3 \cos \theta}{r^2} - \frac{a^3 b^3 \cos \theta_1}{c^3 r_1^2} + b \cos \theta' \right\} \cos \theta' \sin \theta' \, d\theta',$$

taken between proper limits. Putting $b \cos \theta' = x$, we have

$$r \cos \theta = c + x, \quad r_1 \cos \theta_1 = f + x,$$
$$r^2 = b^2 + c^2 + 2cx, \quad r_1^2 = b^2 + f^2 + 2fx.$$

Expressing $\cos \theta$, $\cos \theta_1$, $\cos \theta'$ in terms of x and r, x and r_1, x, and changing the independent variable, first to x, and then in the first term to r and in the second to r_1, we have for the indefinite integral with sign changed

$$\frac{\pi \rho a^3}{12 c^3} \left\{ r^3 - 6b^2 r + 3 (c^4 - b^4) \frac{1}{r} \right\}$$
$$- \frac{\pi \rho a^3 b^3}{12 c^3 f^3} \left\{ r_1^3 - 6b^2 r_1 + 3 (f^4 - b^4) \frac{1}{r_1} \right\} + \frac{\pi \rho a^3}{3},$$

which is to be taken between the limits $r = a$ to $r = c + b$, $r_1 = ab/c$ to $f + b$, $x = -f$ to b. The part of M' due to the integral over the exposed part of the surface of S will be got from the above by interchanging; and on adding the two expressions together, and putting $f = b^2/c$, $c = \sqrt{(a^2 + b^2)}$, we get for the final result

$$M' = \frac{\pi \rho}{3 c^3} \{ 4c^3 (a^3 + b^3) - 2a^6 - 3a^4 b^2 - 6a^3 b^3 - 3a^2 b^4 - 2b^6 \}.$$

When one of the radii, as b, vanishes, we get $M' = \tfrac{2}{3} \pi \rho a^3$ as it ought to be.

[From the *Transactions of the Cambridge Philosophical Society*,
Vol. VIII. p. 533.]

ON THE CRITICAL VALUES OF THE SUMS OF PERIODIC SERIES.

[Read December 6, 1847.]

THERE are a great many problems in Heat, Electricity, Fluid
Motion, &c., the solution of which is effected by developing an
arbitrary function, either in a series or in an integral, by means of
functions of known form. The first example of the systematic
employment of this method is to be found in Fourier's *Theory
of Heat*. The theory of such developements has since become an
important branch of pure mathematics.

Among the various series by which an arbitrary function $f(x)$
can be expressed within certain limits, as 0 and a, of the variable
x, may particularly be mentioned the series which proceeds according to sines of $\pi x/a$ and its multiples, and that which proceeds
according to cosines of the same angles. It has been rigorously
demonstrated that an arbitrary, but finite function of x, $f(x)$, may
be expanded in either of these series. The function is not
restricted to be continuous in the interval, that is to say, it may
pass abruptly from one finite value to another; nor is either the
function or its derivative restricted to vanish at the limits 0 and a.
Although however the *possibility* of the expansion of an arbitrary
function in a series of sines, for instance, when the function does
not vanish at the limits 0 and a, cannot but have been contemplated, the *utility* of this form of expansion has hitherto, so far as
I am aware, been considered to depend on the actual evanescence
of the function at those limits. In fact, if the conditions of the
problem require that $f(0)$ and $f(a)$ be equal to zero, it has been

considered that these conditions were satisfied by selecting the form of expansion referred to. The chief object of the following paper is to develope the principles according to which the expansion of an arbitrary function is to be treated when the conditions at the limits which determine the particular form of the expansion are apparently violated; and to shew, by examples, the advantage that frequently results from the employment of the series in such cases.

In Section I. I have begun by proving the possibility of the expansion of an arbitrary function in a series of sines. Two methods have been principally employed, at least in the simpler cases, in demonstrating the possibility of such expansions. One, which is that employed by Poisson, consists in considering the series as the limit of another formed from it by multiplying its terms by the ascending powers of a quantity infinitely little less than 1; the other consists in summing the series to n terms, that is, expressing the sum by a definite integral, and then considering the limit to which the sum tends when n becomes infinite. The latter method certainly appears the more direct, whenever the summation to n terms can be effected, which however is not always the case; but the former has this in its favour, that it is thus that the series present themselves in physical problems. The former is the method which I have followed, as being that which I employed when I first began the following investigations, and accordingly that which best harmonizes with the rest of the paper. I should hardly have ventured to bring a somewhat modified proof of a well-known theorem before the notice of this Society, were it not for the doubts which some mathematicians seem to feel on this subject, and because there are some points which Poisson does not seem to have treated with sufficient detail.

I have next shewn how the existence and nature of the discontinuity of $f(x)$ and its derivatives may be ascertained merely from the series, whether of sines or cosines, in which $f(x)$ is developed, even though the summation of the series cannot be effected. I have also given formulæ for obtaining the developements of the derivatives of $f(x)$ from that of $f(x)$ itself. These developements cannot in general be obtained by the immediate differentiation of the several terms of the developement of $f(x)$, or in other words by differentiating under the sign of summation.

It is usual to restrict the expanded function to be finite. This restriction however is not necessary, as is shewn towards the end of the section. It is sufficient that the integral of the function be finite.

Section II. contains formulæ applicable to the integrals which replace the series considered in Section I. when the extent a of the variable throughout which the function is considered is supposed to become infinite.

Section III. contains some general considerations respecting series and integrals, with reference especially to the discontinuity of the functions which they express. Some of the results obtained in this section are referred to by anticipation in Sections I. and II. They could not well be introduced in their place without too much interrupting the continuity of the subject.

Section IV. consists of examples of the application of the preceding results. These examples are all taken from physical problems, which in fact afford the best illustrations of the application of periodic series and integrals. Some of the problems considered are interesting on their own account, others, only as applications of mathematical processes. It would be unnecessary here to enumerate these problems, which will be found in their proper place. It will be sufficient to make one or two remarks.

The problem considered in Art. 52, which is that of determining the potential due to an electrical point in the interior of a hollow conducting rectangular parallelepiped, and to the electricity induced on the surface, is given more for the sake of the artifice by which it is solved than as illustrating the methods of this paper. The more obvious mode of solving this problem would lead to a very complicated result.

The problem solved in Art. 54 affords perhaps the best example of the utility of the methods given in this paper. The problem consists in determining the motion of a fluid within the sector of a cylinder, which is made to oscillate about its axis, or a line parallel to its axis. The expression for the moment of inertia of the fluid which would be obtained by the methods generally employed in the solution of such problems is a definite integral, the numerical calculation of which would be very laborious; whereas the expression obtained by the method of this paper is an infinite series which may be summed, to a sufficient degree of approximation, without much trouble.

The series for the developement of an arbitrary function considered in this paper are two, a series of sines and a series of
cosines, together with the corresponding integrals ; but similar
methods may be applied in other cases. I believe that the following statement will be found to embrace the cases to which the
method will apply.

Let u be a continuous function of any number of independent
variables, which is considered for values of the variables lying
within certain limits. For facility of explanation, suppose u a
function of the rectangular co-ordinates x, y, z, or of x, y, z and t,
where t is the time, and suppose that u is considered for values of
x, y, z, t lying between 0 and a, 0 and b, 0 and c, 0 and T, respectively. For such values suppose that u satisfies a linear partial
differential equation, and suppose it to satisfy certain linear equations of condition for the limiting values of the variables. Let
$U = 0$, $U' = 0$ be two of the equations of condition, corresponding
to the two limiting values of one of the variables, as x. Then
the expansion of u to which these equations lead may be applied
to the more general problem which leads to the corresponding
equations of condition $U = F$, $U' = F'$, where F and F' are any
functions of all the variables except x, or of any number of
them.

SECTION I.

*Mode of ascertaining the nature of the discontinuity of a function
which is expanded in a series of sines or cosines, and of obtaining the developements of the derived functions.*

1. By the term *function* I understand in this paper a quantity
whose value depends in any manner on the value of the variable,
or on the values of the several variables of which it is composed.
Thus the functions considered need not be such as admit of being
expressed by any combination of algebraical symbols, even between
limits of the variables ever so close. I shall assume the ordinary
rules of the differential and integral calculus as applicable to such
functions, supposing those rules to have been established by the
method of limits, which does not in the least require the
possibility of the algebraical expression of the functions considered.

The term *discontinuous*, as applied to a function of a single variable, has been used in two totally different senses. Sometimes a function is called discontinuous when its algebraical expression for values of the variable lying between certain limits is different from its algebraical expression for values of the variable lying between other limits. Sometimes a function of x, $f(x)$, is called continuous when, for all values of x, the difference between $f(x)$ and $f(x \pm h)$ can be made smaller than any assignable quantity by sufficiently diminishing h, and in the contrary case discontinuous. If $f(x)$ can become infinite for a finite value of x, it will be convenient to consider it as discontinuous according to the second definition. It is easy to see that a function may be discontinuous in the first sense and continuous in the second, and *vice versâ*. The second is the sense in which the term *discontinuous* is I believe generally employed in treatises on the differential calculus which proceed according to the method of limits, and is the sense in which I shall use the term in this paper. The terms continuous and discontinuous might be applied in either of the above senses to functions of two or more independent variables. If I have occasion to employ them as applied to such a function, I shall employ them in the second sense; but for the present I shall consider only functions of one independent variable.

In the case of the functions considered in this paper, the value of the variable is usually supposed to be restricted to lie within certain limits, as will presently appear. I exclude from consideration all functions which either become infinite themselves, or have any of their differential coefficients of the orders considered becoming infinite, within the limits of the variable within which the function is considered, or at the limits themselves, except when the contrary is expressly stated. Thus in an investigation into which $f(x)$ and its first n differential coefficients enter, and in which $f(x)$ is considered between the limits $x = 0$ and $x = a$, those functions are excluded, at least at first, which are such that any one of the quantities $f(x)$, $f'(x) \dots f^n(x)$ is infinite for a value of x lying between 0 and a, or for $x = 0$ or $x = a$; but the differential coefficients of the higher orders may become infinite. The quantities $f(x)$, $f'(x) \dots f^n(x)$ may however alter discontinuously between the limits $x = 0$ and $x = a$, but I exclude from consideration all functions which are such that any one of the above quantities alters discontinuously an infinite number of times between the limits within which x is supposed to lie.

The terms *convergent* and *divergent*, as applied to infinite series, will be used in this paper in their usual sense; that is to say, a series will be called convergent when the sum to n terms approaches a finite and unique limit as n increases beyond all limit, and divergent in the contrary case. Series such as

$$1 - 1 + 1 - ..., \quad \sin x + \sin 2x + \sin 3x + ...,$$

(where x is supposed not to be 0 or a multiple of π,) will come under the class divergent; for, although the sum to n terms does not increase beyond all limit, it does not approach a unique limit as n increases beyond all limit. Of course the first n terms of a divergent series may be the limits of those of a convergent series: nor does it appear possible to invent a series so rapidly divergent that it shall not be possible to find a convergent series which shall have for the limits of its first n terms the first n terms respectively of the divergent series. Of course we may employ a divergent series merely as an abbreviated mode of expressing the limit of the sum of a convergent series. Whenever a divergent series is employed in this way in the present paper, it will be expressly stated that the series is so regarded.

Convergent series may be divided into two classes, according as the series resulting from taking all the terms of the given series positively is convergent or divergent. It will be convenient for the purposes of the present paper to have names for these two classes. I shall accordingly call series belonging to the first class *essentially convergent*, and series belonging to the second *accidentally convergent*, while the term *convergent*, simply, will be used to include both classes. Thus, according to the definitions which will be employed in this paper, the series

$$x + \tfrac{1}{2} x^2 + \tfrac{1}{3} x^3 + ...$$

is essentially convergent so long as $x^2 < 1$; it is divergent when $x^2 > 1$, and when $x = 1$; and it is accidentally convergent when $x = -1$.

The same definitions may be applied to integrals, when one at least of the limits of integration is ∞. Thus, if $a > 0$, $\displaystyle\int_a^\infty x^{-2}\,dx$

s.   16

is essentially convergent at the limit ∞, while $\int_a^\infty \dfrac{\sin x}{x}\,dx$ is only accidentally convergent, and $\int_a^\infty \sin x\,dx$, not being convergent, comes under the class of divergent integrals. These definitions may be applied also to integrals taken between finite limits, when the quantity under the integral sign becomes infinite within the limits of integration, or at one of the limits. Thus $\int_0^a \log x\,dx$ is convergent, but $\int_0^a \dfrac{dx}{x}$ divergent, at the limit 0.

2. Let $f(x)$ be a function of x which is only considered between the limits $x=0$ and $x=a$, and which can be expanded between those limits in a convergent series of sines of $\pi x/a$ and its multiples, so that

$$f(x) = A_1 \sin \frac{\pi x}{a} + A_2 \sin \frac{2\pi x}{a} \ldots + A_n \sin \frac{n\pi x}{a} + \ldots (1).$$

To determine A_n, multiply both sides of (1) by $\sin n\pi x/a\,.\,dx$ and integrate from $x=0$ to $x=a$. Since the series in (1) is convergent, and $\sin n\pi x/a$ does not become infinite for any real value of x, we may first multiply each term by $\sin n\pi x/a\,.\,dx$ and integrate, and then sum, instead of first summing and then integrating[*]. But each term of the series in (1) except the nth will produce in the new series a term equal to zero, and the nth will produce $\frac{1}{2}aA_n$. Hence

$$A_n = \frac{2}{a}\int_0^a f(x)\sin\frac{n\pi x}{a}\,dx,$$

and therefore

$$f(x) = \frac{2}{a}\Sigma\int_0^a f(x)\sin\frac{n\pi x}{a}\,dx\,.\,\sin\frac{n\pi x}{a} \ldots\ldots\ldots (2).$$

3. Hence, whenever $f(x)$ can be expanded in the convergent series which forms the right-hand side of (1), the value of A_n can be very readily found, and the expansion performed. But this leaves us quite in the dark as to the degree of generality that a function which can be so expanded admits of. In considering this

[*] Moigno, *Leçons de Calcul Différentiel*, &c. Tom. II. p. 70.

question it will be convenient, instead of endeavouring to develope $f(x)$, to seek the value of the infinite series

$$\frac{2}{a} \Sigma \int_0^a f(x') \sin \frac{n\pi x'}{a} dx' . \sin \frac{n\pi x}{a} \dots\dots\dots\dots (3),$$

provided the series be convergent; for it is only in that case that we can, without further definition, speak of the sum of the series at all. Now if we had only a finite number n of terms in the series (3) we might of course replace the series by

$$\frac{2}{a} \int_0^a f(x') \left\{ \sin \frac{\pi x'}{a} \sin \frac{\pi x}{a} + \sin \frac{2\pi x'}{a} \sin \frac{2\pi x}{a} \dots \right.$$
$$\left. + \sin \frac{n\pi x'}{a} \sin \frac{n\pi x}{a} \right\} dx' \dots\dots\dots (4).$$

As it is however this transformation cannot be made, because, the series within brackets in the expression which would replace (4) not being convergent, the expression would be a mere symbol without any meaning. If however the series (3) is essentially convergent, its sum is equal to the limit of the sum of the following essentially convergent series

$$\frac{2}{a} \Sigma g^n \int_0^a f(x') \sin \frac{n\pi x'}{a} dx' . \sin \frac{n\pi x}{a} \dots\dots\dots (5),$$

when g from having been less than 1 becomes in the limit 1. It will be observed that if (3) were only accidentally convergent, we could not with certainty affirm the sum of (3) to be the limit of the sum of (5). For it is conceivable, or at least not at present proved to be impossible, that the mode of the mutual destruction of the terms of (3) in the infinitely remote part of the series should be altered by the introduction of the factor g^n, however little g might differ from 1. Let us now, instead of seeking the sum of (3) in those cases in which the series is convergent, seek the limit to which the sum of (5) approaches as g approaches to 1 as its limit.

4. The transformation already referred to, which could not be effected on the series (3), may be effected on (5), that is to say, instead of first integrating the several terms and then summing, we may first sum and then integrate. We have thus, for the value of the series,

$$\frac{2}{a} \int_0^a f(x') \left\{ \Sigma g^n \sin \frac{n\pi x'}{a} \sin \frac{n\pi x}{a} \right\} dx' \dots\dots\dots (6).$$

16—2

The convergent series within brackets can easily be summed.   The expression (6) thus becomes

$$\frac{1}{2a}\int_0^a f(x')\left\{\frac{1-g^2}{1-2g\cos\pi\,(x'-x)/a+g^2}\right.$$
$$\left.-\frac{1-g^2}{1-2g\cos\pi\,(x'+x)/a+g^2}\right\}dx' \ldots\ldots (7).$$

Now since the quantity under the integral sign vanishes when $g=1$, provided $\cos\pi\,(x'\pm x)/a$ be not $=1$, the limit of (7) when $g=1$ will not be altered if we replace the limits 0 and a of x' by any other limits or groups of limits as close as we please, provided they contain the values of x' which render $x'\pm x$ equal to zero or any multiple of $2a$.   Let us first suppose that we are considering a value of x lying between 0 and a, and in the neighbourhood of which $f(x)$ alters continuously.   Then, since $x'+x$ never becomes equal to zero or any multiple of $2a$ within the limits of integration, we may omit the second term within brackets in (7); and since $x'-x$ never becomes equal to any multiple of $2a$, and vanishes only when $x'=x$, we may take for the limits of x' two quantities lying as close as we please to x, and therefore so close as to exclude all values of x' for which $f(x')$ alters discontinuously. Let $g=1-h$, $x'=x+\xi$, expand $\cos\pi\xi/a$ by the ordinary formula, and put $f(x')=f(x)+R$.   Then the limit of (7) will be the same as that of

$$\frac{2-h}{2a}\int\{f(x)+R\}\,\frac{hd\xi}{h^2+g\,(\pi^2\xi^2/a^2-\ldots)}\ldots\ldots\ldots (8),$$

the limits of ξ being as small as we please, the first negative and the second positive.   Let now

$$g\,(\pi^2\xi^2/a^2-\ldots)=\xi'^2,$$

so that $d\xi/d\xi'$ is ultimately equal to a/π, that is to say when g is first made equal to 1, and then the limits of ξ, and therefore those of ξ', are made to coalesce.   Let now G, L be respectively the greatest and least values of $(1-\tfrac{1}{2}h)\,a^{-1}\,d\xi/d\xi'\,\{f(x)+R\}$ within the limits of integration.   Then if we observe that

$$\int\frac{hd\xi'}{h^2+\xi'^2}=\tan^{-1}\xi'/h+C,$$

were $\tan^{-1}$ denotes an angle lying between $-\pi/2$ and $\pi/2$, putting

$-\xi_1$, ξ_2 for the limits of ξ', we shall see that the value of the integral (8) lies between

$$G\left(\tan^{-1}\xi_1/h + \tan^{-1}\xi_2/h\right) \text{ and } L\left(\tan^{-1}\xi_1/h + \tan^{-1}\xi_2/h\right):$$

but in the limit, that is to say, when we first suppose h to vanish and then ξ_1 and ξ_2, G and L become equal to each other and to $\pi^{-1}f(x)$, and $\tan^{-1}\xi_1/h + \tan^{-1}\xi_2/h$ becomes equal to π. Hence, $f(x)$ is the limit of (7).

Next, suppose that the value of x which we are considering lies between 0 and a, and that as x' passes through it $f(x')$ alters suddenly from M to N. Then the reasoning will be exactly as before, except that we must integrate separately for positive and negative values of ξ', replacing $f(x) + R$ by $M + R$ in the latter case, and by $N + R'$ in the former. Hence, the limit of (7) will be

$$\tfrac{1}{2}(M+N).$$

Lastly, if we are considering the extreme values $x = 0$ and $x = a$, it follows at once from the form of (7) that its limiting value is zero.

Hence the limit to which the sum of the convergent series (5) tends as g tends to 1 as its limit is $f(x)$ for values of x lying between 0 and a, for which $f(x)$ alters continuously, it is $\tfrac{1}{2}(M+N)$ for values of x for which $f(x)$ alters suddenly from M to N, and it is zero for the extreme values 0 and a.

5. Of course the limiting value of the series (5) is $f(0)$ and not zero, if we suppose that g first becomes 1 and then x passes from a positive value to zero. In the same way, if $f(x)$ alters abruptly from M to N as x increases through x_1, the limiting value of (5) will be M if we suppose that g first becomes 1 and then x increases to x_1, and it will be N if we suppose that g first becomes 1 and then x decreases to x_1. It would be futile to argue that the limiting value of (5) for $x = 0$ is zero rather than $f(0)$, or $f(0)$ rather than zero, since that entirely depends on the sense in which we employ the expression *limiting value*. Whichever sense we please to adopt, no error can possibly result, provided we are only consistent, and do not in the course of the same investigation change the meaning of our words.

It is a principle of great importance in these investigations, that a function of two independent variables which becomes indeterminate for particular values of the variables may have different limiting values according to the order in which we suppose the variables to assume their particular values, or according to the nature of the arbitrary relation which we conceive imposed on them as they approach those values together.

I would here make one remark on the subject of consistency. We may speak of the sum of an infinite series which is not convergent, if we define it to mean the limit of the sum of a convergent series of which the first n terms become in the limit the same as those of the divergent series. According to this definition, it appears quite conceivable that the same divergent series should have a different sum according as it is regarded as the limit of one convergent series or of another. If however we are careful in the same investigation always to regard the same divergent series, and the series derived from it, as the limits of the same convergent series and the series derived from it, it does not appear possible to fall into error, assuming of course that we always reason correctly. For example, we may employ the series (3), and the series derived from it by differentiation, &c., without fear, provided we always regard these series when divergent, or only accidentally convergent, as the limits of the *particular* convergent series formed by multiplying their n^{th} terms by g^n.

6. We may now consider the convergency of the series (3), in order to find whether we may employ it directly, or whether we must regard it as the limit of (5).

By integrating by parts in the nth term of (3), we have

$$\frac{2}{a}\int f(x') \sin\frac{n\pi x'}{a}\, dx' = -\frac{2}{n\pi}f(x') \cos\frac{n\pi x'}{a}$$

$$+ \frac{2a}{n^2\pi^2}f'(x') \sin\frac{n\pi x'}{a} - \frac{2a}{n^2\pi^2}\int f''(x') \sin\frac{n\pi x'}{a}\, dx' \dots(9).$$

Suppose that $f(x)$ does not necessarily vanish at the limits $x = 0$ and $x = a$, and that it alters discontinuously any finite number of times between those limits, passing abruptly from M_1 to N_1 when x increases through α_1, from M_2 to N_2 when x increases through α_2, and so on. Then, if we put S for the sign of summation referring to the discontinuous values of $f(x')$, on taking the integrals in (9) from $x = 0$ to $x = a$, we shall get for the part of the integral corresponding to the first term at the right-hand side of the equation

$$\frac{2}{n\pi}\left\{f(0)-(-)^n f(a)+S(N-M)\cos\frac{n\pi a}{a}\right\} \dots\dots(10).$$

It is easily seen that the last two terms in (9) will give a part of the integral taken from 0 to a, which is numerically inferior to L/n^2, where L is a constant properly chosen. As far as regards these terms therefore the series (3) will be essentially convergent, and its sum will therefore be the limit of the sum of (5).

Hence, in examining the convergency or divergency of the series (3), we have only got to consider the part of the coefficient of $\sin n\pi x/a$ of which (10) is the expression. The terms $f(0)$, $f(a)$ in this expression may be included under the sign S if we put for the first $a=0$, $M=0$, $N=f(0)$, and for the second $a=a$, $M=f(a)$, $N=0$. We have thus got a set of series to consider of which the type is

$$\frac{2}{\pi}(N-M)\Sigma\frac{1}{n}\cos\frac{n\pi z}{a}\sin\frac{n\pi x}{a} \dots\dots(11).$$

If we replace the product of the sine and cosine in this expression by the sum of two sines, by means of the ordinary formula, and omit unnecessary constants, we shall have for the series to consider

$$\Sigma\frac{1}{n}\sin nz \dots\dots(12).$$

Let now

$$u=\sin z+\tfrac{1}{2}\sin 2z \dots+\frac{1}{n}\sin nz \dots\dots(13),$$

then

$$\frac{du}{dz}=\cos z+\cos 2z \dots+\cos nz=\frac{\sin(n+\tfrac{1}{2})z}{2\sin\tfrac{1}{2}z}-\tfrac{1}{2};$$

and since u vanishes with z, in which case $\dfrac{\sin(n+\tfrac{1}{2})z}{\sin\tfrac{1}{2}z}$ is finite, we shall have, supposing z to lie between -2π and $+2\pi$, so that the quantity under the integral sign does not become infinite within the limits of integration,

$$u=\tfrac{1}{2}\int_0^z\frac{\sin(n+\tfrac{1}{2})z}{\sin\tfrac{1}{2}z}\,dz-\frac{z}{2} \dots\dots(14);$$

and we have to find whether the integral contained in this equation approaches a finite limit as n increases beyond all limit, and if so what that limit is. Since u changes sign with z, we need not consider the negative values of z.

First suppose the superior limit z to lie between 0 and 2π; and to simplify the integral write $2z$ for z, n for $2n+1$, so that the superior limit of the new integral lies between 0 and π; then the integral

$$= \int_0^z \frac{\sin nz}{\sin z}\, dz = \int_0^z \frac{\sin nz}{z} \cdot \frac{z}{\sin z}\, dz = \int_0^z \frac{\sin nz}{z}\,(1 + Rz)\, dz,$$

where $R = \dfrac{z - \sin z}{z \sin z}$, a quantity which does not become infinite within the limits of integration. Hence, as is known, the limit of $\int_0^z \sin nz \,.\, R dz$ when n increases beyond all limit is zero. Hence, if I be the limit of the integral,

$$I = \text{limit of } \int_0^z \frac{\sin nz}{z}\, dz = \text{limit of } \int_0^{nz} \frac{\sin \zeta}{\zeta}\, d\zeta.$$

Now, z being given, the limit of nz is ∞, and therefore

$$I = \int_0^\infty \frac{\sin \zeta}{\zeta}\, d\zeta = \frac{\pi}{2}.$$

Secondly, suppose z in (14) to be equal to 0. Then it follows directly from this equation, or in fact at once from (13), that $u = 0$, and consequently the limit of $u = 0$.

The value of u in all other cases, if required, may be at once obtained from the consideration that the values of u recur when z is increased or diminished by 2π.

Hence, the series (12) is in all cases convergent, and has for its sum 0 when $z = 0$, and $\frac{1}{2}(\pi - z)$ when z lies between 0 and 2π.

Now, if in the theorem of Article 4, we write z for x, and put $a = \pi$, $f(z) = \frac{1}{2}(\pi - z)$, we find, for values of z lying between 0 and π, and for $z = \pi$,

$$\text{limit of } \Sigma \frac{1}{n}\, g^n \sin nz = \frac{1}{2}(\pi - z);$$

and evidently

$$\text{limit of } \Sigma \frac{1}{n}\, g^n \sin nz = 0, \text{ when } z = 0,$$

that is of course supposing z first to vanish and then g to become 1. Also the limit of $\Sigma n^{-1} g^n \sin nz$ changes sign with z, and recurs when z is increased or diminished by 2π. Hence, the series (12), which has been proved to be convergent, is in all cases the limit to which the sum of the convergent series $\Sigma n^{-1} g^n \sin nz$ tends as g tends to 1 as its limit. Now the series (11) may be decomposed into two series of the form just discussed, whence it follows that the series (3) is always convergent, and its sum for all values of x, critical as well as general, is the limit of the sum of the series (5), when g becomes equal to 1.

The examination of the convergency of the series (3) in the only doubtful case, that is to say, the case in which $f(x)$ is discontinuous, or does not vanish for $x = 0$ and for $x = a$, is more curious than important. For in the analytical applications of the series (3) it would be sufficient to regard it as the limit of the series (5); and in the case in which (3) is only accidentally convergent, we should hardly think of employing it in the numerical computation of $f(x)$ if we could possibly help it, and it will immediately appear that in all the cases which are most important to consider we can get rid of the troublesome terms without knowing the sum of the series.

The proof of the convergency of the series (3) which has just been given, though in some respects I believe new, is certainly rather circuitous, and it has the disadvantage of not applying to the case in which $f'(x)$ is infinite*, an objection which does not apply to the proof given by M. Dirichlet†. It has been supposed moreover that $f''(x)$ is not infinite. The latter restriction however may easily be removed, as in the end of the next article.

7. Let $f(x)$ be a function of x which is expanded between the limits $x = 0$ and $x = a$ in the series (3). Let the series be

$$A_1 \sin \frac{\pi x}{a} + A_2 \sin \frac{2\pi x}{a} \ldots + A_n \sin \frac{n\pi x}{a} + \ldots, \ldots (15),$$

and suppose that we have given the coefficients $A_1, A_2 \ldots$, but do not know the sum of the series $f(x)$. We may for all that find the values of $f(0)$ and $f(a)$, and likewise the values of x

---

* This restriction may however be dispensed with by what is proved in Art. 20.
† Crelle's *Journal*, Tom. IV. p. 157.

for which $f(x)$ is discontinuous, and the quantity by which $f(x)$ is increased as x increases through each of these critical values.

For from (9) and (10)

$$nA_n = \frac{2}{\pi}\left\{f(0) - (-1)^n f(a) + S(N-M)\cos\frac{n\pi a}{a}\right\} + \frac{R}{n},$$

R being a quantity which does not become infinite with n. **If** then we use the term *limit* in an extended sense, so as to include quantities of the form $C\cos n\gamma$, [of course $C(-1)^n$ is a particular case,] or the sum of any finite number of such quantities, we shall have for $n = \infty$,

$$\text{limit of } nA_n = \frac{2}{\pi}\left\{f(0) - (-1)^n f(a) + S(N-M)\cos\frac{n\pi x}{a}\right\}\dots(16).$$

Let then the limit of nA_n be found. It will appear under the form

$$C_0 + C_1(-1)^n + SC\cos n\gamma \dots\dots(17).$$

Comparing this expression with (16), we shall have

$$f(0) = \frac{\pi}{2}C_0, \quad f(a) = -\frac{\pi}{2}C_1;$$

and for each term of the series denoted by S we shall have

$$\alpha = \frac{a\gamma}{\pi}, \quad N - M = \frac{\pi}{2}C.$$

In particular, if $f(x)$ is continuous, and if the limit of nA_n is L_0 or L_e according as n is odd or even, we shall have

$$L_0 = \frac{2}{\pi}\{f(0) + f(a)\}, \quad L_e = \frac{2}{\pi}\{f(0) - f(a)\};$$

whence

$$f(0) = \frac{\pi}{4}(L_0 + L_e), \quad f(a) = \frac{\pi}{4}(L_0 - L_e)\dots(18).$$

If $f(x)$ were discontinuous for an infinite number of values of x lying between 0 and a, it is conceivable that the infinite series coming under the sign S might be divergent, or if convergent might have a sum from which n might wholly or partially disappear, in which case the limit of nA_n might not come out under the form (17). It was for this reason among others, that in Art. 1, I excluded such functions from consideration.

If $f(x)$ be expressible algebraically between the limits $x = 0$ and $x = a$, or if it admit of different algebraical expressions within different portions into which that interval may be divided, A_n will be an algebraical function of n, and the limit of nA_n may be found by the ordinary methods. Under the term *algebraical function*, I here include transcendental functions, using the term *algebraical function* in opposition to what has been sometimes called an *empirical function*, or a *general function*, that is, a function in the sense in which the ordinate of a curve traced *liberâ manu* is a function of the abscissa. Of course, in applying the theorem in this article to general functions, it must be taken as a postulate that the limit of nA_n can be found, and put under the form (17).

The theorem in question has been proved by means of equation (9), in which it is supposed that $f''(x)$ does not become infinite within the limits of integration. The theorem is however true independently of this restriction. To prove it we have only got to integrate by parts once instead of twice, and we thus get for the quantity which replaces $\dfrac{R}{n}$ the integral

$$\frac{2}{\pi} \int_0^a f'(x') \cos \frac{n\pi x'}{a} \, dx',$$

which by the principle of fluctuation* vanishes when n becomes infinite. There is however this difference between the two cases. When the series (15) has been cleared of the part for which the limit of nA_n is finite, by the method which will be explained in the next article, the part which remains will be at least as convergent in the former case as the series $\dfrac{1}{1^2} + \dfrac{1}{2^2} \ldots + \dfrac{1}{n^2} + \ldots,$ whereas we cannot affirm this to be true, and in fact it may be proved that it is not true, in the case in which $f''(x)$ becomes infinite. Observing that the same remark will apply when we come to consider the critical values of the differential coefficients

---

* I borrow this term from a paper by Sir William R. Hamilton *On Fluctuating Functions.* (*Transactions of the Royal Irish Academy*, Vol. XIX. p. 264.) Had I been earlier acquainted with this paper, and that of M. Dirichlet already referred to, I would probably have adopted the second of the methods mentioned in the introduction for establishing equation (2) for any function, or rather, would have begun with Art. 7, taking that equation as established. I have retained Arts (2)— (6), first, because I thought the reader would enter more readily into the spirit of the paper if these articles were retained, and secondly, because I thought that Section III, which is adapted to this mode of viewing the subject, might be found useful.

of $f(x)$, I shall suppose the functions and derived functions employed in each investigation not to become infinite, according to what has already been stated in Art. 1.

8. After having found the several values of α, and the corresponding values of $N - M$, we may subtract the expression (10) from A_n, provided we subtract from the sum of the series (15) the sums of the several series such as (11). Now if X be the sum of the series (11),

$$X = \frac{1}{\pi} (N - M) \left\{ \Sigma \frac{1}{n} \sin \frac{n\pi (x + \alpha)}{a} + \Sigma \frac{1}{n} \sin \frac{n\pi (x - \alpha)}{a} \right\} \dots (19).$$

But it has been already shewn that $\Sigma \dfrac{1}{n} \sin nz = \frac{1}{2} (\pi - z)$ when z lies between 0 and 2π, $= 0$ when $z = 0$, and $= -\frac{1}{2} (\pi + z)$ when z lies between 0 and -2π. Now when x lies between 0 and α, $\pi (x + \alpha)/a$ lies between 0 and 2π, and $\pi (x - \alpha)/a$ lies between -2π and 0 ; and when x lies between α and a, $\pi (x + \alpha)/a$ still lies between 0 and 2π, and $\pi (x - \alpha)/a$ now lies between the same limits. Hence

$$\left. \begin{aligned} X &= - (N - M) \frac{x}{a}, \text{ when } x \text{ lies between 0 and } \alpha \\ &= (N - M) \frac{a - x}{a}, \text{ when } x \text{ lies between } \alpha \text{ and } a \end{aligned} \right\} \dots (20).$$

We need not trouble ourselves with the singular values of the sum of the series (15), since we have seen that a singular value is always the arithmetic mean of the values of the sum for values of x immediately above and below the critical value. This rule will apply to the extreme cases in which $x = 0$ and $x = a$, if we consider the sum of the series for values of x lying beyond those limits. The rule applies to the series in (19), which is only a particular case of (15), and consequently will apply to any combination of series having this property, formed by way of addition or subtraction; since, when we increase or diminish any two quantities M_0, N_0 by any other two M, N respectively, we increase or diminish the arithmetic mean of the two former by the arithmetic mean of the two latter.

It has been already stated that we may, with a certain convention, include quantities referring to the limits $x = 0$ and $x = a$ under the sign of summation S. If we do so, and put Ξ for the sum of the series (15), and B_n for the remainder arising from subtracting the expression (10) from A_n, we shall have

$$\Xi - SX = \Sigma B_n \sin \frac{n\pi x}{a},$$

and the sum of the series forming the right-hand side of this equation will be a continuous function of x. As to SX, the value of each series contained in it is given by equation (20).

To illustrate this, suppose Ξ the ordinate of a curve of which x is the abscissa. Let OG be the axis of x; OA, MB, ND, Gb right lines perpendicular to it, and let $OG = a$. Let the curve of

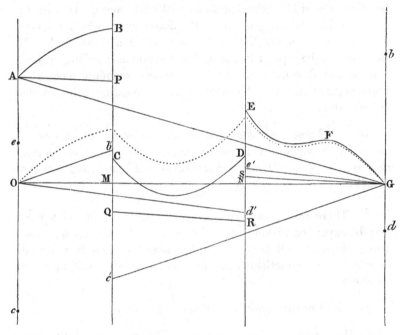

which Ξ is the ordinate be the discontinuous curve AB, CD, EFG. Take Gb equal to BC, and on the positive or negative side of the axis of x according as the ordinate decreases or increases as x increases through OM, and from O measure an equal length Oc on the opposite side of the axis. Take Gd, Oe, each equal to DE, and draw the right lines AG, $Ob'b$, $cc'G$, $Od'd$, $ee'G$. Then it will be easily seen that if X_0, X_1, X_2 be the values of X corresponding to the critical values of x, $x = 0$, $x = OM$, $x = ON$, respectively, X_0 will be represented by the right line AG; X_1 by the discontinuous right line Ob', $c'G$; and X_2 by the discontinuous right line Od', $e'G$. Take MP equal to the sum of the ordinates of the points in which the right lines lying between OA and $c'B$ cut the latter line; MQ equal to the sum of the ordinates of the points in which the right lines lying between $c'B$ and $d'E$ cut the former, and so on, the ordinates being taken with their proper signs. Let P, Q, R, S be the points thus found, and join AP, QR, SG. Then SX will be represented by the discontinuous right line AP, QR, SG. Let the ordinates of the discontinuous curve be diminished by those of the discontinuous right line last mentioned, and let the dotted curve be the result. Then $\Xi - SX$ will be represented by the continuous, dotted curve. It will be observed that the two portions of the dotted curve which meet in each of the ordinates MB, NE may meet at a finite angle. If there should be a point in one of the continuous portions, such as AB, of the discontinuous curve where two tangents meet at a finite angle, there will of course be a corresponding point in the dotted curve.

If we choose to take account of the conjugate points of the curve of which SX is the ordinate, it will be observed that they are situated at O, and midway between P and Q, and between R and S.

9.　　There are a great many series, similar to (3), in which $f(x)$ may be expanded within certain limits of x. I shall consider one other, which as well as (3) is of great use, observing that almost exactly the same methods and the same reasoning will apply in other cases.

The limit of the sum of the series

$$\frac{1}{a}\int_0^a f(x')\,dx' + \frac{2}{a}\Sigma g^n \int_0^a f(x')\cos\frac{n\pi x'}{a}\,dx'.\cos\frac{n\pi x}{a}\ \dots(21),$$

when g from having been less than 1 becomes 1, is $f(x)$, x being supposed not to lie beyond the limits 0 and a. For values, however, of x for which $f(x)$ alters discontinuously, the limit of the sum is the arithmetic mean of the values of $f(x)$ for values of x immediately above and below the critical value. I assume this as being well known, observing that it may be demonstrated just as a similar theorem has been demonstrated in Art. 4.

10. Let us now consider the series

$$\frac{1}{a}\int_0^a f(x')\,dx' + \frac{2}{a}\,\Sigma\int_0^a f(x')\cos\frac{n\pi x'}{a}\,dx' \, . \cos\frac{n\pi x}{a}\ \ldots\text{(22)}.$$

We have by integration by parts

$$\frac{2}{a}\int f(x')\cos\frac{n\pi x'}{a}\,dx'$$

$$=\frac{2}{n\pi}f(x')\sin\frac{n\pi x'}{a} + \frac{2a}{n^2\pi^2}f'(x')\cos\frac{n\pi x'}{a} - \frac{2a}{n^2\pi^2}\int f''(x')\cos\frac{n\pi x'}{a}\,dx';$$

and now, taking the limits properly, and employing the letters M, N, a and S in the same sense as before, we have

$$\frac{2}{a}\int_0^a f(x')\cos\frac{n\pi x'}{a}\,dx' = -\frac{2}{n\pi}\,S\,(N-M)\sin\frac{n\pi a}{a} + \frac{R}{n^2}\ldots\text{(23)},$$

R being a quantity which does not become infinite with n. It follows from (23), that the series (22) is in all cases convergent, and its sum for all values of x, critical as well as general, is the limit of the sum of (21).

It will be observed that if $f(x)$ is a continuous function the series (22) is at least as convergent as the series $\Sigma\,1/n^2$. This is not the case with the series (3), unless $f(0)=f(a)=0$.

If the constant term and the coefficient of $\cos n\pi x/a$ in the general term of (22) are given, $f(x)$ itself not being known, except by its developement, we may as before find the values of x for which $f(x)$ is discontinuous, and the quantity by which $f(x)$ is suddenly increased as x increases through each critical value. We may also, if we please, clear the series (22) of the slowly convergent part corresponding to the discontinuous values of $f(x)$.

11. Since the series (3) is convergent, if we have occasion to integrate $f(x)$ we may, instead of first summing the series and then integrating, first integrate the general term and then sum. More generally, if $\phi(x)$ be any function of x which does not become infinite between the limits $x=0$ and $x=a$, we shall have

$$\int_0^x f(x)\,\phi(x)\,dx = \frac{2}{a}\,\Sigma\int_0^a f(x')\sin\frac{n\pi x'}{a}\,dx' \, . \int_0^x \phi(x)\sin\frac{n\pi x}{a}\,dx,$$

the superior limit x of the integrals being supposed not to lie beyond the limits 0 and a; and the series at the second side of the

above equation will be convergent. In fact, even in the case in
which $f(x)$ is discontinuous the series will be as convergent as the
series $\Sigma\, 1/n^2$. A second integration would give a series still
more rapidly convergent, and so on. Hence, the resulting series
may be employed directly, and not merely when regarded as limits
of converging series. The same remarks apply in all respects to
the series (22) as to the series (3).

12. But the series resulting from differentiating (3) or (22)
once, twice, or any number of times would not in general be convergent, and could not be employed directly, but only as limits of the
convergent series which would be formed by inserting the factor g^n
in the general term. This mode of treating the subject however
appears very inconvenient, except in the case in which the series
are only temporarily divergent, being rendered convergent again
by new integrations; and even then it requires great caution.
The series in question may however be rendered convergent by
means of transformations to which I now proceed, and which,
with their applications, form the principal object of this paper.

The most important case to consider is that in which $f(x)$ and
its derivatives are continuous, so that the divergency arises from
what takes place at the limits 0 and a. I shall suppose then, for
the present, that $f(x)$ and its derivatives of the orders considered
are continuous, except the last, which will only appear under the
sign of integration, and which may be discontinuous.

Consider first the series of sines. Suppose that $f(x)$ is not
given in finite terms, but only by its developement

$$f(x) = \Sigma A_n \sin \frac{n\pi x}{a} \ldots\ldots\ldots\ldots\ldots(24),$$

where A_n is supposed to be given, and where the developement of
$f(x)$ is supposed to be that which would result from the formula
(3). I shall call the expansions of $f(x)$ which are obtained, or
which are to be looked on as obtained from the formulæ (3) and
(22) *direct* expansions, as distinguished from other expansions
which may be obtained by differentiation, and which, being divergent, cannot be directly employed. Let us consider first the even
differential coefficients of $f(x)$, and let A_n'', A_n^4 ... be the coefficients of $\sin n\pi x/a$ in the direct expansions of $f''(x)$, $f^4(x)$ ...
The coefficient of $\sin n\pi x/a$ in the series which would be obtained

by differentiating twice the several terms in the series in (24) would be $-n^2\pi^2/a^2 . A_n$. Now

$$A_n = \frac{2}{a} \int_0^a f(x') \sin \frac{n\pi x'}{a} \, dx';$$

and we have by integrating by parts

$$-\frac{2n^2\pi^2}{a^3} \int f(x') \sin \frac{n\pi x'}{a} \, dx' = \frac{2n\pi}{a^2} f(x') \cos \frac{n\pi x'}{a} - \frac{2}{a} f'(x') \sin \frac{n\pi x'}{a}$$

$$+ \frac{2}{a} \int f''(x') \sin \frac{n\pi x'}{a} \, dx'.$$

Taking now the limits, remembering the expression for A_n'', and transposing, we get

$$A_n'' = \frac{2n\pi}{a^2} \{f(0) - (-1)^n f(a)\} - \frac{n^2\pi^2}{a^2} A_n \ldots\ldots\ldots (25).$$

Any even differential coefficient may be treated in the same way. We thus get, μ being even,

$$(-1)^{\frac{\mu}{2}} A_n^\mu = \left(\frac{n\pi}{a}\right)^\mu A_n - \frac{2}{a} \left(\frac{n\pi}{a}\right)^{\mu-1} \{f(0) - (-1)^n f(a)\}$$

$$+ \frac{2}{a} \left(\frac{n\pi}{a}\right)^{\mu-3} \{f''(0) - (-1)^n f''(a)\} - \ldots$$

$$+ (-1)^{\frac{\mu}{2}} \frac{2}{a} \cdot \frac{n\pi}{a} \{f^{\mu-2}(0) - (-1)^n f^{\mu-2}(a)\} \ldots\ldots\ldots (26).$$

13. In the applications of these equations which I have principally in view, $f(0)$, $f(a)$, $f''(0)\ldots$ are given, and A_1, A_2, $A_3 \ldots$ are indeterminate coefficients. If however A_1, $A_2 \ldots A_n \ldots$ are given, and $f(0)$, $f(a) \ldots$ unknown, we must first find $f(0)$, $f(a) \ldots$, and then we shall be able to substitute in (25) and (26). This may be effected in the following manner.

We get by integrating by parts

$$\int f(x') \sin \frac{n\pi x'}{a} \, dx' = -\frac{a}{n\pi} f(x') \cos \frac{n\pi x'}{a}$$

$$+ \left(\frac{a}{n\pi}\right)^2 f'(x') \sin \frac{n\pi x'}{a} + \left(\frac{a}{n\pi}\right)^3 f''(x') \cos \frac{n\pi x'}{a} - \ldots$$

Multiplying now both sides by $2/a$, and taking the limits of the integrals, we get

$$A_n = \frac{2}{a} \cdot \frac{a}{n\pi} \{f(0) - (-1)^n f(a)\}$$

$$- \frac{2}{a} \cdot \left(\frac{a}{n\pi}\right)^3 \{f''(0) - (-1)^n f''(a)\} + \dots, \quad \dots\dots\dots(27).$$

Hence, if n be always odd or always even, A_n can be expanded, at least to a certain number of terms, in a series according to descending powers of n, the powers being odd, and the first of them -1. The number of terms to which the expansion in this form is possible will depend on the number of differential coefficients of $f(x)$ which remain finite and continuous between the limits $x = 0$ and $x = a$. Let the expansion be performed, and let the result be

$$\left. \begin{aligned} A_n &= O_0 \frac{1}{n} + O_2 \frac{1}{n^3} + O_4 \frac{1}{n^5} + \dots \text{ when } n \text{ is odd,} \\ A_n &= E_0 \frac{1}{n} + E_2 \frac{1}{n^3} + E_4 \frac{1}{n^5} + \dots \text{ when } n \text{ is even} \end{aligned} \right\} \dots(28).$$

Comparing (27) and (28), we shall have

$$\left. \begin{aligned} f(0) &= \frac{\pi}{4}(O_0 + E_0), \quad f(a) = \frac{\pi}{4}(O_0 - E_0), \\ f''(0) &= -\frac{\pi^3}{4a^2}(O_2 + E_2), \quad f''(a) = -\frac{\pi^3}{4a^2}(O_2 - E_2), \\ f^4(0) &= \frac{\pi^5}{4a^4}(O_4 + E_4), \quad f^4(a) = \frac{\pi^5}{4a^4}(O_4 - E_4) \end{aligned} \right\} ..(29),$$

and so on. The first two of these equations agree with (18).

If we conceive the value of A_n given by (27) substituted in (26), we shall arrive at a very simple rule for finding the direct expansion of $f^\mu(x)$. It will only be necessary to expand A_n as far as $1/n^{\mu-1}$, admitting $(-1)^n$ into the expansion as if it were a constant coefficient, and then, subtracting from A_n the sum of the terms thus found, employ the series which would be obtained by differentiating the equation (24) μ times. It will be necessary to assure ourselves that the term in $1/n^\mu$ vanishes in the expansion of A_n, since otherwise $f^\mu(x)$ might be infinite, or $f^{\mu-1}(x)$ discontinuous without our being aware of it. It will be seen however presently (Art. 20) that the former circumstance would not vitiate the result, nor introduce a term involving $1/n^\mu$.

Should A_n already appear under such a form as

$$\frac{1}{n} + c^n; \quad (-1)^n \frac{1}{n^3} + n^2 c^n, \text{ &c.,}$$

where $c^2 < 1$, it will be sufficient to differentiate equation (24) μ times, and leave out the part of the series which becomes divergent. For it will be observed that the terms c^n, $n^2 c^n$, in the examples chosen, decrease with $1/n$ faster than any inverse power of n.

14. Let us now consider the odd differential coefficients of $f(x)$, supposing $f(x)$ to be expanded in a series of cosines, so that

$$f(x) = B_0 + \Sigma B_n \cos \frac{n\pi x}{a} \dots\dots\dots\dots(30).$$

Let A'_n, $A'''_n \dots$ be the coefficients of $\sin n\pi x/a$ in the direct expansions of $f'(x)$, $f'''(x)\dots$ in series of sines. If we were to differentiate (30) once we should have $-n\pi/a . B_n$ for the coefficient of $\sin n\pi x/a$. Now

$$-\frac{n\pi}{a} \cdot \frac{2}{a}\int f(x') \cos \frac{n\pi x'}{a} dx'$$

$$= -\frac{2}{a} f(x') \sin \frac{n\pi x'}{a} + \frac{2}{a}\int f'(x') \sin \frac{n\pi x'}{a} dx';$$

and taking the limits of the integrals, and introducing B_n and A'_n, we get

$$A'_n = -\frac{n\pi}{a} B_n \dots\dots\dots\dots\dots(31).$$

Hence, the series arising from differentiating (30) once gives the direct expansion of $f(x)$ in a series of sines.

The coefficient of $\sin n\pi x/a$ in the series which would be obtained by differentiating (30) μ times, μ being odd, would be $(-1)^{(\mu+1)/2} (n\pi/a)^\mu B_n$. By proceeding just as in the last article we obtain

$$(-1)^{\frac{\mu+1}{2}} A_n^\mu = \left(\frac{n\pi}{a}\right)^\mu B_n + \frac{2}{a}\left(\frac{n\pi}{a}\right)^{\mu-2}\{f'(0) - (-1)^n f'(a)\}$$

$$-\frac{2}{a}\left(\frac{n\pi}{a}\right)^{\mu-4}\{f'''(0) - (-1)^n f'''(a)\} + \dots$$

$$+ (-1)^{\frac{\mu+1}{2}}\frac{2}{a} \cdot \frac{n\pi}{a}\{f^{\mu-2}(0) - (-1)^n f^{\mu-2}(a)\}\dots(32).$$

17—2

When $f'(0)$, $f'(a)$, &c., are known, this series enables us to develope $f^\mu(x)$ in a direct series of sines, the direct developement of $f(x)$ in a series of cosines being given.

15. If we treat the expression for B_n by integration by parts, just as the expression for A_n was treated, going on till we arrive at the integral which gives $A_n{}^\mu$, and observe that the very same process is used in deducing the value of $A_n{}^\mu$ from that of B_n as in expanding the latter according to inverse powers of n, and that the index of n in the coefficient of $A_n{}^\mu$ is $-\mu$, and that A_n vanishes when n becomes infinite, we shall see that in order to obtain the direct expansion of $f^\mu(x)$ we have only got to expand B_n as far as $1/n^\mu$, (the coefficient of $1/n^\mu$ will vanish,) and subtract from B_n these terms of the expansion, and then differentiate (30) μ times.

The expansion of B_n, at least to a certain number of terms, will proceed according to even powers of $1/n$, beginning with $1/n^2$. If we suppose that

$$B_n = O_1 \frac{1}{n^2} + O_3 \frac{1}{n^4} + O_5 \frac{1}{n^6} + \dots \text{ when } n \text{ is odd,} \left.\begin{matrix} \\ \\ \end{matrix}\right\} \dots (33),$$
$$B_n = E_1 \frac{1}{n^2} + E_3 \frac{1}{n^4} + E_5 \frac{1}{n^6} + \dots \text{ when } n \text{ is even}$$

and compare these expansions with that given by integration by parts, we shall have

$$f'(0) = -\frac{\pi^2}{4a}(O_1 + E_1), \quad f'(a) = -\frac{\pi^2}{4a}(O_1 - E_1), \left.\begin{matrix} \\ \\ \end{matrix}\right\} \dots (34),$$
$$f'''(0) = \frac{\pi^4}{4a^3}(O_3 + E_3), \quad f'''(a) = \frac{\pi^4}{4a^3}(O_3 - E_3)$$

and so on, the signs of the coefficients being alternately $+$ and $-$, and the index of π/a increasing by 2 each time.

16. The values of $f^\mu(0)$ and $f^\mu(a)$ when $f^\mu(x)$ is expanded in a series of sines and μ is odd, or when $f(x)$ is expanded in a series of cosines and μ is even, will be expressed by infinite series. To find these values we should first have to obtain the direct expansion of $f^\mu(x)$, which would be got by differentiating the equation (24) or (30) μ times, expanding A_n or B_n according to powers of $1/n$, and rejecting the terms which would render the series contained in the μth derived equation divergent. The reason of this is the same as before.

17. The direct expansions of the derivatives of $f(x)$ may be obtained in a similar manner in the cases in which $f(x)$ itself, or any one of its derivatives is discontinuous. In what follows, a will be taken to denote a value of x for which $f(x)$ or any one of its derivatives of the orders considered is discontinuous; Q, $Q_1, \ldots Q_\mu$ will denote the quantities by which $f(x)$, $f'(x)$, $\ldots f^\mu(x)$ are suddenly increased as x increases through a; S will be used for the sign of summation relative to the different values of a, and will be supposed to include the extreme values 0 and a, under the convention already mentioned in Art. 6. Of course $f(x)$ may be discontinuous for a particular value of x while $f^\mu(x)$ is continuous, and *vice versâ*. In this case one of the two Q, Q_μ will be zero while the other is finite.

The method of proceeding is precisely the same as before, except that each term such as $f(x) \cos n\pi x/a$ in the indefinite integral arising from the integration by parts will give rise to a series such as $- SQ \cos n\pi a/a$ in the integral taken between limits. We thus get in the case of the even derivatives of $f(x)$, when $f(x)$ is expanded in a series of sines,

$$(-1)^{\frac{\mu}{2}} A_n{}^\mu = \left(\frac{n\pi}{a}\right)^\mu A_n - \frac{2}{a} \cdot \left(\frac{n\pi}{a}\right)^{\mu-1} SQ \cos \frac{n\pi a}{a}$$

$$+ \frac{2}{a} \cdot \left(\frac{n\pi}{a}\right)^{\mu-2} SQ_1 \sin \frac{n\pi a}{a} + \frac{2}{a} \cdot \left(\frac{n\pi}{a}\right)^{\mu-2} SQ_2 \cos \frac{n\pi a}{a} - \ldots$$

$$+ (-1)^{\frac{\mu}{2}+1} \cdot \frac{2}{a} \cdot SQ_{\mu-1} \sin \frac{n\pi a}{a} \ldots\ldots\ldots\ldots(35).$$

In the case of the odd derivatives of $f(x)$, when $f(x)$ is expanded in a series of cosines, we get

$$(-1)^{\frac{\mu+1}{2}} A_n{}^\mu = \left(\frac{n\pi}{a}\right)^\mu B_n + \frac{2}{a}\left(\frac{n\pi}{a}\right)^{\mu-1} SQ \sin \frac{n\pi a}{a}$$

$$+ \frac{2}{a}\left(\frac{n\pi}{a}\right)^{\mu-2} SQ_1 \cos \frac{n\pi a}{a} - \ldots + (-1)^{\frac{\mu-1}{2}} \frac{2}{a} SQ_{\mu-1} \sin \frac{n\pi a}{a} \ldots(36).$$

When the several values of a, Q, $Q_1 \ldots$ are given, these equations enable us to find the direct expansion of $f^\mu(x)$. The series corresponding to the odd derivatives in the first case and the even in the second might easily be found.

If we wish to find the direct expansion of $f^\mu(x)$ in the case in which A_n or B_n is given, we have only to expand A_n or B_n in a series according to descending powers of n, regarding $\cos n\gamma$ or sin $n\gamma$, as well as $(-1)^n$, as constant coefficients, and then reject from the series obtained by the immediate differentiation of (24) or (30) those terms which would render it divergent. This readily follows as in Art. 15, from the consideration of the mode in which $A_n{}^\mu$ is obtained from A_n or B_n.

The equations (35) and (36) contain as particular cases (26) and (32) respectively. It was convenient however to have the latter equations, on account of their utility, expressed in a form which requires no transformation.

18.   If we transform A_n and B_n by integration by parts, we get

$$A_n = \frac{2}{n\pi} SQ \cos \frac{n\pi\alpha}{a} - \frac{2a}{n^2\pi^2} SQ_1 \sin \frac{n\pi\alpha}{a}$$
$$- \frac{2a^2}{n^3\pi^3} SQ_2 \cos \frac{n\pi\alpha}{a} + ..., \quad ....(37),$$

$$B_n = -\frac{2}{n\pi} SQ \sin \frac{n\pi\alpha}{a} - \frac{2a}{n^2\pi^2} SQ_1 \cos \frac{n\pi\alpha}{a}$$
$$+ \frac{2a^2}{n^3\pi^3} SQ_2 \sin \frac{n\pi\alpha}{a} + ..., \quad ....(38),$$

where the law of the series is evident, if we only observe that two signs of the same kind are always followed by two of the opposite kind. The equations (37), (38) may be at once obtained from (35), (36). The former equations give the true expansions of A_n and B_n according to powers of $1/n$; because when we stop after any number of integrations by parts the last integral with its proper coefficient always vanishes compared with the coefficient of the preceding term.

Hence A_n and B_n admit of expansion according to powers of $1/n$, if we regard cos $n\gamma$ or sin $n\gamma$ as a constant coefficient in the expansion. Moreover quantities such as cos $n\gamma$, sin $n\gamma$ will occur alternately in each expansion, the one kind going along with odd powers of $1/n$ and the other along with even. If we suppose the value of A_n or B_n, as the case may be, given, and the expansion performed, so that

$$A_n = SF \cos n\gamma . \frac{1}{n} + SF_1 \sin n\gamma . \frac{1}{n^2} + SF_2 \cos n\gamma . \frac{1}{n^3} + ..., \quad ...(39),$$

$$B_n = SG \sin n\gamma . \frac{1}{n} + SG_1 \cos n\gamma . \frac{1}{n^2} + SG_2 \sin n\gamma . \frac{1}{n^3} + ..., \quad ...(40),$$

and compare these expansions with (37) or (38), we shall get the several values of α, and the corresponding values of Q, Q_1, Q_2... We may thus, without being able to sum the series in equation (24) or (30), find the values of x for which $f(x)$ itself or any one of its derivatives is discontinuous, and likewise the quantity by which the function or derivative is suddenly increased. This remark will apply to the extreme values 0 and a of x if we continue to denote the sum of the series by $f(x)$ when x is outside of the limits 0 and a.

19. Having found the values of α, Q, Q_1..., we may if we please clear the series in (24) or (30) of the terms which render $f(x)$ itself, or any one of its derivatives, discontinuous. If we wish the function which remains expressed by an infinite series and its first μ derivatives to be continuous, we have only to subtract from A_n or B_n the terms at the commencement of its expansion, ending with the term containing $1/n^{\mu+1}$, and from $f(x)$ itself the sums of the series corresponding to the terms subtracted from A_n or B_n. These sums will be obtained by transforming products of sines and cosines into sums or differences, and then employing known formulæ such as

$$\frac{\cos z}{1^2} + \frac{\cos 3z}{3^2} + \ldots = \frac{\pi^2}{8} - \frac{\pi z}{4}, \text{ from } z = 0 \text{ to } z = \pi \ldots\ldots (41),$$

which are obtained by integrating several times the equation

$$\sin z + \tfrac{1}{2}\sin 2z + \tfrac{1}{3}\sin 3z + \ldots = \tfrac{1}{2}(\pi - z), \text{ from } z = 0 \text{ to } z = 2\pi,$$

or the equation deduced from it by writing $\pi - z$ for z, and taking the semi-sum of the results. It will be observed that in the several series to be summed we shall always have sines coming with odd powers of n and cosines with even. Of course, by clearing the series in (24) or (30) in the way just mentioned we shall increase the convergency of the infinite series in which a part of $f(x)$ still remains developed.

When A_n or B_n decreases faster than any inverse power of n as n increases, (as is the case for instance when it is the nth term of a geometric series with a ratio less than 1,) all the terms of its expansion in a series according to inverse powers of n vanish. In this case, then, $f(x)$ and its derivatives of all orders are continuous.

20. In establishing the several theorems contained in this Section, it has been supposed that none of the derivatives of $f(x)$ which enter into the investigation are infinite. It should be observed, however, that if $f^\mu(x)$ is the last derivative employed, which only appears under the sign of integration, it is allowable to suppose that $f^\mu(x)$ becomes infinite any finite number of times within the limits of integration. To shew this, we have only got to prove that

$$\int_0^a f^\mu(x) \sin \nu x \, dx \text{ or } \int_0^a f^\mu(x) \cos \nu x \, dx$$

approaches zero as its limit as ν increases beyond all limit. Let us consider the former of these integrals, and suppose that $f^\mu(x)$ becomes infinite only once, namely, when $x = \alpha$, within the limits of integration. Let the interval from 0 to a be divided into these four intervals 0 to $\alpha - \zeta$, $\alpha - \zeta$ to α, α to $\alpha + \zeta'$, $\alpha + \zeta'$ to a, where ζ and ζ' are supposed to be taken sufficiently small to exclude all values of x lying between the limits $\alpha - \zeta$ and $\alpha + \zeta'$ for which $f^{\mu-1}(x)$ alters discontinuously, or for which $f^\mu(x)$ changes sign, unless it be the value α. For the first and fourth intervals $f^\mu(x)$ is not infinite, and therefore, as it is known, the corresponding parts of the integral vanish for $\nu = \infty$. Since $\sin \nu x$ cannot lie beyond the limits $+1$ and -1, and is only equal to either limit for particular values of x, it is evident that the second and third portions of the integral are together numerically inferior to I, where

$$I = \{f^{\mu-1}(\alpha - \epsilon) \sim f^{\mu-1}(\alpha - \zeta)\} + \{f^{\mu-1}(\alpha + \zeta') \sim f^{\mu-1}(\alpha + \epsilon)\},$$

the symbol $A \sim B$ denoting the arithmetical difference of A and B, and ϵ being an infinitely small quantity, so that $f(\alpha - \epsilon)$, $f(\alpha + \epsilon)$ denote the limits to which $f(x)$ tends as x tends to the limit α by increasing and decreasing respectively. Hence the limit of the integral first considered, for $\nu = \infty$, must be less than I. But I may be made as small as we please by diminishing ζ and ζ', and therefore the limit required is zero.

The same proof applies to the integral containing $\cos \nu x$, and there is no difficulty in extending it to the case in which $f^\mu(x)$ is infinite more than once within the limits of integration, or at one of the limits.

21. It has hitherto been supposed that the function expanded in the series (3) or (22) does not become infinite; but the expansions will still be correct even if $f(x)$ becomes infinite any finite

number of times, provided that $\int f(x)\,dx$ be essentially convergent. Suppose that $f(x)$ becomes infinite only when $x = a$. Then it is evident that we may find a function of x, $F(x)$, which shall be equal to $f(x)$ except when x lies between the limits $a - \zeta$ and $a + \zeta'$, which shall remain finite from $x = a - \zeta$ to $x = a + \zeta'$, and which shall be such that $\int_{a-\zeta}^{a+\zeta'} F(x)\,dx = \int_{a-\zeta}^{a+\zeta'} f(x)\,dx$. Suppose that we are considering the series (3). Then, if C_n be the coefficient of $\sin n\pi x/a$ in the expansion of $F(x)$ in a series of the form (3), it is evident that C_n will approach the finite limit A_n when ζ and ζ' vanish, where $A_n = \dfrac{2}{a}\int_0^a f(x)\sin\dfrac{n\pi x}{a}\,dx$. But so long as ζ and ζ' differ from zero the series $\Sigma\,C_n\sin n\pi x/a$ is convergent, and has $F(x)$ for its sum, and $F(x)$ becomes equal to $f(x)$ when ζ and ζ' vanish, for any value of x except a. We might therefore be disposed to conclude at once that the series (3) is convergent, and has $f(x)$ for its sum, unless it be for the particular value $x = a$; but this point will require examination, since we might conceive that the series (3) became divergent, or if it remained convergent that it had a sum different from $f(x)$, when ζ and ζ' were made to vanish before the summation was performed. If we agree not to consider the series (3) directly, but only the limit of the series (5) when g becomes 1, it follows at once from (7) that for values of x different from a that limit is the same as in Art. 4. For $x = a$ the limit required is that of $\frac{1}{2}\{f(a - \epsilon) + f(a + \epsilon)\}$ when ϵ vanishes. If $f(x)$ does not change sign as x passes through a the limit required is therefore positive or negative infinity, according as $f(x)$ is positive or negative; but if $f(x)$ changes sign in passing through ∞ the limit required may be zero, a finite quantity, or infinity. The expression just given for the limit may be proved without difficulty. In fact, according to the method of Art. 4, we are led to examine an integral of the form

$$\frac{1}{\pi}\int_0^\zeta \{f(a - \xi) + f(a + \xi)\}\frac{h\,d\xi}{h^2 + \xi^2},$$

where ζ is a constant quantity which may be taken as small as we please, and supposed to vanish after h. Now by a known property of integrals the above integral is equal to

$$\frac{1}{\pi}\{f(a - \xi_1) + f(a + \xi_1)\}\int_0^\zeta \frac{h\,d\xi}{h^2 + \xi^2}, \text{ where } \xi_1 \text{ lies between 0 and } \zeta.$$

But $\int_0^\zeta \dfrac{h\,d\xi}{h^2+\xi^2}$, which is equal to $\tan^{-1}\dfrac{\zeta}{h}$, becomes equal to $\dfrac{\pi}{2}$ when h vanishes, and the limit of ξ_1 when h vanishes must be zero, since it cannot be greater than ζ, and ζ may be made to vanish after h.

22. The same thing may be proved by the method which consists in summing the series $\Sigma \sin n\pi x/a \,.\, \sin n\pi x'/a$ to n terms. If we adopt this method, then so long as we are considering a value of x different from α it will be found that the only peculiarity in the investigation is, that the quantity under the integral sign in the integrals we have to consider becomes infinite for one value of the variable; and it may be proved just as in Art. 20, that this circumstance has no effect on the result. If we are considering the value $x = \alpha$, it will be found that the integral we shall have to consider will be

$$\frac{1}{\pi}\int_0^\zeta \frac{\sin \nu\xi}{\xi}\left\{ f\left(\alpha+\frac{2a}{\pi}\xi\right) + f\left(\alpha-\frac{2a}{\pi}\xi\right)\right\} d\xi \,\ldots\ldots (42),$$

where ν is first to be made infinite, and then ζ may be supposed to vanish. If $f(\alpha+\epsilon)+f(\alpha-\epsilon)$ approaches a finite limit, or zero, when ϵ vanishes, as may be the case if $f(x)$ changes sign in passing through ∞, it may be proved, just as in the case in which $f(x)$ does not become infinite, that the above integral approaches the same limit as $\frac{1}{2}\{f(\alpha+\epsilon)+f(\alpha-\epsilon)\}$. In all cases however in which $f(x)$ does not change sign in passing through ∞, and in some cases in which it does change sign, $f(\alpha+\epsilon)+f(\alpha-\epsilon)$ becomes infinite when ϵ vanishes.

In such cases put for shortness

$$f\left(\alpha+\frac{2a}{\pi}\xi\right) + f\left(\alpha-\frac{2a}{\pi}\xi\right) = F(\xi),$$

and let the numerical values of the integral $\int \dfrac{\sin \nu\xi}{\xi}\,d\xi$ taken from 0 to π/ν, from π/ν to $2\pi/\nu$ ... or which is the same those of $\int \dfrac{\sin \xi}{\xi}\,d\xi$ taken from 0 to π, from π to 2π ... be denoted by $I_1,\ I_2\ldots$ Then evidently $I_1 > I_2 > I_3 \ldots$ Also, if ζ be sufficiently small, $F(\xi)$ will decrease from $\xi = 0$ to $\xi = \zeta$, if we suppose, as we may, $F(\xi)$ to be positive. Hence the integral (42), which is equal to

$$\frac{1}{\pi}\{I_1\,F(\xi_1) - I_2\,F(\xi_2) + I_3\,F(\xi_3) - \ldots\}\,\ldots\ldots\ldots(43),$$

where ξ_1, ξ_2 ... are quantities lying between 0 and π/ν, π/ν and $2\pi/\nu$ ... is greater than

$$\frac{1}{\pi}\{I_1 F(\xi_1) - I_2 F(\xi_2)\},$$

if we neglect the incomplete pair of terms which may occur at the end of the series (43), and which need not be considered, since they vanish when $\nu = \infty$. Hence, the integral (42) is *a fortiori* $> \pi^{-1}(I_1 - I_2) F(\xi_1)$. But ξ_1 vanishes and $F(\xi_1)$ becomes infinite when ν becomes infinite; and therefore for the particular value $x = a$ the sum of the first n terms of the series (3) increases indefinitely with n.

If α coincides with one of the extreme values 0 and a of x, the sum of the series (3) vanishes for $x = a$. This comes under the formula given above if we consider the sum of the series for values of x lying beyond the limits 0 and a. The same proof as that given in the present and last article will evidently apply if $f(x)$ become infinite for several values of x, or if the series considered be (22) instead of (3). In this case, the sum of the series becomes infinite for $x = \alpha$ when $\alpha = 0$ or $= a$.

23. Hence it appears that $f(x)$ may be expanded in a series of the form (3) or (22), provided only $\int f(x) \, dx$ be continuous. It should be observed however that functions like $(\sin c/x)^{-\frac{2}{3}}$, which become infinite or discontinuous an infinite number of times within the limits of the variable within which they are considered, have been excluded from the previous reasoning.

Hence, we may employ the formulæ such as (26), (35), &c., to obtain the direct developement of $f^\mu(x)$, without enquiring whether it becomes infinite or not within the limits of the variable for which it is considered. All that is necessary is that $f(x)$ and its derivatives up to the $(\mu - 1)^{\text{th}}$ inclusive should not be infinite within those limits, although they may be discontinuous.

24. In obtaining the formulæ of Arts. 7 and 13, and generally the formulæ which apply to the case in which A_n or B_n is given, and $f(x)$ is unknown, it has hitherto been supposed that we knew *a priori* that $f(x)$ was a function of the class proposed in Art. 1 for consideration, or at least of that class with the extension mentioned

in the preceding article. Suppose now that we have simply presented to us the series (3) or (22), namely

$$\Sigma A_n \sin \frac{n\pi x}{a}, \ \text{ or } B_0 + \Sigma B_n \cos \frac{n\pi x}{a},$$

where A_n or B_n is supposed given and want to know, *first*, whether the series is convergent, *secondly*, whether if it be convergent it is the direct developement of its sum $f(x)$, and *thirdly*, whether we may directly employ the formulæ already obtained, trusting to the formulæ themselves to give notice of the cases to which they do not apply by leading to processes which cannot be effected.

25. If the series ΣA_n or ΣB_n is essentially convergent, it is evident *a fortiori* that the series (3) or (22) is convergent.

If $A_n = S\dfrac{c}{n}\cos n\gamma + C_n$, or if $B_n = S\dfrac{c}{n}\sin n\gamma + C_n$, where ΣC_n is essentially convergent, the given series will be convergent, as is proved in Art. 6.

In either of these cases let $f(x)$ be the sum of the given series. Suppose that it is the series of sines which we are considering. Let E_n be the coefficient of $\sin n\pi x/a$ in the direct developement of $f(x)$. Then we have

$$f(x) = \Sigma A_n \sin \frac{n\pi x}{a} = \Sigma E_n \sin \frac{n\pi x}{a};$$

and since both series are convergent, if we multiply by any finite function of x, $\phi(x)$, and integrate, we may first integrate each term, and then sum, instead of first summing and then integrating. Taking $\phi(x) = \sin n\pi x/a$, and integrating from $x = 0$ to $x = a$, we get $E_n = A_n$, so that the given series is the direct developement of its sum $f(x)$. The proof is the same for the series of cosines.

26. Consider now the more general case in which the series $\Sigma 1/n \,.\, A_n$ is essentially convergent. The reasoning which is about to be offered can hardly be regarded as absolutely rigorous; nevertheless the proposition which it is endeavoured to establish seems worthy of attention. Let u_n be the sum of the first n terms of the given series, and $F(n, x)$ the sum of the first n terms of the series $\Sigma - a/n\pi \,.\, A_n \cos n\pi x/a$. Then we have

$$\int (u_{n+m} - u_n)\, dx = F(n+m, x) - F(n, x) = \psi(n, x), \text{ suppose} \dots (44).$$

Now by hypothesis the series $\Sigma 1/n \,.\, A_n$ is essentially convergent, and therefore *a fortiori* the series $\Sigma - a/n\pi \,.\, A_n \cos n\pi x/a$ is convergent, and therefore $\psi(\infty, x) = 0$, whatever be the value of m. Let the limits of x in (44) be x and $x + \Delta x$, and divide by Δx, and we get

$$\frac{1}{\Delta x} \int_x^{x+\Delta x} (u_{n+m} - u_n)\, dx = \frac{\Delta \psi(n, x)}{\Delta x}:$$

and as we have seen the limit of the second side of this equation when we suppose n first to become infinite and then Δx to vanish is zero. But *for general values of x* the limit will remain the same if we first suppose Δx to vanish and then n to become infinite; and on this supposition we have

$$\text{limit of } (u_{n+m} - u_n) = 0, \text{ for } n = \infty;$$

so that for general values of x the series considered is convergent.

To illustrate the assumption here made that for general values of x the order in which n and Δx assume their limiting values is immaterial, let $\psi(y, x)$ be a continuous function of x which becomes equal to $\psi(n, x)$ when y is a positive integer; and consider the surface whose equation is $z = \psi(y, x)$. Since

$$\psi(\infty, x) = 0$$

for integral values of y, the surface approaches indefinitely to the plane xy when y becomes infinite; or rather, among the infinite number of admissible forms of $\psi(y, x)$ we may evidently choose an infinite number for which that is the case. Now the assertion made comes to this; that if we cut the surface by a plane parallel to the plane xz, and at a distance n from it, the tangent at the point of the section corresponding to any given value of x will ultimately lie in the plane xy when n becomes infinite, except in the case of singular, isolated values of x, whose number is finite between $x = 0$ and $x = a$. For such values the sum $f(x)$ of the infinite series may become infinite, while $\int f(x)\, dx$ remains finite. The assumption just made appears evident unless A_n be a function of n whose complexity increases indefinitely with its rank, *i.e.* with the value of n.

Since the integral of $f(x)$ is continuous, $f(x)$ may be expanded by the formula in a series of sines. Let E_n be the coefficient of $\sin n\pi x/a$ in its direct expansion; so that,

$$\left.\begin{aligned}
f(x) &= \Sigma A_n \sin \frac{n\pi x}{a}, \\
f(x) &= \Sigma E_n \sin \frac{n\pi x}{a},
\end{aligned}\right\} \quad \ldots\ldots\ldots\ldots\ldots (45),$$

where both series are convergent, except it be for isolated values of x. Consequently, we have, in a series which is convergent, at least for general values of x,

$$0 = \Sigma\,(A_n - E_n)\,\sin\frac{n\pi x}{a}\dots\dots\dots\dots\dots(46).$$

The series (45) may become divergent for isolated values of x, and are in fact divergent for values of x which render $f(x)$ infinite. But the first side of (46) being constantly zero, and the series at the second side being convergent for general values of x, it does not seem that it can become divergent for isolated values. Hence according to the preceding article the second side of the equation is the direct developement of the first side, i.e. of zero; and therefore $E_n = A_n$, or the given series is the direct developement of its sum, which is what it was required to prove. The same reasoning applies to the series of cosines.

It may be observed that the well known series,

$$\tfrac{1}{2} + \cos x + \cos 2x + \cos 3x\dots\dots\dots\dots\dots(47),$$

forms no exception to the preceding observation. This series is in fact divergent for general values of x, that is to say not convergent, and in that respect it totally differs from the series in (46). When it is asserted that the sum of the series (47) is zero except for $x = 0$ or any multiple of 2π, when it is infinite, all that is meant is that the limit to which the sum of the convergent series $\tfrac{1}{2} + \Sigma g^n \cos nx$ approaches when g becomes 1 is zero, except for $x = 0$ or any multiple of 2π, in which case it is infinity.

27. It follows from the preceding article that even without knowing *a priori* the nature of the function $f(x)$ we may employ the formulæ such as (35), provided that if $n^{-\mu}$ be the highest power of $1/n$ required by the formula, and $n^{-\mu}C_n$ the remainder in the expansion of A_n, the series $\Sigma n^{-1}C_n$ be essentially convergent. For let G_n be the sum of the terms as far as that containing $n^{-\mu}$ in the expansion of A_n, those terms having the form assigned by (35), that is to say cosines like $\cos n\gamma$ coming along with odd powers of $1/n$, and sines along with even powers. Then

$$A_n = G_n + n^{-\mu}C_n.$$

Let
$$\Sigma\,G_n \sin\frac{n\pi x}{a} = F(x)\,;$$

then $$f(x) - F(x) = \Sigma n^{-\mu} C_n \sin \frac{n\pi x}{a} \ldots\ldots\ldots(48).$$

Now if $\phi(x) = \Sigma u_n$, where the series Σu_n, $\Sigma du_n/dx$ are both convergent, we may find $\phi'(x)$ by differentiating under the sign of summation. This is evident, since by the theorem referred to in Art. 2 (note), we may find $\int \Sigma \dfrac{du_n}{dx} dx$ by integrating under the sign of summation. Consequently we have from (48)

$$f^{\mu-1}(x) - F^{\mu-1}(x) = \pm \left(\frac{\pi}{a}\right)^{\mu-1} \Sigma \frac{1}{n} C_{n\,\cos}^{\,\sin} \frac{n\pi x}{a} \ldots..(49);$$

and since the series $\Sigma n^{-1} C_n$ is essentially convergent, the convergency of the series forming the right-hand side of (49) cannot become infinitely slow (see Sect. III.), and therefore, the nth term being a continuous function of x, the sum is also a continuous function of x, and therefore $f^\mu(x) - F^\mu(x)$ is a function which by Art. 23 can be expanded in a series of sines or cosines. But $F_\mu(x)$ is also such a function, being in fact a constant, and therefore $f^\mu(x)$ is a function of the kind considered in Art. 23, which is what is assumed in obtaining the formula (35).

It may be observed that these results do not require the assumptions of Art. 26 in the case in which the series ΣC_n is essentially convergent, or composed of an essentially convergent series and of a series of the form $\Sigma Scn^{-1} \sin n\gamma$ or $\Sigma Scn^{-1} \cos n\gamma$, according as C_n is the coefficient of a cosine or of a sine.

SECTION II.

Mode of ascertaining the nature of the discontinuity of the integrals which are analogous to the series considered in Section I., and of obtaining the developements of the derivatives of the expanded functions.

28. Let us consider the following integral, which is analogous to the series in (1),

$$\int_0^\infty \phi(\beta) \sin \beta x \, d\beta \ldots\ldots\ldots\ldots (50),$$

where $$\phi(\beta) = \frac{2}{\pi} \int_0^a f(x') \sin \beta x' \, dx' \ldots\ldots\ldots(51).$$

Although the integral (50) may be written as a double integral,

$$\frac{2}{\pi} \int_0^\infty \int_0^a f(x') \sin \beta x \sin \beta x' \, d\beta dx' \ldots\ldots\ldots(52),$$

the integration with respect to x' must be performed first, because, the integral of $\sin \beta x \sin \beta x' \, d\beta$ not being convergent at the limit ∞, $\int_0^\infty \sin \beta x \sin \beta x' d\beta$ would have no meaning. Suppose, however, that instead of (52) we consider the integral,

$$\frac{2}{\pi} \int_0^\infty \int_0^a f(x') \, \epsilon^{-h\beta} \sin \beta x \sin \beta x' \, d\beta dx' \ldots\ldots\ldots(53),$$

where h is a positive constant, and ϵ is the base of the Napierian logarithms. It is easy to see that at least in the case in which the integral (50) is essentially convergent its value is also the limit to which the integral (53) tends when h tends to zero as its limit. It is well known that the limit of (53) when h vanishes is in general $f(x)$; but when $x = 0$ the limit is zero; when $x = a$ the limit is $\frac{1}{2} f(a)$; and when $f(x)$ is discontinuous it is the arithmetic mean of the values of $f(x)$ for two values of x infinitely little greater and less respectively than the critical value. When $x > a$ it is zero, and in all cases it is the same, except as to sign, for negative as for positive values of x.

We may always speak of (53), but we cannot speak of the integral (50) till we assure ourselves that it is convergent. Now we get by integration by parts,

$$\int f(x') \sin \beta x' \, dx' = -\frac{1}{\beta} f(x') \cos \beta x'$$

$$+ \frac{1}{\beta^2} f'(x') \sin \beta x' - \frac{1}{\beta^2} \int f''(x') \sin \beta x' \, dx' \ldots\ldots\ldots(54).$$

When this integral is taken between limits, the first term will furnish a set of terms of the form $C/\beta \cdot \cos \beta a$, where a may be zero, and the last two terms will give a result numerically less than L/β^2, where L is a constant properly chosen. Now whether a be zero or not, $\int \cos \beta a \sin \beta x \cdot \beta^{-1} d\beta$ is convergent at the limit ∞, and moreover its value taken from any finite value of β to $\beta = \infty$ is the limit to which the integral deduced from it by inserting the factor $\epsilon^{-h\beta}$ tends when h vanishes. The remaining part of the integral (50) is essentially convergent at the limit ∞. Hence the integral (50) is convergent, and its value for all values of x, both critical and general, is the limit to which the value of the integral (53) tends when h vanishes.

29. Suppose that we want to find $f''(x)$, knowing nothing about $f(x)$, at least for general values of x, except that it is the value of the integral (50), and that it is not a function of the class excluded from consideration in Art. 1. We cannot differentiate under the integral sign, because the resulting integral would, usually at least, be divergent at the limit ∞. We may however find $f''(x)$ provided we know the values of x for which $f(x)$ and $f'(x)$ are discontinuous, and the quantities by which $f(x)$ and $f'(x)$ are suddenly increased as x increases through each critical value, supposing the extreme values included among those for which $f(x)$ or $f'(x)$ is discontinuous, under the same convention as in Art. 6. Let α be any one of the critical values of x; Q, Q_1 the quantities by which $f(x)$, $f'(x)$ are suddenly increased as x increases through α; S the sign of summation referring to the critical values of x; $\phi_2(\beta)$ the coefficient of $\sin \beta x$ in the direct developement of $f''(x)$ in a definite integral of the form (50). Then taking the integrals in (54) between limits, and applying the formula (51) to $f''(x)$, we get

$$\phi_2(\beta) = -\beta^2 \phi(\beta) + \frac{2}{\pi} \beta S Q \cos \beta \alpha - \frac{2}{\pi} S Q_1 \sin \beta \alpha.$$

We may find $\phi_\mu(\beta)$ in a similar manner. We get thus when μ is even

$$(-1)^{\frac{\mu}{2}} \phi_\mu(\beta) = \beta^\mu \phi(\beta) - \frac{2}{\pi} \beta^{\mu-1} S Q \cos \beta \alpha + \frac{2}{\pi} \beta^{\mu-2} S Q_1 \sin \beta \alpha + \dots$$

$$+ (-1)^{\frac{\mu}{2}+1} \frac{2}{\pi} S Q_{\mu-1} \sin \beta \alpha \dots \dots (55),$$

where sines and cosines occur alternately, and two signs of the same kind are always followed by two of the opposite. The expression for $\phi^\mu(\beta)$ when μ is odd might be found in a similar manner. These formulæ enable us to express $f^\mu(x)$ when $\phi(\beta)$ is an arbitrary function which has to be determined, and $f(0)$, &c. are given.

30. If however $\phi(\beta)$ should be given, and $f(0)$, &c. be unknown, $\phi(\beta)$ will admit of expansion according to powers of β^{-1}, beginning with the first, provided we treat $\sin \beta x$ or $\cos \beta x$ as if it

s. 18

were a constant coefficient; and sin βx, cos βx will occur with even and odd powers of β respectively. The possibility of the expansion of $\phi(\beta)$ in this form depends of course on the circumstance that $\phi(x)$ is a function of the class which it is proposed in Art. 1 to consider, or at least with the extension mentioned in Art. 23. It appears from (55) that in order to express $f^\mu(x)$ as a definite integral of the form (50) we have only got to expand $\phi(\beta)$, to differentiate (50) μ times with respect to x, differentiating under the integral sign, and to reject those terms which appear under the integral sign with positive powers of β or with the power 0. The same rule applies whether μ be odd or even.

31. If we have given $\phi(x)$, but are not able to evaluate the integral (50), we may notwithstanding that find the values of x which render $f(x)$ or any of its derivatives discontinuous, and the quantities by which the function considered is suddenly increased. For this purpose it is only necessary to compare the expansion of $\phi(\beta)$ with the expansion

$$\phi(\beta) = \frac{2}{\pi\beta} S Q \cos \beta a - \frac{2}{\pi\beta^2} S Q_1 \sin \beta a - \ldots\ldots\ldots \quad (56),$$

given by (55), just as in the case of series.

We may easily if we please clear the function $\phi(\beta)$ of the part for which $f(x)$ or any one of its derivatives is discontinuous, or does not vanish for $x = 0$ and $x = a$. For this purpose it will be sufficient to take any function $F(x)$ at pleasure, which as well as its derivatives of the orders considered has got the same discontinuity as $f(x)$ and its derivatives, to develope $F(x)$ in a definite integral of the form $\int_0^\infty \Phi(\beta) \sin \beta x \, d\beta$ by the formula (51), and to subtract $F(x)$ from $f(x)$ and $\Phi(\beta)$ from $\phi(\beta)$. It will be convenient to choose such simple functions as $l + mx + nx^2$; $l \sin x + m \cos x$; $l\epsilon^{-x} + m\epsilon^{-kx}$, &c. for the algebraical expressions of $F(x)$ for the several intervals throughout which it is continuous, the functions chosen being such as admit of easy integration when multiplied by $\sin \beta x \, dx$, and which furnish a sufficient number of indeterminate coefficients to allow of the requisite conditions as to discontinuity being satisfied. These conditions are that the several values of Q, Q_1, &c. shall be the same for $F(x)$ as for $f(x)$.

32. Whenever $\int_0^\infty f(x)\, dx$ is essentially convergent, we may at once put $a = \infty$ in the preceding formulæ. For, first, it may be easily proved that in this case, (though not in this case only,) the limit of (53) when h vanishes is $f(x)$; secondly, the limit of (53) is also the value of (52); and, lastly, all the derivatives of $f(x)$ have their integrals, (which are the preceding derivatives,) essentially convergent, and therefore ∞ may be put for a in the developements of the derivatives in definite integrals.

When $f(x)$ tends to zero as its limit as x becomes infinite, and moreover after a finite value of x does not change from decreasing to increasing nor from increasing to decreasing,

$$\int_0^\infty \epsilon^{-hx'} f(x') \sin \beta x'\, dx'$$

will be more convergent than $\int_0^\infty f(x') \sin \beta x'\, dx'$, and the latter integral will be convergent, and its convergency will remain finite* when β vanishes. In this case also we may put $a = \infty$.

Thus if $f(x) = \sin lx\, (b^2 + x^2)^{-1}$, we may put $a = \infty$ because $f(x)$ has its integral essentially convergent: if $f(x) = (b + x)^{-\frac{1}{2}}$, we may put $a = \infty$ because $f(x)$ is always decreasing to zero as its limit. But if $f(x) = \sin lx\, (b + x)^{-\frac{1}{2}}$, the preceding rules will not apply, because $f(x)$, though it has zero for its limit, is sometimes increasing and sometimes decreasing. And in fact in this case the integral in equation (51) will be divergent when $\beta = l$, and $\phi(\beta)$ will become infinite for that value of β. It is true that $f(x)$ is still the limit to which the integral (53) tends when h vanishes; but I do not intend to enter into the consideration of such cases in this paper.

33. When ∞ may be put for a, and $f(x)$ is continuous, we get from (55)

$$(-1)^{\frac{\mu}{2}} \phi_\mu(\beta) = \beta^\mu \phi(\beta) - \frac{2}{\pi} \beta^{\mu-1} f(0) + \frac{2}{\pi} \beta^{\mu-3} f''(0) - \ldots$$

$$+ (-1)^{\frac{\mu}{2}} \frac{2}{\pi} \beta f^{\mu-2}(0) \ldots\ldots\ldots(57).$$

In this case $\phi(\beta)$ will admit of expansion, at least to a certain

* See next Section.

18—2

number of terms, according to odd negative powers of β.   If
we suppose $\phi\,(\beta)$ known, and the expansion performed, so that

$$\phi\,(\beta) = H_0\beta^{-1} + H_2\beta^{-3} + H_4\beta^{-5} + \ldots$$

and compare the result (49), we shall get

$$f\,(0) = \frac{\pi}{2}\,H_0\,; \quad f''\,(0) = -\frac{\pi}{2}\,H_2\,; \quad f^4\,(0) = \frac{\pi}{2}\,H_4\,; \quad \&\text{c}\ldots\ldots(58).$$

34.   The integral

$$\int_0^\infty \psi\,(\beta)\cos\beta x\,d\beta \ldots\ldots\ldots\ldots\ldots(59),$$

where

$$\psi(\beta) = \frac{2}{\pi}\int_0^a f(x')\cos\beta x'\,dx' \quad\ldots\ldots\ldots(60),$$

which is analogous to the series (22), is another in which it is sometimes useful to develope a function or conceive it developed.   For
positive values of x the value of (59) is the same as that of (50).
When $x = 0$ the value is $f\,(0)$; and for negative values of x it is
the same as for positive.   It is supposed here that the integral (59)
is convergent, which it may be proved to be in the same manner
as the integral (50) was proved to be convergent.

Suppose that we wish to find, in terms of $\psi\,(\beta)$, the developement of $f^\mu\,(x)$ in a definite integral of the form (50) or (59),
according as μ is odd or even.   We cannot differentiate under the
integral sign, because the resulting integral would be divergent.
We may however obtain the required developement by transform-
ing the expression $\psi\,(\beta)$ by integration by parts, just as before.   We
thus get for the case in which μ is odd

$$(-1)^{\frac{\mu+1}{2}}\,\phi_\mu\,(\beta) = \beta^\mu\psi(\beta) + \frac{2}{\pi}\,\beta^{\mu-1}\,SQ\,\sin\beta x + \frac{2}{\pi}\,\beta^{\mu-2}SQ_1\cos\beta x - \ldots$$

$$+ (-1)^{\frac{\mu-1}{2}}\,\frac{2}{\pi}\,SQ_{\mu-1}\sin\beta x\ldots\ldots\ldots(61),$$

where $\phi_\mu\,(\beta)$ is the value of $\phi\,(\beta)$ in the direct developement of
$f^\mu\,(x)$ in the integral (50).   In the same way we may get the value
of $\psi_\mu\,(\beta)$ when μ is even, $\psi_\mu(\beta)$ being the value of $\psi\,(\beta)$ in the
direct developement of $f^\mu\,(x)$ by the formulæ (59), (60).

The equation (61) is applicable to the case in which $\psi\,(\beta)$ is an
arbitrary function, and a, Q, &c., are given.   If however $\psi(\beta)$

should be given, we may find $\phi_\mu(\beta)$ or $\psi_\mu(\beta)$ by the same rule as before.

In the case in which $\psi(\beta)$ is given, we may find the values of a, Q, &c., without being able to evaluate the integral (59). For this purpose it is sufficient to expand $\psi(\beta)$ according to negative powers of β, and compare the expansion with that furnished by equation (61).

35. The same remarks as to the cases in which we are at liberty to put ∞ for a apply to (60) as to (51), with one exception. In the case in which $f(x)$ approaches zero as its limit, and is at last always decreasing numerically, or at least never increasing, as x increases, while $\int f(x)\,dx$ is divergent at the limit ∞, it has been observed that $\phi(\beta)$ remains finite when β vanishes. This however is not the case with $\psi(\beta)$, at least in general. I say *in general*, because, although $\int_0^x f(x)\,dx$ increases indefinitely with its superior limit, we are not entitled at once to conclude from thence that $\int_0^\infty \cos\beta x\, f(x)\,dx$ becomes infinite when β vanishes, as will appear in Section III. It may be shewn from the known value of $\int_0^\infty x^{-n}\cos\beta x\, dx$, where $1 > n > 0$, that if $f(x) = F(x) + Cx^{-n}$, where $F(x)$ is such that $\int F(x)\,dx$ is convergent at the limit ∞, $\psi(\beta)$ becomes infinite when β vanishes; and the same would be true if there were any finite number of terms of the form Cx^{-n}. There is no occasion however to enquire whether $\psi(\beta)$ *always* becomes infinite: the point to consider is whether the integral (59) is always convergent at the limit zero.

In considering this question, we may evidently begin the integration relative to x' at any value x_0 that we please. Suppose first we integrate from $x' = x_0$ to $x' = X$, and let $\varpi(\beta)$ be the result so that

$$\varpi(\beta) = \frac{2}{\pi}\int_{x_0}^X f(x')\cos\beta x'\,dx'.$$

Let $\varpi_{,}(\beta)$ be the indefinite integral of $\varpi(\beta)\,d\beta$: then, c being a positive quantity, we get from the above equation

$$\varpi_{,}(\beta) - \varpi_{,}(c) = \frac{2}{\pi}\int_{x_0}^X f(x')\left\{\sin\beta x' - \sin cx'\right\}\frac{dx'}{x'}.$$

Now put $X = \infty$. Then since $\int_{x_0}^{\infty} f(x') \dfrac{\sin \beta x'}{x'} \, dx'$ is a convergent integral, and its convergency remains finite (Art. 39) when β vanishes, as may be proved without much difficulty, its value can-. not become infinite, and therefore $\varpi_{,}(\beta)$ does not become infinite when β vanishes. Now

$$\int \varpi(\beta) \cos \beta x \, d\beta = \varpi_{,}(\beta) \cos \beta x + x \int \varpi_{,}(\beta) \sin \beta x \, d\beta \ldots\ldots(62),$$

when x is positive ; and when $x = 0$,

$$\int \varpi(\beta) \, d(\beta) = \varpi_{,}(\beta) :$$

hence in either case $\int \varpi(\beta) \cos \beta x \, d\beta$ is convergent at the limit zero. Now the quantity by which $\varpi(\beta)$ differs from $\psi(\beta)$ evidently cannot render (59) divergent, and therefore in the case considered the integral (59) is convergent at the limit zero.

By treating $\displaystyle\int_{0}^{\infty} \varpi(\beta) \, \epsilon^{-h\beta} \cos \beta x \, d\beta$ in the manner in which $\int \varpi(\beta) \cos \beta x \, d\beta$ is treated in (62), it may be shewn that the convergency of the former integral remains finite when h vanishes. Hence, not only is the integral (59) convergent, but its value is the limit to which the integral similar to (53) tends when h vanishes.

When $f(x)$ is continuous, and ∞ may be put for a, we have from (61), μ being odd,

$$(-1)^{\frac{\mu+1}{2}} \phi_{\mu}(\beta) = \beta^{\mu} \psi(\mu) + \frac{2}{\pi} \beta^{\mu-2} f'(0) - \frac{2}{\pi} \beta^{\mu-4} f'''(0) + \ldots$$

$$+ (-1)^{\frac{\mu+1}{2}} \frac{2}{\pi} \beta f^{\mu-2}(0) \ldots\ldots(63).$$

If $\psi(\beta)$ be given we can find the values of $f'(0)$, $f'''(0) \ldots$ just as before.

36.   The integral

$$\frac{1}{\pi} \int_{0}^{\infty} \int_{-a_{,}}^{a} \cos \beta \, (x' - x) f(x') \, d\beta \, dx' \ldots\ldots\ldots(64),$$

in which the integration with respect to x' is supposed to be performed before that with respect to β, so that the integral has the form

$$\int_{0}^{\infty} \chi(\beta) \cos \beta x \, d\beta + \int_{0}^{\infty} \sigma(\beta) \sin \beta x \, d\beta \ldots\ldots(65),$$

may be treated just as the integral (59); and it may be shewn that in the same circumstances we may replace the limits $-a$, and a by $-\infty, +\infty$ respectively. If we suppose $\chi(\beta)$ and $\sigma(\beta)$ known, we may find as before the values of x for which $f(x)$, $f'(x) \ldots$ are discontinuous, and the quantities by which those functions are suddenly increased. We may also find the direct developement of $f'(x), f''(x) \ldots$ in two integrals of the form (65); and we may if we please clear the integrals (65) of the part which renders $f(x), f'(x) \ldots$ discontinuous.

37. In the developement of $f(x)$ in an integral of the form (50) or (59), or in two integrals of the form (65), it has hitherto been supposed that $f(x)$ is not infinite. It may be observed however that it is allowable to suppose $f(x)$ to become infinite any finite number of times, provided $\int f(x)\, dx$ be essentially convergent about the values of x which render $f(x)$ infinite. This may be shewn just as in the case of series. Hence, the formulæ such as (55) which give the developement of $f^\mu(x)$ are true even when $f^\mu(x)$ is infinite, $f^{\mu-1}(x)$ being finite.

SECTION III.

On the discontinuity of the sums of infinite series, and of the values of integrals taken between infinite limits.

38. LET

$$u_1 + u_2 \ldots + u_n + \ldots \quad\quad\quad\quad (66),$$

be a convergent infinite series having U for its sum. Let

$$v_1 + v_2 \ldots + v_n + \ldots \quad\quad\quad\quad (67),$$

be another infinite series of which the general term v_n is a function of the positive variable h, and becomes equal to u_n when h vanishes. Suppose that for a sufficiently small value of h and all inferior values the series (67) is convergent, and has V for its sum. It might at first sight be supposed that the limit of V for $h = 0$ was necessarily equal to U. This however is not true. For let the sum to n terms of the series (67) be denoted by $f(n, h)$: then the limit of V is the limit of $f(n, h)$ when n first becomes infinite and then h vanishes, whereas U is the limit of $f(n, h)$ when h first vanishes and then n becomes infinite, and these limits may be different. Whenever a discontinuous function is developed in a periodic series like (15) or (30) we have an instance of this ; but it is easy to form two series, having nothing to to with periodic series, in which the same happens. For this purpose it is only requisite to take for $f(n, h) - U_n$, (U_n being the sum of the first n terms of (66),) a quantity which has different limiting values according to the order in which n and h are supposed to assume their limiting values, and which has for its finite difference a quantity which vanishes when n becomes infinite, whether h be a positive quantity sufficiently small or be actually zero.

For example, let

$$f(n, h) - U_n = \frac{2nh}{nh + 1} \quad \dots\dots\dots\dots\dots(68),$$

which vanishes when $n = 0$.   Then

$$\Delta \{f(n, h) - U_n\} = v_{n+1} - u_{n+1} = \frac{2h}{(nh + 1)(nh + h + 1)}.$$

Assume

$$U_n = 1 - \frac{1}{n+1}, \quad \text{so that } u_n = \Delta U_{n-1} = \frac{1}{n(n+1)},$$

and we get the series

$$\frac{1}{1.2} + \frac{1}{2.3} \dots + \frac{1}{n(n+1)} + \dots \quad \dots\dots\dots\dots(69),$$

$$\frac{1 + 5h}{2(1 + h)} \dots + \frac{h(h+2)n^2 + h(4-h)n + 1 - h}{n(n+1)\{(n-1)h + 1\}(nh + 1)} + \dots \quad \dots\dots(70),$$

which are both convergent, and of which the general terms become the same when h vanishes.   Yet the sum of the first is 1, whereas the sum of the second is 3.

If the numerator of the fraction on the right-hand side of (68) had been pnh instead of $2nh$, the sum of the series (70) would have been $p + 1$, and therefore the limit to which the sum approaches when h vanishes would have been $p + 1$.   Hence we can form as many series as we please like (67) having different quantities for the limits of their sums when h vanishes, and yet all having their n^{th} terms becoming equal to u_n when h vanishes.   This is equally true whether the series (66) be convergent or divergent, the series like (67) of course being always supposed to be convergent for all positive values of h however small.

39. It is important for the purposes of the present paper to have a ready mode of ascertaining in what cases we may replace the limit of (67) by (66). Now it follows from the following theorem that this substitution may at once be made in an extensive class of cases.

THEOREM. The limit of V can never differ from U unless the convergency of the series (67) becomes infinitely slow when h vanishes.

The convergency of the series is here said to become infinitely slow when, if n be the number of terms which must be taken in order to render the sum of the neglected terms numerically less than a given quantity e which may be as small as we please, n increases beyond all limit as h decreases beyond all limit.

DEMONSTRATION. If the convergency do not become infinitely slow, it will be possible to find a number n_1 so great that for the value of h we begin with and for all inferior values greater than zero the sum of the neglected terms shall be numerically less than e. Now the limit of the sum of the first n_1 terms of (67), when h vanishes is the sum of the first n_1 terms of (66). Hence if e' be the numerical value of the sum of the terms after the n_1^{th} of the series (66), U and the limit of V cannot differ by a quantity so great as $e + e'$. But e and e' may be made smaller than any assignable quantities, and therefore U is equal to the limit of V.

COR. 1. If the series (66) is essentially convergent, and if, either from the very beginning, or after a certain term whose rank does not depend upon h, the terms of (67) are numerically less than the corresponding terms of (66), the limit of V is equal to U.

For in this case the series (67) is more rapidly convergent than (66), and therefore its convergency remains finite.

COR. 2. If the series (66) is essentially convergent, and if the terms of (67) are derived from those of (66) by multiplying them by the ascending powers of a quantity g which is less than 1, and which becomes 1 in the limit, the limit of V is equal to U.

It may be observed that when the convergency of (67) does not become infinitely slow when h vanishes there is no occasion to prove the convergency of (66), since it follows from that of (67). In fact, let V_n be the sum of the first n terms of (67), U_n the same for (66), V_0 the value of V for $h = 0$. Then by hypothesis we may find a finite value of n such that $V - V_n$ shall be numerically less than e, however small h may be; so that

$$V = V_n + \text{ a quantity always numerically less than } e.$$

Now let h vanish: then V becomes V_0 and V_n becomes U_n. Also e may be made as small as we please by taking n sufficiently great. Hence U_n approaches a finite limit when n becomes infinite, and that limit is V_0.

Conversely, if (66) is convergent, and if $U = V_0$, the convergency of the series (67) cannot become infinitely slow when h vanishes.

For if U_n', V_n' represent the sums of the terms after the nth in the series (66), (67) respectively, we have

$$V = V_n + V_n', \quad U = U_n + U_n';$$

whence
$$V_n' = V - U - (V_n - U_n) + U_n'.$$

Now $V - U$, $V_n - U_n$ vanish with h, and U_n' vanishes when n becomes infinite. Hence for a sufficiently small value of h and all inferior values, together with a value of n sufficiently large, and independent of h, the value of V_n' may be made numerically less than any given quantity e however small; and therefore, by definition, the convergency of the series (67) does not become infinitely slow when h vanishes.

On the whole, then, when the convergency of the series (67) does not become infinitely slow when h vanishes, the series (66) is necessarily convergent, and has V_0 for its sum: but in the contrary case there must necessarily be a discontinuity of some kind. Either V must become infinite when h vanishes, or the series (66) must be divergent, or, if (66) is convergent as well as (67), U must be different from V_0.

When a finite function of x, $f(x)$, which passes suddenly from M to N as x increases through a, where $a > a > 0$, is expanded in the series (15) or (30), we have seen that the series is always convergent, and its sum for all values of x except critical values is $f(x)$, and for $x = a$ its sum is $\frac{1}{2}(M + N)$. Hence the convergency of the series necessarily becomes infinitely slow when $a - x$ vanishes. In applying the preceding reasoning to this case it will be observed that h is $a - x$, V_0 is M, and U is $\frac{1}{2}(M + N)$, if we are considering values of x a little less than a; but h is $x - a$ and V_0 is N, if we are considering values of x a little greater than a.

When the series (66) is convergent as well as (67), it may be
easily proved that in all cases

$$U = V_0 - L,$$

where L is the limit of V_n' when h is first made to vanish and then
n to become infinite.

40. Reasoning exactly similar to that contained in the preceding article may be applied to integrals, and the same definitions
may be used. Thus if $\int_a^\infty F(x, h)\, dx$ is a convergent integral, we
may say that the convergency becomes infinitely slow when h
vanishes, when, if X be the superior limit to which we must integrate in order that the neglected part of the integral, or

$$\int_X^\infty F(x, h)\, dx,$$

may be numerically less than a given constant e which may be as
small as we please, X increases beyond all limit when h vanishes.

The reasoning of the preceding article leads to the following
theorems.

If $V = \int_a^\infty F(x, h)\, dx$, if V_0 be the limit of V when $h = 0$, and if
$F(x, 0) = f(x)$; then, if the convergency of the integral V do not
become infinitely slow when h vanishes, $\int_a^\infty f(x)\, dx$ must be convergent, and its value must be V_0. But in the contrary case either
V must become infinite when h vanishes, or the integral

$$\int_a^\infty f(x)\, dx$$

must be divergent, or if it be convergent its value must differ
from V_0.

When the integral $\int_a^\infty f(x)\, dx$ is convergent, if we denote its
value by U, we shall have in all cases

$$U = V_0 - L,$$

when L is the limit to which $\int_X^\infty F(x, h)\, dx$ approaches when h is
first made to vanish and then X to become infinite.

The same remarks which have been made with reference to the convergency of series such as (15) or (30) for values of x near critical values will apply to the convergency of integrals such as (50), (59) or (65).

The question of the convergency or divergency of an integral might arise, not from one of the limits of integration being ∞, but from the circumstance that the quantity under the integral sign becomes infinite within the limits of integration. The reasoning of the preceding article will apply, with no material alteration, to this case also.

41. It may not be uninteresting to consider the bearing of the reasoning contained in this Section on a method frequently given of determining the values of two definite integrals, more especially as the values assigned to the integrals have recently been called into question, on account of their discontinuity.

Consider first the integral

$$u = \int_0^\infty \frac{\sin ax}{x}\, dx \,\ldots\ldots\ldots\ldots\ldots\ldots\ldots (71),$$

where a is supposed positive. Consider also the integral

$$v = \int_0^\infty \epsilon^{-hx} \frac{\sin ax}{x}\, dx.$$

It is easy to prove that the integral v is convergent, and that its convergency does not become infinitely slow when h vanishes. Consequently the integral u is also convergent, (as might also be proved directly in the same way as in the case of v,) and its value is the limit of u for $h = 0$. But we have

$$\frac{dv}{dh} = -\int_0^\infty \epsilon^{-hx} \sin ax\, dx = -\frac{a}{a^2 + h^2};$$

whence

$$v = C - \tan^{-1} \frac{h}{a};$$

and since v evidently vanishes when $h = \infty$, we have $C = \pi/2$, whence

$$v = \frac{\pi}{2} - \tan^{-1} \frac{h}{a}, \quad u = \frac{\pi}{2}.$$

Also $u = 0$ when $a = 0$, and $u = -\pi/2$ when a is negative, since u changes sign with a. By the value of u for $a = 0$, which is asserted to be 0, is of course meant the limit of $\int_0^X \frac{\sin ax}{x}\, dx$ when a is *first* made to vanish and *then* X made infinite.

It is easily proved that the convergency of the integral u becomes infinitely slow when a vanishes. In fact if

$$u' = \int_X^\infty \frac{\sin ax}{x}\, dx,$$

we get by changing the independent variable

$$u' = \int_{aX}^\infty \frac{\sin x}{x}\, dx:$$

but for any given value of X, however great, the value of u' becomes when a vanishes $\int_0^\infty \frac{\sin x}{x}\, dx$, an integral which might have been very easily proved to be greater than zero even had we been unable to find its value. It readily follows from the above that if u' has to be less than e the value of X increases indefinitely as a approaches to zero.

42. Consider next the integrals

$$u = \int_0^\infty \frac{\cos ax\, dx}{1+x^2}, \quad v = \int_0^\infty \epsilon^{-hx} \frac{\cos ax\, dx}{1+x^2} \dots\dots(72).$$

It is easily proved that the convergency of the integral v does not become infinitely slow when h vanishes, whatever be the value of a. Consequently u is in all cases the limit of v for $h=0$. Now v satisfies the equation

$$\frac{d^2 v}{da^2} - v = -\int_0^\infty \epsilon^{-hx} \cos ax\, dx = -\frac{h}{h^2+a^2} \dots\dots(73).$$

It is not however necessary to find the general value of v; for if we put $h=0$ we see that u satisfies the equation

$$\frac{d^2 u}{da^2} - u = 0 \dots\dots\dots (74),$$

so long as a is kept always positive or always negative : but we cannot pass from the value of u found for positive values of a to the value which belongs to negative values of a by merely writing $-a$ for a in the algebraical expression obtained. For although u is a continuous function of a, it readily follows from (73) that $\frac{du}{da}$ is discontinuous. In fact, we have from this equation

$$\left(\frac{dv}{da}\right)_{a=\lambda} - \left(\frac{dv}{da}\right)_{a=-\lambda} = \int_{-\lambda}^\lambda v\, da - 2\tan^{-1}\frac{\lambda}{h}.$$

Now let h first vanish and then λ. Then v becomes u, and $\int_{-\lambda}^{\lambda} v\,da$ vanishes, since v does not become infinite for $a = 0$, whether h be finite or be zero. Therefore du/da is suddenly decreased by π as a increases through zero, as might have been easily proved from the expression for u by means of the known integral (71), even had we been unable to find the value of u in (72). The equation (74) gives, a being supposed positive,

$$u = C\epsilon^{-a} + C'\,\epsilon^{a}.$$

But u evidently does not increase indefinitely with a, and

$$u = \int_0 \frac{dx}{1+x^2} = \frac{\pi}{2} \text{ when } a = 0;$$

whence $C' = 0$, $C = \pi/2$, $u = \pi/2 \,.\, \epsilon^{-a}$. Also, since the numerical value u is unaltered when the sign of a is changed, we have $u = \pi/2 \,.\, \epsilon^{a}$ when a is negative.

It may be observed that if the form of the integral u had been such that we could not have inferred its value for a negative from its value for a positive, nor even known that u is not infinite for $a = -\infty$, we might yet have found its value for a negative by means of the known continuity of u and discontinuity of du/da when a vanishes. For it follows from (74) that $u = C_1 \epsilon^{a} + C_2 \epsilon^{-a}$ for a negative; and knowing already that $u = \pi/2 \,.\, \epsilon^{-a}$ for a positive, we have

$$\frac{\pi}{2} = C_1 + C_2, \quad -\frac{\pi}{2} = C_1 - C_2 - \pi;$$

whence $C_1 = \pi/2$, $C_2 = 0$, $u = \pi/2 \,.\, \epsilon^{a}$, for a negative.

Of course the easiest way of verifying the result $u = \pi/2 \,.\, \epsilon^{-a}$ for a positive is to develope ϵ^{-x} for x positive in a definite integral of the form (59), by means of the formula (60).

SECTION IV.

Examples of the application of the formulæ proved in the preceding Sections.

43. Before proceeding with the consideration of particular examples, it will be convenient to write down the formulæ which

will have to be employed. Some of these formulæ have been proved, and others only alluded to, in the preceding Sections.

In the following formulæ, when series are considered, $f(x)$ is supposed to be a function of x which, as well as each of its derivatives up to the $(\mu - 1)^{\text{th}}$ order inclusive, is continuous between the limits $x = 0$ and $x = a$, and which is expanded between those limits in a series either of sines or of cosines of $\pi x/a$ and its multiples. A_n denotes the coefficient of $\sin n\pi x/a$ when the series is one of sines, B_n the coefficient of $\cos n\pi x/a$ when the series is one of cosines, A_n^{μ} or B_n^{μ} the coefficient of $\sin n\pi x/a$ or $\cos n\pi x/a$ in the expansion of the μth derivative. When integrals are considered $f(x)$ and its first $\mu - 1$ derivatives are supposed to be functions of the same nature as before, which are considered between the limits $x = 0$ and $x = \infty$; and it is moreover supposed that $f(x)$ decreases as x increases to ∞, sufficiently fast to allow $\int f(x)\, dx$ to be essentially convergent at the limit ∞, or else that $f(x)$ vanishes when $x = \infty$, and after a finite value of x never changes from increasing to decreasing nor from decreasing to increasing. $\phi(\beta)$ or $\psi(\beta)$ denotes the coefficient of $\sin \beta x$ or $\cos \beta x$ in the developement of $f(x)$ in a definite integral of the form $\int_0^{\infty} \phi(\beta) \sin \beta x\, dx$ or $\int_0^{\infty} \psi(\beta) \cos \beta x\, dx$, $\phi_\mu(\beta)$ or $\psi_\mu(\beta)$ denotes the coefficient of $\sin \beta x$ or $\cos \beta x$ in the developement of the μth derivative of $f(x)$. The formulæ are

$$(-1)^{\frac{\mu-1}{2}} B_n^{\mu} = \left(\frac{n\pi}{a}\right)^{\mu} A_n - \frac{2}{a}\left(\frac{n\pi}{a}\right)^{\mu-1}\{f(0) - (-1)^n f(a)\}$$

$$+ \frac{2}{a}\left(\frac{n\pi}{a}\right)^{\mu-3}\{f''(0) - (-1)^n f''(a)\} - \dots (\mu \text{ odd})\dots(A),$$

$$(-1)^{\frac{\mu}{2}} A_n^{\mu} = \left(\frac{n\pi}{a}\right)^{\mu} A_n$$

$$- \frac{2}{a}\left(\frac{n\pi}{a}\right)^{\mu-1}\{f(0) - (-1)^n f(a)\} + \dots (\mu \text{ even})\dots(B),$$

$$(-1)^{\frac{\mu+1}{2}} A_n^{\mu} = \left(\frac{n\pi}{a}\right)^{\mu} B_n$$

$$+ \frac{2}{a}\left(\frac{n\pi}{a}\right)^{\mu-2}\{f'(0) - (-1)^n f'(a)\} - \dots (\mu \text{ odd})\dots(C),$$

$$(-1)^{\frac{\mu}{2}} B_n{}^\mu = \left(\frac{n\pi}{a}\right)^\mu B_n$$

$$+ \frac{2}{a} \left(\frac{n\pi}{a}\right)^{\mu-2} \{f'(0) - (-1)^n f'(a)\} - \dots \ (\mu \text{ even})\dots..(D),$$

except when $n = 0$, in which case we have always

$$B_0{}^\mu = \frac{1}{a} \{ f^{\mu-1}(a) - f^{\mu-1}(0) \},$$

B_0 being the constant term in the expansion of $f^\mu(x)$ in a series of cosines. In the formulæ (A), (B), (C), (D) we must stop when we have written the term containing the power 1 or 0, (as the case may be,) of $n\pi/a$.

The formulæ for integrals are

$$(-1)^{\frac{\mu-1}{2}} \psi_\mu(\beta) = \beta^\mu \phi(\beta) - \frac{2}{\pi} \beta^{\mu-1} f(0)$$

$$+ \frac{2}{\pi} \beta^{\mu-3} f''(0) - \dots \ (\mu \text{ odd})\dots\dots\dots(a),$$

$$(-1)^{\frac{\mu}{2}} \phi_\mu(\beta) = \beta^\mu \phi(\beta) - \frac{2}{\pi} \beta^{\mu-1} f(0)$$

$$+ \frac{2}{\pi} \beta^{\mu-3} f''(0) - \dots \ (\mu \text{ even})\dots\dots\dots..(b),$$

$$(-1)^{\frac{\mu+1}{2}} \phi_\mu(\beta) = \beta^\mu \psi(\beta) + \frac{2}{\pi} \beta^{\mu-2} f'(0)$$

$$- \frac{2}{\pi} \beta^{\mu-4} f'''(0) + \dots \ (\mu \ \text{ odd}) \dots\dots..(c),$$

$$(-1)^{\frac{\mu}{2}} \psi_\mu(\beta) = \beta^\mu \psi(\beta) + \frac{2}{\pi} \beta^{\mu-2} f'(0)$$

$$- \frac{2}{\pi} \beta^{\mu-4} f'''(0) + \dots \ (\mu \text{ even})\dots\dots\dots(d),$$

where we must stop with the last term involving a positive power of β or the power zero.

44. As a first example of the application of the principles contained in Sections I. and II. suppose that we have to determine the value of ϕ for values of x lying between 0 and a, 0 and b respectively, from the equation

$$\frac{d^2\phi}{dx^2} + \frac{d^2\phi}{dy^2} = 0 \dots\dots\dots \dots\dots\dots(75),$$

with the particular conditions

$$\frac{d\phi}{dy} = \omega\,(x - \tfrac{1}{2}a), \quad \text{when } y = 0 \text{ or } = b \ldots\ldots\ldots(76),$$

$$\frac{d\phi}{dx} = -\omega\,(y - \tfrac{1}{2}b), \quad \text{when } x = 0 \text{ or } = a \ldots\ldots\ldots(77).$$

This is the problem in pure analysis to which we are led in seeking to determine the motion of a liquid within a closed rectangular box which is made to oscillate.

For a given value of y, the value of ϕ can be expanded in a convergent series of cosines of $\pi x/a$ and its multiples; for another value of y, ϕ can be expanded in a similar series with different coefficients, and so on. Hence, in general, ϕ can be expanded in a convergent series of the form

$$\Sigma Y_n \cos \frac{n\pi x}{a}\,,\ldots\ldots\ldots\ldots\ldots\ldots\ldots\ldots(78),$$

where Y_n is a certain function of y, which has to be determined.

In the first place the value of ϕ given by (78) must satisfy (75). Now the direct developement of $d^2\phi/dy^2$ in a series of cosines will be obtained from (78) by differentiating under the sign of summation; the direct developement of $d^2\phi/dx^2$ will be given by the formula (D). We thus get

$$\Sigma\left[\frac{d^2 Y_n}{dy^2} - \frac{n^2\pi^2}{a^2} Y_n + \frac{2\omega}{a}\{1 - (-1)^n\}(y - \tfrac{1}{2}b)\right]\cos\frac{n\pi x}{a} = 0\,;$$

and the left-hand member of this equation being the result of directly developing the right-hand member in a series of cosines, we have

$$\frac{d^2 Y_n}{dy^2} - \frac{n^2\pi^2}{a^2} Y_n = -\frac{4\omega}{a}(y - \tfrac{1}{2}b) \text{ or } = 0,$$

according as n is odd or even. This equation is easily integrated, and the integral contains two arbitrary constants, C_n, D_n, suppose. It only remains to satisfy (76). Now the direct developement of dY_n/dy will be obtained by differentiating under the sign of summation, and the direct developement of $\omega\,(x - \tfrac{1}{2}a)$ is easily found to be $-\Sigma_0\,4\omega a/\pi^2 n^2 \cdot \cos n\pi x/a$, the sign Σ_0 denoting that odd values only of n are to be taken. We have then, both for

s.   19

$y = 0$ and for $y = b$,

$$\frac{dY_n}{dy} = -\frac{4\omega a}{\pi^2 n^2} \text{ or } = 0,$$

according as n is odd or even, which determines C_n and D_n.

It is unnecessary to write down the result, because I have already given it in a former paper*, where it is obtained by considerations applicable to this particular problem. The result is contained in equation (4) of that paper. The only step of the process which I have just indicated which requires notice is, that the term containing $(x - \tfrac{1}{2}a)(y - \tfrac{1}{2}b)$ at first appears as an infinite series, which may be summed by the formula (41). The present example is a good one for shewing the utility of the methods contained in the present paper, inasmuch as in the Supplement referred to I have pointed out the advantage of the formula contained in equation (6), with respect to facility of numerical calculation, over one which I had previously arrived at by using developements, in series of cosines, of functions whose derivatives vanish for the limiting values of the variable.

45. Let it be required to determine the permanent state of temperature in a rectangle which has two of its opposite edges kept up to given temperatures, varying from point to point, while the other edges radiate into a space at a temperature zero. The rectangle is understood to be a section of a rectangular bar of infinite length, which has all the points situated in the same line parallel to the axis at the same temperature, so that the propagation of heat takes place in two dimensions.

Let the rectangle be referred to the rectangular axes of x, y, the axis of y coinciding with one of the edges whose temperature is given, and the origin being in the middle point of the edge. Let the unit of length be so chosen that the length of either edge parallel to the axis of x shall be π, and let 2β be the length of each of the other edges. Let u be the temperature at the point (x, y), h the ratio of the exterior, to the interior conductivity. Then we have

$$\frac{d^2u}{dx^2} + \frac{d^2u}{dy^2} = 0 \dots\dots\dots(79),$$

$$\frac{du}{dy} - hu = 0, \text{ when } y = -\beta \dots\dots\dots(80),$$

---

* Supplement to a Memoir 'On some Cases of Fluid Motion,' p. 409 of the present Volume [*Ante*, p. 188].

$$\frac{du}{dy} + hu = 0, \text{ when } y = \beta \dots\dots(81),$$

$$u = f(y), \text{ when } x = 0 \dots \dots\dots(82),$$

$$u = F(y), \text{ when } x = a \dots\dots\dots(83),$$

$f(y)$, $F(y)$ being the given temperatures of two of the edges.

According to the method by which Fourier has solved a similar problem, we should first take a particular function $Y\epsilon^{\lambda x}$, where Y is a function of y, and restrict it to satisfy (79). This gives $Y = A \cos \lambda y + B \sin \lambda y$, A and B being arbitrary constants. We may of course take, still satisfying (79), the sum of any number of such functions. It will be convenient to take together the functions belonging to two values of λ which differ only in sign. We may therefore take, by altering the arbitrary constants,

$$u = \Sigma \left\{ A \left(\epsilon^{\lambda(\pi-x)} - \epsilon^{-\lambda(\pi-x)}\right) + B \left(\epsilon^{\lambda x} - \epsilon^{-\lambda x}\right)\right\} \cos \lambda y,$$

$$+ \Sigma \left\{ C \left(\epsilon^{\lambda(\pi-x)} - \epsilon^{-\lambda(\pi-x)}\right) + D \left(\epsilon^{\lambda x} - \epsilon^{-\lambda x}\right)\right\} \sin \lambda y \dots..(84),$$

in which expression it will be sufficient to take only one of two values of λ which differ only by sign, so that λ, if real, may be taken positive. Substituting now in (80) and (81) the value of u given by (84), we get either $C = 0$, $D = 0$, and

$$\lambda\beta \cdot \tan \lambda\beta = h\beta \dots\dots\dots(85),$$

or else $A = 0$, $B = 0$, and

$$\lambda\beta \cdot \cot \lambda\beta = - h\beta \dots\dots\dots(86).$$

It is easy to prove that the equation (85), in which $\lambda\beta$ is regarded as the unknown quantity, has an infinite number of real positive roots lying between each even multiple of $\pi/2$, including zero, and the next odd multiple. The equation (86) has also an infinite number of real positive roots lying between each odd multiple of $\pi/2$ and the next even multiple. The negative roots of (85) and (86) need not be considered, since the several negative roots have their numerical values equal to those of the positive roots; and it may be proved that the equations do not admit of imaginary roots. The values of λ in (84) must now be restricted to be those given by (85) for the first line, and those given by (86) for the second. It remains to satisfy (82) and (83). Now let

$$f(y) + f(-y) = 2f_1(y), \quad f(y) - f(-y) = 2f_2(y),$$
$$F(y) + F(-y) = 2F_1(y), \quad F(y) - F(-y) = 2F_2(y):$$

19—2

then we must have for all values of y from 0 to β, and therefore for all values from $-\beta$ to 0,

$$\Sigma AL \cos \lambda y = f_1(y), \quad \Sigma BL \cos \lambda y = F_1(y)\ldots\ldots\ldots(87),$$

$$\Sigma CM \sin \mu y = f_2(y), \quad \Sigma DM \sin \mu y = F_2(y)\ldots\ldots\ldots(88),$$

where $\qquad L = \epsilon^{\lambda\pi} - \epsilon^{-\lambda\pi}, \quad M = \epsilon^{\mu\pi} - \epsilon^{-\mu\pi},$

μ denoting one of the roots of the equation

$$\mu\beta \, . \, \cot \mu\beta = -h\beta \ldots\ldots\ldots\ldots\ldots\ldots(89),$$

and the two signs Σ extending to all the positive roots of the equations (85), (89), respectively. To determine A and B, multiply both sides of each of the equations (87) by $\cos \lambda' y \, dy$, λ' being any root of (85), and integrate from $y = 0$ to $y = \beta$. The integral at the first side will vanish, by virtue of (85), except when $\lambda' = \lambda$, in which case it will become $1/4\lambda \, . \, (2\lambda\beta + \sin 2\lambda\beta)$, whence A and B will be known. C and D may be determined in a similar manner by multiplying both sides of each of the equations (88) by $\sin \mu' y \, dy$, μ' being any root of (89), integrating from $y = 0$ to $y = \beta$, and employing (89). We shall thus have finally

$$u = 4\Sigma\lambda \, (2\lambda\beta + \sin 2\lambda\beta)^{-1} \, (\epsilon^{\lambda\pi} - \epsilon^{-\lambda\pi})^{-1} \left\{ (\epsilon^{\lambda(\pi-x)} - \epsilon^{-\lambda(\pi-x)}) \int_0^\beta f_1(y) \right.$$

$$\cos \lambda y \, dy + (\epsilon^{\lambda x} - \epsilon^{-\lambda x}) \int_0^\beta F_1(y) \cos \lambda y \, dy \bigg\} \cos \lambda y,$$

$$+ 4\Sigma\mu \, (2\mu\beta - \sin 2\mu\beta)^{-1} \, (\epsilon^{\mu\pi} - \epsilon^{-\mu\pi})^{-1} \left\{ (\epsilon^{\mu(\pi-x)} - \epsilon^{-\mu(\pi-x)}) \int_0^\beta f_2(y) \right.$$

$$\sin \mu y \, dy + (\epsilon^{\mu x} - \epsilon^{-\mu x}) \int_0^\beta F_2(y) \sin \mu y \, dy \bigg\} \sin \mu y \ldots(90).$$

46.   Such is the solution obtained by a method similar to that employed by Fourier. A solution very different in appearance may be obtained by expanding u in a series $\Sigma Y \sin nx$, and employing the formula (B). We thus get from the equation (79)

$$\frac{d^2 Y}{dy^2} - n^2 Y + \frac{2n}{\pi} \left\{ f(y) - (-1)^n F(y) \right\} = 0,$$

which gives

$$Y = A\epsilon^{ny} + B\epsilon^{-ny} - \frac{1}{\pi} \int_0^y \left\{ f(y') - (-1)^n F(y') \right\} (\epsilon^{n(y-y')} - \epsilon^{-n(y-y')}) \, dy';$$

whence, $du/dy = \Sigma\, Y' \sin nx$, where

$$Y' = nA\epsilon^{ny} - nB\epsilon^{-ny} - \frac{n}{\pi}\int_0^y \{f'(y') - (-1)^n F'(y')\}\, (\epsilon^{n(y-y')}$$
$$+ \epsilon^{-n(y-y')})\, dy'.$$

The values of A and B are to be determined by (80) and (81), which require that

$$\frac{dY}{dy} \pm hY = 0 \text{ when } y = \pm\,\beta.$$

We thus get

$$(n+h)\,\epsilon^{n\beta}\, A - (n-h)\,\epsilon^{-n\beta}\, B - \frac{1}{\pi}\int_0^\beta \{f(y') - (-1)^n F(y')\}$$
$$\{(n+h)\,\epsilon^{n(\beta-y')} + (n-h)\,\epsilon^{-n(\beta-y')}\}\, dy' = 0,$$

and the equation derived from this by changing the signs of h and β; whence the values of A and B may be found. We get finally

$$u = \Sigma\, Y \sin nx \dots\dots\dots\dots\dots\dots(91),$$

where

$$Y = \frac{1}{\pi}\{(n+h)\,\epsilon^{n\beta} - (n-h)\,\epsilon^{-n\beta}\}^{-1}\,(\epsilon^{ny} + \epsilon^{-ny})\int_0^\beta \{(n+h)\,\epsilon^{n(\beta-y')}$$
$$+ (n-h)\,\epsilon^{-n(\beta-y')}\}\,\{f_1(y') - (-1)^n F_1(y')\}\, dy'$$
$$- \frac{1}{\pi}\int_0^y (\epsilon^{n(y-y')} - \epsilon^{-n(y-y')})\,\{f_1(y') - (-1)^n F_1(y')\}\, dy'$$
$$+ \frac{1}{\pi}\{(n+h)\,\epsilon^{n\beta} + (n-h)\,\epsilon^{-n\beta}\}^{-1}(\epsilon^{ny} - \epsilon^{-ny})\int_0^\beta \{(n+h)\,\epsilon^{n(\beta-y')}$$
$$+ (n-h)\,\epsilon^{-n(\beta-y')}\}\,\{f_2(y') - (-1)^n F_2(y')\}\, dy'$$
$$- \frac{1}{\pi}\int_0^y (\epsilon^{n(y-y')} - \epsilon^{-n(y-y')})\,\{f_2(y') - (-1)^n F_2(y')\}\, dy'\dots(92).$$

47. The two expressions for u given, one by (90), and the other by (91) and (92), are necessarily equal for values of x and y lying between the limits 0 and π, $-\beta$ and β respectively. They are also equal for the limiting values $y = -\beta$ and $y = \beta$, but not for the limiting values $x = 0$ and $x = \pi$, since for these values (91) fails; that is to say, in order to find from this series the value of u for $x = 0$ or $x = \pi$, we should have *first* to sum the series, and *then* put $x = 0$ or $x = \pi$.

The comparison of these expressions leads to two remarkable formulæ. In the first place it will be observed that the first and second portions of the right-hand side of (92) are unchanged when y changes sign, while the third and fourth portions change sign with y. This is obvious with respect to the first and third portions, and may be easily proved with respect to the second and fourth by taking $-y'$ instead of y' for the variable with respect to which the integration is performed, and remembering that $f_1(y)$, $F_1(y)$ are unchanged, and $f_2(y)$, $F_2(y)$ change sign, when y changes sign. Consequently the part of u corresponding to the first two portions of (92) is equal to the part expressed by the first two lines in (90), and the part corresponding to the last two portions of (92) equal to the part expressed by the last two lines in (90). Hence the equation obtained by equating the two expressions for u splits into two; and each of the new equations will again split into two in consequence of the independence of the functions f, F, which are arbitrary from $y = 0$ to $y = \beta$. As far however as anything peculiar in the transformations is concerned, it is evident that we may suppress one of the functions f, F, suppose F, and consider only an element of the integral by which f is developed, or, which is the same, suppose $f_1(y')$ or $f_2(y')$ to be zero except for values of the variable infinitely close to a particular value y', and divide both sides of the equation by

$$\int f_1(y')\, dy' \quad \text{or} \quad \int f_2(y')\, dy'.$$

We get thus from the first two lines of (90) and the first two portions of (92), supposing y and y' positive, and y' the greater of the two,

$$\Sigma \frac{4\lambda}{2\lambda\beta + \sin 2\lambda\beta} \frac{\epsilon^{\lambda(\pi-x)} - \epsilon^{-\lambda(\pi-x)}}{\epsilon^{\lambda\pi} - \epsilon^{-\lambda\pi}} \cos \lambda y \cos \lambda y'$$

$$= \frac{1}{\pi} \Sigma \frac{(\epsilon^{ny} + \epsilon^{-ny}) \{(n+h)\, \epsilon^{n(\beta-y')} + (n-h)\, \epsilon^{-n(\beta-y')}\}}{(n+h)\, \epsilon^{n\beta} - (n-h)\, \epsilon^{-n\beta}} \sin nx \dots (93),$$

where the first Σ refers to the positive roots of (85), and the second to positive integral values of n from 1 to ∞.

Of course, if y become greater than y', y and y' will have to change places in the second side of (93). This is in accordance with the formula (92), since now the second line does not vanish; and it will easily be found that the first and second lines together give the same result as if we had at once made y and y' change places. Although y has been supposed positive in (93), it is easily seen that it may be supposed negative, provided it be numerically less than y'.

The other formula above alluded to is obtained in a manner exactly similar by comparing the last two portions of (92) with the last two lines in (90). It is

$$\Sigma \frac{4\mu}{2\mu\beta - \sin 2\mu\beta} \frac{\epsilon^{\mu(\pi-x)} - \epsilon^{-\mu(\pi-x)}}{\epsilon^{\mu\pi} - \epsilon^{-\mu\pi}} \sin \mu y \sin \mu y'$$

$$= \frac{1}{\pi} \Sigma \frac{(\epsilon^{ny} - \epsilon^{-ny})\{(n+h)\,\epsilon^{n(\beta-y')} + (n+h)\,\epsilon^{-n(\beta-y')}\}}{(n+h)\,\epsilon^{n\beta} + (n-h)\,\epsilon^{-n\beta}} \sin nx \dots (94),$$

where the first Σ refers to the positive roots of (89), the second to positive integral values of n, and where x is supposed to lie between 0 and π, y' between 0 and β, y between 0 and y', or, it may be, between $-y'$ and y'. Although x has been supposed less than π, it may be observed that the formulæ (93), (94) hold good so long as x, being positive, is less than 2π.

48. Let it be required to determine the permanent state of temperature in a homogeneous rectangular parallelepiped, supposing the surface kept up to a given temperature, which varies from point to point.

Let the origin be in one corner of the parallelepiped, and let the adjacent edges be taken for the axes of x, y, z. Let a, b, c be the lengths of the edges; $f_1(y, z)$, $F_1(y, z)$, the given temperatures of the faces for which $x = 0$ and $x = a$ respectively; $f_2(z, x)$, $F_2(z, x)$ the same for the faces perpendicular to the axis of y; $f_3(x, y)$, $F_3(x, y)$ the same for those perpendicular to the axis of z. Then if we put for shortness ∇ to denote the operation otherwise denoted by

$$\frac{d^2}{dx^2} + \frac{d^2}{dy^2} + \frac{d^2}{dz^2},$$

as will be done in the rest of this paper, and write only the characteristics of the functions, we shall have, to determine the temperature u, the general equation $\nabla u = 0$ with the particular conditions

$$u = f_1, \text{ when } x = 0; \quad u = F_1, \text{ when } x = a \dots\dots (95);$$

$$u = f_2, \text{ when } y = 0; \quad u = F_2, \text{ when } y = b \dots\dots (96);$$

$$u = f_3, \text{ when } z = 0; \quad u = F_3, \text{ when } z = c \dots\dots (97);$$

It is evident that u is the sum of three temperatures u_1, u_2, u_3, where u_1 satisfies the conditions (95), and vanishes at the four remaining faces, and u_2, u_3 are related to the axes of y, z as u_1 is related to that of x, each of the quantities u_1, u_2, u_3 representing a possible permanent temperature. Now u_3 may be expanded in a double series $\Sigma\Sigma Z_{mn} \sin m\pi x/a \cdot \sin n\pi y/b$, where Z_{mn} is a function of z which has to be determined. Let for shortness

$$\frac{m\pi}{a} = \mu, \quad \frac{n\pi}{b} = \nu, \quad \frac{p\pi}{c} = \varpi;$$

then the substitution of the above value of u_3 in the equation $\nabla u_3 = 0$ leads to the equation

$$\frac{d^2 Z_{mn}}{dz^2} - q^2 Z_{mn} = 0,$$

where $q^2 = \mu^2 + \nu^2$, which gives $Z_{mn} = A_{mn}\epsilon^{qz} + B_{mn}\epsilon^{-qz}$; and the constants A_{mn}, B_{mn} are easily determined by the condition (97). We may find u_1 and u_2 in a similar manner, and the sum of the results gives u. It is thus that such problems are usually solved.

We may, however, expand u in a series of the form

$$\Sigma\Sigma Z_{mn} \sin \mu x \sin \nu y,$$

even though it does not vanish for $x = 0$ and $x = a$, and for $y = 0$ and $y = b$; and the formulæ proved in Section I. enable us to make use of this expansion.

Let then $u = \Sigma\Sigma Z \sin \mu x \sin \nu y,$

the suffixes of Z being omitted for the sake of simplicity. We have by the formula (B)

$$\frac{d^2 u}{dx^2} = \Sigma\left\{-\mu^2 \Sigma Z \sin \nu y + \frac{2\mu}{a}[f_1 - (-1)^m F_1]\right\} \sin \mu x.$$

Let $f_1(y, z) - (-1)^m F_1(y, z)$ be expanded in the series $\Sigma Q \sin \nu y$ by the formula (3), so that Q will be a known function of z, m, and n. Then

$$\frac{d^2 u}{dx^2} = \Sigma\Sigma\left\{-\mu^2 Z + \frac{2u}{a}Q\right\} \sin \mu x \sin \nu y.$$

The value of $d^2 u/dy^2$ may be expressed in a similar manner, and that of $d^2 u/dz^2$ is found by direct differentiation. We have thus, for the direct developement of ∇u, the double series

$$\Sigma\Sigma\left\{\frac{d^2 Z}{dz^2} - (\mu^2 + \nu^2)Z + \frac{2\nu}{b}P + \frac{2\mu}{a}Q\right\} \sin \mu x \sin \nu y,$$

where P is for x what Q is for y. The above series being the direct developement of ∇u, and ∇u being equal to zero, each coefficient must be equal to zero, which gives

$$\frac{d^2Z}{dz^2} - q^2Z + \frac{2\nu}{b}P + \frac{2\mu}{a}Q = 0 \dots\dots\dots\dots(98),$$

where q means the same as before. The integral of the equation (98) is

$$Z = A\epsilon^{qz} + B\epsilon^{-qz} - \frac{1}{q}\epsilon^{qz}\int_0^z \epsilon^{-qz}T\,dz + \frac{1}{q}\epsilon^{-qz}\int_0^z \epsilon^{qz}T\,dz,$$

$2T$ denoting the sum of the last two terms of (98). It only remains to satisfy (97). If the known functions $f_3(x, y)$, $F_3(x, y)$ be developed in the double series $\Sigma\Sigma G \sin \mu x \sin \nu y$, $\Sigma\Sigma H \sin \mu x \sin \nu y$, we shall have from (97)

$$A + B = G,$$

$$A\epsilon^{qc} + B\epsilon^{-qc} - \frac{1}{q}\epsilon^{qc}\int_0^c \epsilon^{-qz}T\,dz + \frac{1}{q}\epsilon^{-qc}\int_0^c \epsilon^{qz}T\,dz = H.$$

A and B may be easily found from these equations, and we shall have finally

$$(\epsilon^{qc} - \epsilon^{-qc})Z = G(\epsilon^{q(c-z)} - \epsilon^{-q(c-z)}) + H(\epsilon^{qz} - \epsilon^{-qz})$$

$$+ \frac{1}{q}(\epsilon^{q(c-z)} - \epsilon^{-q(c-z)})\int_0^z (\epsilon^{qz'} - \epsilon^{-qz'})\,T'\,dz'$$

$$+ \frac{1}{q}(\epsilon^{qz} - \epsilon^{-qz})\int_z^c (\epsilon^{q(c-z')} - \epsilon^{-q(c-z')})\,T'\,dz',$$

T' being the value of T when $z = -z'$. It will be observed that the letters Z, P, Q, T, A, B, G, H ought properly to be affected with the double suffix mn. It would be useless to write down the expression for u in terms of the known quantities $f_1(y, z)$, &c.

It will be observed that u might equally have been expressed by means of the double series $\Sigma\Sigma X_{np} \sin \nu y \sin \varpi z$, or $\Sigma\Sigma Y_{mp} \sin \mu x \sin \varpi z$, where p is any integer. We should thus have three different expressions for the same quantity u within the limits $x = 0$ and $x = a$, $y = 0$ and $y = b$, $z = 0$ and $z = c$. The comparison of these three expressions when particular values are assigned to the known functions $f_1(y, z)$ &c. would lead to remarkable transformations. The expressions differ however in one respect which deserves notice. Their numerical values are the same for values of the variables lying within the limits 0 and a, 0 and b, 0 and c. The first expression holds good for the extreme values of z, but fails for those of x and y: in other words, in order to find from the series the value of u for the face considered, instead of first giving x or y its extreme value and then summing, which would lead to a result zero, we should first have to sum with respect to m or n, or conceive the summation performed, and then give x or y its extreme value. The same remarks apply, *mutatis mutandis*, to the second and third expressions; so that the three expressions are not equivalent if we take in the extreme values of the variables.

49.   Many other remarkable transformations might be obtained from those already referred to by differentiation and integration. We might for instance compare the three expressions which would be obtained for $\int_0^a \int_0^b \int_0^c u\,dx\,dy\,dz$, and we should thus have three different expressions for the same function of the three independent variables a, b, c, which are supposed to be positive, but may be of any magnitudes.   Some examples of the results of transformations of this kind may be seen by comparing the formulæ obtained in the Supplement alluded to in Art. 44 with the corresponding formulæ contained in the Memoir itself to which the Supplement has been added.   Such transformations, however, when separated from physical problems, are more curious than useful.   Nevertheless, it may be worth while to exhibit in its simplest shape the formula from which they all flow, so long as we restrict ourselves to a function u satisfying the equation $\nabla u = 0$, and expanded between the limits $x = 0$ and $x = a$, &c. in a double series of sines.

The functions $f_1(y, z)$ &c., which are supposed known, are arbitrary, and enter into the expression for u under the sign of double integration.   Consequently we shall not lose generality, so far as anything peculiar in the transformations is concerned, by considering only one element of the integrals by which one of the functions is developed.   Let then all the functions be zero except f_3; and since in the process f_3 has to be developed in the double series

$$\frac{4}{ab} \Sigma\Sigma \int_0^a \int_0^b f_3(x', y') \sin \mu x' \sin \nu y' dx' dy' . \sin \mu x \sin \nu y,$$

consider only the element $f_3(x', y') \sin \mu x' \sin \nu y' dx' dy'$ of the double integral, omit the $dx' dy'$, and put $f_3(x', y') = 1$ for the sake of simplicity. If we adopt the first expansion of u, and put q^2 for $\mu^2 + \nu^2$, we shall have

$$Z = A\left(\epsilon^{q(c-z)} - \epsilon^{-q(c-z)}\right), \quad \left(\epsilon^{qc} - \epsilon^{-qc}\right)A = \frac{4}{ab}\sin \mu x' \sin \nu y';$$

whence

$$u = \frac{4}{ab}\Sigma\Sigma\,\frac{\epsilon^{q(c-z)} - \epsilon^{-q(c-z)}}{\epsilon^{qc} - \epsilon^{-qc}}\sin \mu x' \sin \nu y' \sin \mu x \sin \nu y \ldots (99).$$

By expanding u in the double series $\Sigma\Sigma\,Y\sin \mu x \sin \varpi z$ we should get

$$u = \frac{2}{ac}\Sigma\Sigma\,\frac{\varpi}{s}\,\frac{\left(\epsilon^{sy} - \epsilon^{-sy}\right)\left(\epsilon^{s(b-y')} - \epsilon^{-s(b-y')}\right)}{\epsilon^{sb} - \epsilon^{-sb}}\sin \mu x' \sin \mu x \sin \varpi z$$
$$\ldots\ldots(100),$$

where $s^2 = \mu^2 + \varpi^2$, and y' is the greater of the two y, y'. The third expansion would be derived from the second by interchanging the requisite quantities. In these formulæ z may have any positive value less than $2c$.

We should get in a similar manner in the case of two variables x, y

$$u = \frac{2}{b}\Sigma\,\frac{\epsilon^{\nu(a-x)} - \epsilon^{-\nu(a-x)}}{\epsilon^{\nu a} - \epsilon^{-\nu a}}\sin \nu y' \sin \nu y$$

$$= \frac{1}{a}\Sigma\,\frac{\left(\epsilon^{\mu y} - \epsilon^{-\mu y}\right)\left(\epsilon^{\mu(b-y')} - \epsilon^{-\mu(b-y')}\right)}{\epsilon^{\mu b} - \epsilon^{-\mu b}}\sin \mu x \ldots\ldots\ldots\ldots(101),$$

where x is supposed to lie between 0 and a, y' between 0 and b, and y between 0 and y'. This formula is however true so long as x lies between 0 and $2a$, and y between $-y'$ and y'.

If we compare the two expressions for $\int_0^b\int_0^b\int_0^a u\,dy\,dy'\,dx$ obtained from (101), taking Σ_0 for the sign of summation corresponding to odd values of n from 1 to ∞, putting $a = rb$, and replacing $\Sigma_0\,1/n^2$ by its value $\pi^2/8$, we shall get the formula

$$\frac{1}{r}\Sigma_0\,\frac{1}{n^3}\,\frac{1 - \epsilon^{-n\pi r}}{1 + \epsilon^{-n\pi r}} + r\Sigma_0\,\frac{1}{n^3}\,\frac{1 - \epsilon^{-n\pi/r}}{1 + \epsilon^{-n\pi/r}} = \frac{\pi^3}{16}\ \ldots\ldots(102),$$

which is true for all positive values of r, and likewise for all negative values, since the left-hand side of (102) is not changed when $-r$ is put for r. In integrating the second side of (101), supposing that we integrate for y before integrating for y', we must integrate separately from $y = 0$ to $y = y'$, and from $y = y'$

to $y = b$, since the algebraical expression of the quantity to be integrated changes when y passes the value y'.

It would be useless to go on with these transformations, which may be multiplied to any extent, and which cease to be useful when they are separated from physical problems to which they relate, and of which we wish to obtain solutions.

It may be observed that instead of supposing, in the case of the parallelepiped, the value of u known for all points of the surface, we might have supposed the value of the flux known, subject of course to the condition that the total flux shall be zero. This would correspond to the following problem in fluid motion, u taking the place of the quantity usually denoted by ϕ, "To determine the initial motion at any point of a homogeneous incompressible fluid contained in a closed vessel of the form of a rectangular parallelepiped, which it completely fills, supposing the several points of the surface of the vessel suddenly moved in any manner consistent with the condition that the volume be not changed." In this case we should expand u in a series of cosines instead of sines, and employ the formula (D) instead of (B). We might, again, suppose the value of u known for the faces perpendicular to one or two of the axes, and the value of the flux known for the remaining faces. In this case we should employ sines involving the co-ordinates perpendicular to the first set of faces, and cosines involving the others.

The formulæ would also be modified by supposing some one or more of the faces to move off to an infinite distance. In this case some of the series would be replaced by integrals. Thus, in the case in which the value of u at the surface is known, if we supposed a to become infinite we should employ the integral (50) instead of the series (3), as far as relates to the variable x, and the formula (b) instead of (B). If we were considering a rectangular bar infinitely extended both ways we should employ the integral (65). Of course, if we had already obtained the result for the case of the parallelepiped, the shortest way would be thence to deduce the result for the case of the bar infinite in one or in both directions, but if we began with considering the bar it would be best to start with the integrals (50) or (65).

50. To give one example of transformations of this kind, let us suppose b to become infinite in (101). Observing that

$\nu = n\pi/b$, $\Delta\nu = \pi/b$, we get on passing to the limit

$$\frac{2}{\pi}\int_0^\infty \frac{\epsilon^{\nu(a-x)} - \epsilon^{-\nu(a-x)}}{\epsilon^{\nu a} - \epsilon^{-\nu a}}\, \sin \nu y'\, \sin \nu y\, d\nu$$

$$= \frac{1}{a}\, \Sigma\, (\epsilon^{\mu y} - \epsilon^{-\mu y})\, \epsilon^{-\mu y'}\, \sin \mu x \dots\dots(103).$$

Multiply both sides of this equation by $dx\, dy$, and integrate from $x = 0$ to $x = a$, and from $y = 0$ to $y = \infty$. With respect to the integration of the second side, it is only necessary to remark that when y becomes greater than y', y and y' must be made to change places in the expression written down in (103). As to the integration of the first side, if we first integrate from $y = 0$ to $y = Y$, we get, putting $f(\nu, x)$ for the fraction involving x,

$$\frac{2}{\pi}\int_0^\infty f(\nu, x)\, \sin \nu y'\, (1 - \cos \nu Y)\, \frac{d\nu}{\nu}.$$

Now let Y become infinite; then the term involving $\cos \nu Y$ may be omitted, not because $\cos \nu Y$ vanishes when Y becomes infinite, which is not true, but because, as may be rigorously proved, the integral in which it occurs vanishes when Y becomes infinite. If we write 1 for a, as we may without loss of generality, we get finally

$$\int_0^\infty \frac{1 - \epsilon^{-\nu}}{1 + \epsilon^{-\nu}}\, \sin \nu y'\, \frac{d\nu}{\nu^2} = \frac{2}{\pi}\, \Sigma_0\, \frac{1}{n^2}\, (1 - \epsilon^{-n\pi y'}) \dots\dots(104).$$

51. Hitherto in satisfying the general equation $\nabla u = 0$, together with the particular conditions at the surface, the value of u has been expanded in a double series involving two of the variables, and the functions of the third variable which enter as coefficients into the double series have been determined by an ordinary differential equation such as (98). We might however expand u in a triple series, and thus satisfy at the same time the equation $\nabla u = 0$ and the conditions at the surface, without using an ordinary differential equation at all, but simply by means of the terms introduced into the series by differentiation, which are given by the formulæ at the beginning of this Section; and then by summing the triple series once, which may be done in any one of three ways, we should arrive at the same results as if we had employed in succession three double series, involving circular functions of x and y, y and z, z and x respectively, and the corresponding ordinary differential equations. I am indebted

for this method to my friend Prof. William Thomson, to whom I shewed the method given in Art. 48.

Let us take the case of the permanent state of temperature in a rectangular parallelepiped, supposing the temperature at the several points of the surface known. For more simplicity suppose the temperature zero at the surface, except infinitely close to the point (x', y') in the face for which $z = 0$, so that all the functions f, &c. are zero, except $f_3(x, y)$, and $f_3(x, y)$ itself zero except for values of x, y infinitely close to x', y' respectively; and let $\iint f_3(x, y)\, dx dy = 1$, provided the limits of integration include the values $x = x'$, $y = y'$. Let u be expanded in the triple series

$$\Sigma\Sigma\Sigma A_{mnp} \sin \mu x \sin \nu y \sin \varpi z \dots\dots\dots\dots(105),$$

where μ, ν, ϖ mean the same as in Art. 48.   Then

$$\frac{d^2 u}{dz^2} = \Sigma_p \left\{ - \Sigma_m \Sigma_n \varpi^2 A_{mnp} \sin \mu x \sin \nu y \right.$$
$$\left. + \frac{2\varpi}{c} f_3(x, y) \right\} \sin \varpi z \dots\dots\dots\dots(106).$$

Now the expansion of $f_3(x, y)$ in a double series is

$$4/ab \cdot \Sigma\Sigma \sin \mu x' \sin \nu y' \sin \mu x \sin \nu y,$$

that is to say with this understanding, that the result is to be substituted in (106); for it would be absurd to speak, except by way of abbreviation, of a quantity which is zero except for particular values of x and y, for which it is infinite. The values of d^2u/dx^2 and d^2u/dy^2 will be obtained by direct differentiation. We have therefore for the direct developement of ∇u in a triple series

$$\nabla u = \Sigma\Sigma\Sigma \left\{ - (\mu^2 + \nu^2 + \varpi^2) A_{mnp} \right.$$
$$\left. + \frac{8\varpi}{abc} \sin \mu x' \sin \nu y' \right\} \sin \mu x \sin \nu y \sin \varpi z.$$

But ∇u being equal to zero, each coefficient will be equal to zero, from whence we get A_{mnp}, and then

$$u = \frac{8}{abc} \Sigma\Sigma\Sigma \frac{\varpi}{\mu^2 + \nu^2 + \varpi^2} \sin \mu x' \sin \nu y' \sin \mu x \sin \nu y \sin \varpi z \dots(107).$$

One of the three summations, whichever we please, may be performed by means of the known formulæ

$$\Sigma \frac{\varpi \sin \varpi z}{\varpi^2 + k^2} = \frac{c}{2} \frac{\epsilon^{k(c-z)} - \epsilon^{-k(c-z)}}{\epsilon^{kc} - \epsilon^{-k}}, \text{ if } 2c > z > 0 \ldots\ldots(108),$$

$$\frac{1}{2k} + \Sigma \frac{k \cos \nu y_{\prime}}{k^2 + \nu^2} = \frac{b}{2} \frac{\epsilon^{k(b-y_{\prime})} + \epsilon^{-k(b-y)}}{\epsilon^{kb} - \epsilon^{-bk}}, \text{ if } 2b > y_{\prime} > 0 \ldots\ldots(109),$$

which may be obtained by developing the second members between the limits $z = 0$ and $z = c$, $y_{\prime} = 0$ and $y_{\prime} = b$ by the formulæ (2), (22), and observing that the expansions hold good within the limits written after the formulæ, since $\epsilon^{k(c-z)} - \epsilon^{-k(c-z)}$ has the same magnitude and opposite signs for values of z equidistant from c, and $\epsilon^{k(b-y_{\prime})} + \epsilon^{-k(b-y_{\prime})}$ has the same magnitude and sign for values of $y_{\prime}$ equidistant from b. If in equation (107) we perform the summation with respect to p, by means of the formula (108), we get the equation (99): if we perform the summation with respect to n, by means of the formula (109), we get the equation (100).

52. The following problem will illustrate some of the ideas contained in this paper, although, in the mode of solution which will be adopted, the formulæ given at the beginning of this Section will not be required.

A hollow conducting rectangular parallelepiped is in communication with the ground: required to express the potential, at any point in the interior, due to a given interior electrical point and to the electricity induced on the surface.

Let the axes be taken as in Art. 48. Let x', y', z' be the co-ordinates of the electrical point, m the electrical mass, v the required potential. Then v is determined *first* by satisfying the equation $\nabla v = 0$, *secondly* by being equal to zero at the surface, *thirdly* by being equal to m/r infinitely close to the electrical point, r being the distance of the points (x, y, z), (x', y', z'), and by being finite and continuous at all other points within the parallelepiped.

Let $v = m/r + v_1$, so that v_1 is the potential due to the electricity induced on the surface. Then v_1 is finite and continuous within the parallelepiped, and is determined by satisfying the general equation $\nabla v_1 = 0$, and by being equal to $- m/r$ at the surface. Consequently v_1 can be determined precisely as u in Art. 48 or 51. This separation however of v into two parts seems to introduce a degree of complexity not inherent in the problem; for v itself vanishes at the surface; and it is when the function expanded vanishes at the limits that the application of the series (2) involves least complexity. On the other hand we cannot immediately expand v in a triple series of the form (105), on account of its becoming infinite at the point (x', y', z').

Suppose, therefore, for the present that the electricity is diffused over a finite space: then it is evident that we may suppose the electrical density, ρ, to change so gradually, and pass so gradually into zero, that the derivatives of v, of as many orders as we please, shall be continuous functions. We may now suppose v expanded in a triple series, so that

$$v = \Sigma\Sigma\Sigma A_{mnp} \sin \mu x \sin \nu y \sin \varpi z:$$

and we shall have

$$\nabla v = - \Sigma\Sigma\Sigma (\mu^2 + \nu^2 + \varpi^2) A_{mnp} \sin \mu x \sin \nu y \sin \varpi z.$$

But we have also, by a well-known theorem, $\nabla v = - 4\pi\rho$; and

$$\rho = \Sigma\Sigma\Sigma R_{mnp} \sin \mu x \sin \nu y \sin \varpi z,$$

where

$$R_{mnp} = \frac{8}{abc} \int_0^a \int_0^b \int_0^c \rho' \sin \mu x' \sin \nu y' \sin \varpi z' \, dx' \, dy' \, dz',$$

ρ' being the same function of x', y', z' that ρ is of x, y, z. We get therefore by comparing the two expansions of ∇v

$$A_{mnp} = 4\pi (\mu^2 + \nu^2 + \varpi^2)^{-1} R_{mnp},$$

whence the value of v is known. We may now, if we like, suppose the electricity condensed into a point, which gives

$$R_{mnp} = \frac{8m}{abc} \sin \mu x' \sin \nu y' \sin \varpi z',$$

$$v = \frac{32\pi m}{abc} \Sigma\Sigma\Sigma (\mu^2 + \nu^2 + \varpi^2)^{-1}$$

$$\sin \mu x' \sin \nu y' \sin \varpi z' \sin \mu x \sin \nu y \sin \varpi z \ldots(110).$$

One of the summations may be performed just as before. We thus get, by summing with respect to p,

$$v = \frac{8\pi m}{ab} \Sigma\Sigma \frac{1}{q} \frac{(\epsilon^{qz} - \epsilon^{-qz})(\epsilon^{q(c-z')} - \epsilon^{-q(c-z')})}{\epsilon^{qc} - \epsilon^{-qc}}$$

$$\sin \mu x' \sin \nu y' \sin \mu x \sin \nu y \ldots\ldots(111),$$

where $q^2 = \mu^2 + \nu^2$, and z is supposed to be the smaller of the two z, z'. If z be greater than z', we have only to make z and z' change places in (111).

53. The equation (110) shows that the potential at the point (x, y, z) due to a unit of electricity at the point (x', y', z') and to the electricity induced on the surface of the parallelepiped is equal to the potential at the point (x', y', z') due to a unit of electricity at the point (x, y, z) and to the electricity induced on the surface. This however is only a particular case of a general theorem proved by Green*.

Of course the parallelepiped includes as particular cases two parallel infinite planes, two parallel infinite planes cut at right angles by a third infinite plane, &c. The value of v being known the density of the induced electricity at any point of the surface is at once obtained, by means of a known theorem.

If we suppose a ball-pendulum to oscillate within a rectangular case, the value of ϕ belonging to the motion of the fluid which is due to the direct motion of the ball and to the motion reflected from the case can be found in nearly the same manner. The expression *reflected motion* is here used in the sense explained in Art. 6 of my paper, "On some Cases of Fluid Motion†." In the present instance we should expand ϕ in a triple series of cosines.

54. Let a hollow cylinder, containing one or more plane partitions reaching from the axis to the curved surface, be filled with homogeneous incompressible fluid, and made to oscillate about its axis, both ends being closed: required to determine the effect of the inertia of the fluid on the motion of the cylinder.

If there be more than one partition, it will evidently be sufficient to consider one of the sectors into which the cylinder is divided, since the solution obtained may be applied to the others. In the present case the motion is such that $udx + vdy + wdz$ (according to the usual notation) is an exact differential $d\phi$. The motion considered is in two dimensions, taking place in planes perpendicular to the axis of the cylinder. Let the fluid be referred to polar co-ordinates r, θ in a plane perpendicular to the axis, r being measured from the axis, and θ from one of the bounding

* *Essay on Electricity*, p. 19.
† See p. 111 of the present Volume. [*Ante*, p. 28.]

partitions of the sector considered, being reckoned positive when measured inwards. Let the radius of the cylinder be taken for the unit of length, and let α be the angle of the sector, and ω the angular velocity of the cylinder at the instant considered. It will be observed that $\alpha = 2\pi$ corresponds to the case of a single partition. Then to determine ϕ we have the general equation

$$\frac{d^2\phi}{dr^2} + \frac{1}{r}\frac{d\phi}{dr} + \frac{1}{r^2}\frac{d^2\phi}{d\theta^2} = 0 \quad\dots\dots\dots\dots(112),$$

with the conditions

$$\frac{1}{r}\frac{d\phi}{d\theta} = \omega r, \text{ when } \theta = 0 \text{ or } = \alpha \dots\dots\dots(113),$$

$$\frac{d\phi}{dr} = 0, \text{ when } r = 1 \quad\dots\dots\dots\dots(114),$$

and, that ϕ shall not become infinite when r vanishes.

Let $r = \epsilon^{-\lambda}$, and take θ, λ for the independent variables; then (112), (113), (114) become

$$\frac{d^2\phi}{d\lambda^2} + \frac{d^2\phi}{d\theta^2} = 0\dots\dots\dots\dots\dots(115),$$

$$\frac{d\phi}{d\theta} = \omega\epsilon^{-2\lambda}, \text{ when } \theta = 0 \text{ or } = \alpha\dots\dots\dots(116),$$

$$\frac{d\phi}{d\lambda} = 0, \text{ when } \lambda = 0 \quad\dots\dots\dots\dots(117).$$

Let ϕ be expanded between the limits $\theta = 0$ and $\theta = \alpha$ in a series of cosines, so that

$$\phi = \Lambda_0 + \Sigma\Lambda_n \cos\frac{n\pi\theta}{\alpha} \dots\dots\dots\dots(118),$$

Λ_0, Λ_n being functions of λ. Then we have by the formula (D) and the condition (116) applied to the general equation (115)

$$\frac{d^2\Lambda_0}{d\lambda^2} = 0,$$

$$\frac{d^2\Lambda_n}{d\lambda^2} - \left(\frac{n\pi}{\alpha}\right)^2\Lambda_n - \frac{2\omega}{\alpha}\{1-(-1)^n\}\epsilon^{-2\lambda} = 0;$$

whence

$$\Lambda_0 = A_0\lambda + B_0,$$

$$\Lambda_n = A_n\epsilon^{\frac{n\pi\lambda}{\alpha}} + B_n\epsilon^{-\frac{n\pi\lambda}{\alpha}} - \frac{2\omega\alpha\{1-(-1)^n\}}{n^2\pi^2 - 4\alpha^2}\epsilon^{-2\lambda}.$$

Since ϕ is not to be infinite when r vanishes, that is when λ becomes infinite, we have in the first place $A_0 = 0$, $A_n = 0$. We have then by the condition (117)

$$B_n = \frac{8\omega\alpha^2}{n\pi\,(n^2\pi^2 - 4\alpha^2)},$$

when n is odd, and $B_n = 0$ when n is even. If then we omit B_0, which is useless, and put for λ its value, we get

$$\phi = 4\omega\alpha\Sigma_0\frac{2\alpha/n\pi\,.\,r^{n\pi/\alpha} - r^2}{n^2\pi^2 - 4\alpha^2}\cos n\pi\theta/\alpha \ldots\ldots(119).$$

The series multiplied by r^2 may be summed. For if we expand $\sin 2\,(\theta - \tfrac{1}{2}\alpha)$ between the limits $\theta = 0$, $\theta = \alpha$ in a series of cosines, we get

$$\sin(2\theta - \alpha) = -\Sigma_0\frac{8\alpha\cos\alpha}{n^2\pi^2 - 4\alpha^2}\cos n\pi\theta/\alpha\,;$$

whence

$$\phi = 8\omega\alpha^2\Sigma_0\frac{r^{n\pi/\alpha}\cos n\pi\theta/\alpha}{n\pi\,(n^2\pi^2 - 4\alpha^2)} + \frac{\omega}{2\cos\alpha}r^2\sin(2\theta - \alpha)\ldots(120).$$

In determining the motion of the cylinder, the only quantity we care to know is the moment of the fluid pressures about the axis. Now if the motion be so small that we may omit the square of the velocity we shall have, putting $\phi = -\omega f(r,\,\theta)$,

$$p = \psi\,(t) + \frac{d\omega}{dt}f(r,\,\theta),$$

where p is the pressure, $\psi\,(t)$ a function of the time t, whose value is not required, and where the density is supposed to be 1, and the pressure due to gravity is omitted, since it may be taken account of separately. The moment of the pressure on the curved surface is zero, since the direction of the pressure at any point passes through the axis. The expression (119) or (120) shows that the moments on the plane faces of the sector are equal, and act in the same direction; so that it will be sufficient to find the moment on one of these faces and double the result. If we consider a portion of the face for which $\theta = 0$ whose length in the direction of the axis is unity, we shall have for the pressure on an element dr of the surface $d\omega/dt\,.\,f(r, 0)\,dr$; and if we denote the whole moment of the pressures by $- C\,d\omega/dt$, reckoned positive when it tends to make the cylinder move in the direction of θ positive, we shall have

$$C = 2 \int_0^1 f(r, 0) \, r \, dr.$$

Taking now the value of $f(r, 0)$ from (120), and performing the integration, we shall have

$$C = \frac{1}{4} \tan \alpha - 16\alpha^3 \, \Sigma_0 \, \frac{1}{(n\pi - 2\alpha) \, n\pi \, (n\pi + 2\alpha)^2} \dots\dots(121).$$

The mass of the portion of fluid considered is $\frac{1}{2}\alpha$; and if we put

$$C = \tfrac{1}{2}\alpha k'^2,$$

and write $s\pi/2$ for α, so that s may have any value from 0 to 4, we shall have

$$k'^2 = \frac{1}{s\pi} \tan \frac{s\pi}{2} - \frac{8s^2}{\pi^2} \Sigma_0 \, \frac{1}{(n - s) \, n \, (n + s)^2} \dots\dots(122).$$

When s is an odd integer, the expression for k'^2 takes the form $\infty - \infty$, and we shall easily find

$$k'^2 = \frac{4}{s^2 \pi^2} - \frac{8s^2}{\pi^2} \Sigma \, \frac{1}{(n - s) \, n \, (n + s)^2} \dots\dots(123),$$

where all odd values of n except s are to be taken.

The quantity k' may be called the *radius of gyration* of the fluid about the axis. It would be easy to prove from general dynamical principles, without calculation, that if k be the corresponding quantity for a parallel axis passing through the centre of gravity of the fluid, h the distance of the axes,

$$k'^2 = k^2 + h^2 \dots\dots\dots (124),$$

in fact, in considering the motion of the cylinder, which is supposed to take place in two dimensions, the fluid may be replaced by a solid having the same mass and centre of gravity as the fluid, but a moment of inertia about an axis passing through the centre of gravity and parallel to the axis of the cylinder different from the moment of inertia of the fluid supposed to be solidified. If K', K be the radii of gyration of the solidified fluid about the axis of the cylinder and a parallel axis passing through the centre of gravity respectively, we shall have

$$K'^2 = \tfrac{1}{2} = K^2 + h^2, \quad h = \frac{4}{3} \frac{\sin \tfrac{1}{2}\alpha}{\alpha} = \frac{8}{3s\pi} \sin \frac{s\pi}{4} \dots (125).$$

If we had restricted the application of the series and the integrals involving cosines to those cases in which the derivative of the expanded function vanishes at the limits, we should have expanded ϕ in the definite integral $\int_0^\infty \zeta(\theta, \beta) \cos \beta \lambda d\beta$, and the equation (115) would have given

$$\zeta(\theta, \beta) = \xi(\beta) \epsilon^{\beta\theta} + \chi(\beta) \epsilon^{-\beta\theta},$$

ξ, χ denoting arbitrary functions, which must be determined by the conditions (116). We should have obtained in this manner

$$\phi = \frac{4\omega}{\pi} \int_0^\infty \frac{\epsilon^{\beta(\theta-\frac{1}{2}a)} - \epsilon^{-\beta(\theta-\frac{1}{2}a)}}{\beta(\beta^2+4)(\epsilon^{\frac{1}{2}\beta a}+\epsilon^{-\frac{1}{2}\beta a})} \cos\left(\beta \log \frac{1}{r}\right) d\beta \ldots (126),$$

$$k'^2 = \frac{32}{\pi\alpha} \int_0^\infty \frac{1-\epsilon^{-\beta a}}{1+\epsilon^{-\beta a}} \frac{d\beta}{\beta(\beta^2+4)^2} \ldots\ldots\ldots(127).$$

It will be seen at once that k'^2 is expressed in a much better form for numerical computation by the series in (122) than by the integral in (127). Although the nature of the problem restricts α to be at most equal to 2π, it will be observed that there is no such restriction in the analytical proof of the equivalence of the two expressions for ϕ, or for k'^2.

In the following table the first column gives the angle of the cylindrical sector, the second the square of the radius of gyration of the fluid about the axis of the cylinder, the radius of the cylinder being taken for the unit of length, the third the square of the radius of gyration of the fluid about a parallel axis passing through the centre of gravity, the fourth and fifth the ratios of the quantities in the second and third to the corresponding quantities for the solidified fluid.

a	k'^2	k^2	$\dfrac{k'^2}{K'^2}$	$\dfrac{k^2}{K^2}$
0°	·50000	·05556	1·0000	1·0000
45°	·45385	·03179	·9077	·4079
90°	·39518	·03492	·7904	·2499
135°	·34775	·07442	·6955	·3283
180°	·31057	·13044	·6211	·4078
225°	·28101	·18261	·5620	·4547
270°	·25703	·21700	·5141	·4718
315°	·23720	·22858	·4744	·4652
360°	·22051	·22051	·4410	·4410

55. When α is greater than π, it will be observed that the expression for the velocity which is obtained from (119) becomes infinite when r vanishes. Of course the velocity cannot really become infinite, but the expression (119) fails for points very near the axis. In fact, in obtaining this expression it has been assumed that the motion of the fluid is continuous, and that a fluid particle at the axis may be considered to belong to either of the plane faces indifferently, so that its velocity in a direction normal to either of the faces is zero. The velocity obtained from (119) satisfies this latter condition so long as α is not greater than π. For when $\alpha < \pi$ the velocity vanishes with r, and when $\alpha = \pi$ the velocity is finite when r vanishes, and is directed along the single plane face which is made up of the two plane faces before considered.

But when α is greater than π the motion which takes place appears to be as follows. Let $OABC$ be a section of the cylindrical sector made by a plane perpendicular to the axis, and cutting it in O. Suppose the cylinder to be turning round O in the direction indicated by the arrow at B. Then the fluid in contact with OA and near O will be flowing, relatively to OA, towards O, as indi-

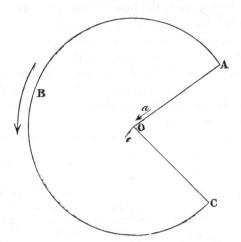

cated by the arrow a. When it gets to O it will shoot past the face OC; so that there will be formed a surface of discontinuity Oe extending some way into the fluid, the fluid to the left of Oe and near O flowing in the direction AO, while the fluid to the right is nearly at rest. Of course, in the case of fluids such as they exist in nature, friction would prevent the velocity in a direction tangential to Oe from altering abruptly as we pass from a particle on one side of Oe to a particle on the other; but I have all along been going on the supposition that the fluid is perfectly smooth, as is usually supposed in Hydrodynamics*. The extent of

[* It may be said that the motion of a perfect fluid which is at first at rest, and is then set in motion by the action of solid bodies of finite curvature in contact with it, is unique and continuous, so that no surface of discontinuity can be formed; and that that being always true will be true in the limit, when we suppose the curvature at a certain point or along a certain line to become infinite, in such a manner as to pass in the limit to a salient conical point or edge intruding into the fluid, and therefore even in this case no surface of discontinuity can be formed.

This may be perfectly true in one sense and yet not in another. A perfect fluid is an ideal abstraction, representing something which does not exist in nature. All actual fluids are more or less viscous, and we arrive at the conception of a perfect fluid by starting with fluids such as we find them, and then in imagination making abstraction of the viscosity. Similarly any edge that we can mechanically form is more or less rounded off; but we have no difficulty in conceiving of an edge perfectly sharp. The motion that belongs to a perfect fluid and perfectly sharp edge may be regarded as the limit, if unique limit there be, of the motion which belongs to a slightly viscous fluid interrupted by a solid presenting a salient, slightly rounded edge, when both the viscosity of the fluid and the radius of curvature of the edge are supposed to vanish. For the sake of clear ideas we may suppose that the mass of fluid we are dealing with is contained in a vessel differing from that mentioned in the text in being bounded on the side towards the centre by a cylindrical surface of very small radius a, coaxial with the outer cylinder. Then we may represent the motion, in a sense which the reader will readily apprehend, by $f(\mu, a)$, where μ denotes the coefficient or index of internal friction. We proposed to contemplate the limit of $f(\mu, a)$ when μ and a vanish. But for anything that appears to the contrary there may be no such unique limit, but the limit $\lim_{a=0} \lim_{\mu=0} f(\mu, a)$ may be one thing, and the limit $\lim_{\mu=0} \lim_{a=0} f(\mu, a)$ a totally different thing; and I am strongly disposed to believe that such is actually the case. When a is finite and $\mu=0$, we pass to a case of motion of a perfect fluid similar to that in the text, and capable of being attacked by a similar analysis, but in which the motion nowhere becomes infinite; and the limit to which this tends as a vanishes does not present a surface of discontinuity, but the velocity near the centre increases indefinitely as a decreases indefinitely. But when on the other hand μ though small remains finite, and a diminishes without limit, the motion which would be investigated by an analysis resembling that in the text would be such that near the centre there would be an enormous gliding, which would call into play a great tangential force, in which work would be consumed, that is, converted into heat. This I believe would be an unstable condition, and what would actually take place would be that so large a local consumption of work would be avoided by the fluid rushing past the corner, somewhat as represented in the figure, carrying with it by adhesion a narrow stratum in which there would be very great molecular rotation, inasmuch as the fluid of which the stratum consists had previously been pent up between the radial wall of the vessel on one side of it, the surface of discontinuity Oe will be the less the smaller be the motion of the cylinder; and although the expression (119) fails for points very near O, that does not prevent it from being sensibly correct for the remainder of the fluid, so that we may calculate k'^2 from (122) without committing a sensible error. In fact, if γ be the angle through which the cylinder oscillates, since the extent of the surface of discontinuity depends upon the first power of γ, the error we should commit would depend upon γ^2. I expect, therefore, that the moment of inertia of the fluid which would be determined by experiment would agree with theory nearly, if not quite, as well when $\alpha > \pi$ as when $\alpha < \pi$, care being taken that the oscillations of the cylinder be very small.

As an instance of the employment of analytical expressions which give infinite values for physical quantities, I may allude to the distribution of electricity on the surfaces of conducting bodies which have sharp edges.

56. The preceding examples will be sufficient to show the utility of the methods contained in this paper. It may be observed that in all cases in which an arbitrary function is expanded between certain limits in a series of quantities whose form is determined by certain conditions to be satisfied at the limits, the expansion can be performed whether the conditions at the limits be satisfied or not, since the expanded function is supposed perfectly arbitrary. Analogy would lead us to conclude that the derivatives of the expanded functions could not be found by direct differentiation, but would have to be obtained from formulæ answering to those at the beginning of this Section. If such expansions should be found useful, the requisite formulæ would probably be obtained without difficulty by integration by parts. This is in fact the case with the only expansion of the kind which I have tried, which is that employed in Art. 45. By means of this expansion and the corresponding formulæ we might determine in a double series the permanent temperature in a homogeneous rectangular parallelepiped which radiates into a medium whose

which had no radial motion and but little in a perpendicular direction, and the rapidly rushing fluid on the other side. The smaller μ is made, the narrower will this stratum be, but not, so far as I can see, the shorter; and a very narrow stratum in which there is intense molecular rotation passes, or may pass, in the limit to a surface of discontinuity.

The above is what was referred to by anticipation in the footnote at p. 99.]

temperature varies in any given manner from point to point; or we might determine in a triple series the variable temperature in such a solid, supposing the temperature of the medium to vary in a given manner with the time as well as with the co-ordinates, and supposing the initial temperature of the parallelepiped given as a function of the co-ordinates. This problem, made a little more general by supposing the exterior conductivity different for the six faces, has been solved in another manner by M. Duhamel in the Fourteenth Volume of the *Journal de l'École Polytechnique*. Of course such a problem is interesting only as an exercise of analysis.

ADDITIONAL NOTE.

If the series by which r^2 is multiplied in (119) had been left without summation, the series which would have been obtained for k'^2 would have been rather simpler in form than the series in (122), although more slowly convergent. One of these series may of course be obtained from the other by means of the development of tan x in a harmonic series. When s is an integer, k'^2 can be expressed in finite terms. The result is

$$k'^2 = 8s^{-1}\pi^{-2}\log_e 2 + 8s^{-1}\pi^{-2}\{2^{-1} + 4^{-1} \ldots$$
$$+ (s-1)^{-1}\} + 4\pi^{-2}\{2^{-2} + 4^{-2} \ldots + (s-1)^{-2}\} - \tfrac{1}{6}, \text{ (s odd,)}$$

$$k'^2 = 8s^{-1}\pi^{-2}\{1^{-1} + 3^{-1} \ldots + (s-1)^{-1}\} + 4\pi^{-2}\{1^{-2} + 3^{-2} \ldots$$
$$+ (s-1)^{-2}\} - \tfrac{1}{2}. \text{ (s even.)}$$

Moreover when $2s$ is an odd integer, or when $\alpha = 45^\circ$, or $= 135^\circ$, &c., k'^2 can be expressed in finite terms if the sum of the series $1^{-2} + 5^{-2} + 9^{-2} + \ldots$ be calculated, and then be regarded as a known transcendental quantity.

[Not before published.   (See page 229.)]

SUPPLEMENT TO A PAPER ON THE THEORY OF OSCILLATORY WAVES.

THE labour of the approximation in proceeding to a high order, when conducted according to the method of the former paper whether we employ the function ϕ or ψ, depends in great measure upon the circumstance that the two equations which have to be satisfied simultaneously at the free surface are both composed in a rather complicated manner of the independent variables, and in the elimination of y the length of the process is still further increased by the necessity of expanding the exponentials in y according to series of powers, giving for each exponential a whole set of terms. This depends upon the circumstance that of the limits of y belonging to the boundaries of the fluid, one instead of being a constant is a function of x, and that too a function which is only known from the solution of the problem.

If we convert the wave motion into steady motion, and refer the fluid to two independent variables of which one is the parameter of the stream lines or a function of the parameter, and the other is x or a quantity which extends with x from $-\infty$ to $+\infty$, we shall ensure constancy of each independent variable at both its limits, but in general the equations obtained will be of great complexity. It occurred to me however that if from among the infinite number of systems of independent variables possessing the above character we were to take the functions ϕ, ψ, where

$$\phi = \int(u\,dx + v\,dy), \qquad \psi = \int(u\,dy - v\,dx),$$

simplicity might be gained in consequence of the immediate relation of these functions to the problem.

We know that ϕ, ψ are conjugate solutions of the equation

$$\frac{d^2\phi}{dx^2} + \frac{d^2\phi}{dy^2} = 0 \dots\dots\dots\dots\dots(1),$$

satisfying the equations

$$\frac{d\phi}{dx} = \frac{d\psi}{dy}, \quad \frac{d\phi}{dy} = -\frac{d\psi}{dx} \dots\dots\dots\dots(2),$$

so that if the form of either be assigned, satisfying of course the equation (1), the other may be deemed known, since it can be obtained by the integration of a perfect differential. If now we take ϕ, ψ for the independent variables, of which x and y are regarded as functions, we get by changing the independent variables in differentiation

$$\frac{d\phi}{dx} = \frac{1}{S}\frac{dy}{d\psi}, \quad \frac{d\phi}{dy} = -\frac{1}{S}\frac{dx}{d\psi}, \quad \frac{d\psi}{dx} = -\frac{1}{S}\frac{dy}{d\phi}, \quad \frac{d\psi}{dy} = \frac{1}{S}\frac{dx}{d\phi} \dots(3),$$

where

$$S = \frac{dx}{d\phi}\frac{dy}{d\psi} - \frac{dx}{d\psi}\frac{dy}{d\phi},$$

whence from (2)

$$\frac{dx}{d\phi} = \frac{dy}{d\psi}, \quad \frac{dx}{d\psi} = -\frac{dy}{d\phi} \dots\dots\dots\dots(4),$$

so that x, y are conjugate solutions of the equation

$$\frac{d^2x}{d\phi^2} + \frac{d^2x}{d\psi^2} = 0 \dots\dots\dots\dots\dots(5).$$

We have also from (4)

$$S = \left(\frac{dx}{d\phi}\right)^2 + \left(\frac{dx}{d\psi}\right)^2 \dots\dots\dots\dots\dots(6).$$

We get from (3), (4) and (6)

$$u^2 + v^2 = \left(\frac{d\phi}{dx}\right)^2 + \left(\frac{d\psi}{dx}\right)^2 = \frac{1}{S^2}\left\{\left(\frac{dx}{d\phi}\right)^2 + \left(\frac{dx}{d\psi}\right)^2\right\} = \frac{1}{S} \dots\dots(7),$$

whence

$$\frac{p}{\rho} = g(y + C) - \frac{1}{2S} \dots\dots\dots\dots\dots(8),$$

where C is an arbitrary constant.

The mode of proceeding is the same in principle whether the depth of the fluid be finite or infinite; but as the formulæ are simpler in the latter case, it may be well to consider it separately in the first instance.

If c be the velocity of propagation, c will be the horizontal velocity at a great depth when the wave motion is converted into steady motion. The difference between ϕ and $-cx$ will be a periodic function of x or of ϕ. We may therefore assume in accordance with equation (5)

$$x = -\frac{\phi}{c} + \Sigma_1^\infty (A_i e^{im\psi/c} + B_i e^{-im\psi/c}) \sin im\phi/c \ldots\ldots(9).$$

No cosines are inserted in this equation because if we take, as we may, the origins of x and of ϕ at a trough or a crest (suppose a trough), x will be an odd function of ϕ, in accordance with what has already been shown at page 212. Corresponding to the above value of x we have

$$y = -\frac{\psi}{c} + \Sigma_1^\infty (A_i e^{im\psi/c} - B_i e^{-im\psi/c}) \cos im\phi/c \ldots..(10),$$

the arbitrary constant being omitted, as may be done provided we leave open the origin of y.

The origin of ψ being arbitrary, we may take, as it will be convenient to do, $\psi = 0$ at the free surface. We see from (10) that ψ increases negatively downwards; and therefore of the two exponentials that with $-im\psi/c$ for index is the one which must be omitted, as expressing a disturbance that increases indefinitely in descending.

We may without loss of generality shorten the formulæ during a rather long approximation by writing 1 for any two of the constants which depend differently on the units of space and time. These constants can easily be reintroduced in the end by rendering the equations homogeneous. We may accordingly put $m = 1$ and $c = 1$. The expressions for x and y as thus shortened become, on retaining only the exponential which decreases downwards,

$$x = -\phi + \Sigma_1^\infty A_i e^{i\psi} \sin i\phi \ldots\ldots\ldots\ldots(11),$$
$$y = -\psi + \Sigma_1^\infty A_i e^{i\psi} \cos i\phi \ldots\ldots\ldots\ldots(12).$$

At the free surface $\psi = 0$, and we must therefore have for $\psi = 0$
$$g (y + C) S - \tfrac{1}{2} = 0,$$
which gives

$$(C + \Sigma A_i \cos i\phi) \{1 - 2\Sigma i A_i \cos i\phi + \Sigma i^2 A_i^2 + 2\Sigma ij A_i A_j \cos [(i-j)\phi]\}$$
$$- \frac{1}{2g} = 0 \ldots\ldots\ldots \ldots..(13),$$

where in the last term within parentheses each different combination of unequal integers i, j is to be taken once.

On account of the complicated form of this equation, we can proceed further only by adopting some system of approximation. The most obvious is that adopted in the former paper, namely to proceed according to powers of the coefficient of the term of the first order. If we multiply out in equation (13), and replace products of cosines by cosines of sums and differences, we may arrange the equation in the form

$$B_0 + B_1 \cos \phi + B_2 \cos 2\phi + \ldots = 0,$$

where the several B's are series of terms involving the coefficients A. And as the equation has to be satisfied independently of ϕ, we must have separately

$$B_0 = 0, \quad B_1 = 0, \quad B_2 = 0, \ \&c.$$

A slight examination of the process will show that A_i is of the order i, and that consequently the product of any number of the A's is of the order marked by the sum of the suffixes, and that B_i is of the order i. In proceeding therefore to any desired order we can see at once what terms need not be written down, as being of a superior order.

Thus in proceeding to the fifth order we must take the six equations $B_0 = 0$, $B_1 = 0, \ldots B_5 = 0$, which when written at length are

$$C(1 + A_1^2 + 4A_2^2) - A_1^2 + 2A_1^2 A_2 - 2A_2^2 - \tfrac{1}{2}g^{-1} = 0,$$

$$C(-2A_1 + 4A_1 A_2 + 12 A_2 A_3) + A_1 + A_1^3 - 3A_1 A_2 + 6A_1 A_2^2$$
$$+ 3A_1^2 A_3 - 5A_2 A_3 = 0,$$

$$C(-4A_2 + 6A_1 A_3) + A_2 - A_1^2 + 3A_1^2 A_2 - 4A_1 A_3 = 0,$$

$$C(-6A_3 + 8A_1 A_4) + A_3 - 3A_1 A_4 + 4A_1^2 A_3 + 2A_1 A_2^2 - 5A_1 A_4 = 0,$$

$$C(-8A_4) + A_4 - 4A_1 A_3 - 2A_2^2 = 0,$$

$$C(-10A_5) + A_5 - 5A_1 A_4 - 5A_2 A_3 = 0.$$

These equations may be looked on as giving, the first, the arbitrary constant C, the second, the velocity of propagation, and the succeeding ones taken in order the values of the constants A_2, A_3, A_4, A_5, respectively. I say "may be looked on as giving", for it is only when we restrict ourselves to the terms of the lowest order in each equation that those quantities are actually given in succession; the equations contain also terms of higher orders; and to get the complete values of the quantities true to the order to which we are working, we must use the method of successive substitutions. As to the second equation, if we take the terms of lowest order in the first two we get $C = \frac{1}{2}g^{-1}$, and then by substitution in the second equation $1 = g$, the constant A_1 dividing out. The equation $1 = g$ becomes on generalizing the units of space and time $c^2 = g/m$, and accordingly gives the velocity of propagation to the lowest order of approximation.

On eliminating the arbitrary constant in the above equations, and writing b for A_1, the results become

$$1 = g\left(1 + b^2 + \tfrac{7}{2}b^4\right)\dots\dots\dots\dots\dots(14),$$

$$x = -\phi + be^\psi \sin\phi - (b^2 + \tfrac{1}{2}b^4)e^{2\psi}\sin 2\phi + (\tfrac{3}{2}b^3 + \tfrac{19}{12}b^5)e^{3\psi}\sin 3\phi$$
$$- \tfrac{8}{3}b^4e^{4\psi}\sin 4\phi + \tfrac{125}{24}b^5e^{5\psi}\sin 5\phi\dots\dots\dots(15),$$

$$y = -\psi + be^\psi \cos\phi - (b^2 + \tfrac{1}{2}b^4)e^{2\psi}\cos 2\phi + (\tfrac{3}{2}b^3 + \tfrac{19}{12}b^5)e^{3\psi}\cos 3\phi$$
$$- \tfrac{8}{3}b^4e^{4\psi}\cos 4\phi + \tfrac{125}{24}b^5e^{5\psi}\cos 5\phi\dots\dots\dots(16).$$

The equation (14) gives to the fifth order the square of the velocity of propagation in the wave motion; and (15), (16) give the point where the parameters ϕ, ψ have given values, and also, by the aid of the formulæ previously given, the components of the velocity, and the pressure, in the steady motion. These same equations (15), (16), if we suppose ψ constant give implicitly the equation of the corresponding stream line, or if we suppose ϕ constant the equation of one of the orthogonal trajectories.

To find implicitly the equation of the surface, we have only to put $\psi = 0$ in (15), (16), which gives

$$x = -\phi + b\sin\phi - (b^2 + \tfrac{1}{2}b^4)\sin 2\phi + (\tfrac{3}{2}b^3 + \tfrac{19}{12}b^5)\sin 3\phi$$
$$- \tfrac{8}{3}b^4 \sin 4\phi + \tfrac{125}{24}b^5 \sin 5\phi\dots\dots\dots\dots(17),$$

$$y = \qquad b\cos\phi - (b^2 + \tfrac{1}{2}b^4)\cos 2\phi + (\tfrac{3}{2}b^3 + \tfrac{19}{12}b^5)\cos 3\phi$$
$$- \tfrac{8}{3}b^4 \cos 4\phi + \tfrac{125}{24}b^5 \cos 5\phi\dots\dots\dots\dots(18).$$

It is not necessary to form the explicit equation, but we can do so if we please, most conveniently by the aid of Lagrange's theorem. The result, carried to the fourth order only, which will suffice for the object more immediately in view, is

$$y + \tfrac{1}{2}b^2 + b^4 = (b + \tfrac{9}{8}b^3)\cos x - (\tfrac{1}{2}b^2 + \tfrac{11}{6}b^4)\cos 2x$$
$$+ \tfrac{3}{8}b^3 \cos 3x - \tfrac{1}{3}b^4 \cos 4x\dots(19).$$

If we put $b + \frac{9}{8} b^3 = a$, we have to the fourth order
$$b = a - \tfrac{9}{8} a^3,$$
and substituting in (19) we get
$$y + \tfrac{1}{2} a^2 - \tfrac{1}{8} a^4 = a \cos x - (\tfrac{1}{2} a^2 + \tfrac{17}{24} a^4) \cos 2x + \tfrac{3}{8} a^3 \cos 3x$$
$$- \tfrac{1}{3} a^4 \cos 4x \ldots\ldots(20).$$

The expression (14) for the square of the velocity of propagation, and the equation of the surface (20), agree with the results previously obtained by the former method (see p. 221) to the degree of approximation to which the latter were carried, as will be seen when we remember that the origins of y are not the same in the two cases; but it would have been much more laborious to obtain the approximation true to the fifth order by the old method.

It has already been remarked (p. 211) that the equation of the profile in deep water agrees with a trochoid to the third order, which is as far as the approximation there proceeded. This is no longer true when we proceed to the fourth order. On shifting the origin of y so as to get rid of the constant term, the equation (20) of the profile becomes

$$y = a \cos x - (\tfrac{1}{2} a^2 + \tfrac{17}{24} a^4) \cos 2x + \tfrac{3}{8} a^3 \cos 3x - \tfrac{1}{3} a^4 \cos 4x \ldots(21).$$

On the other hand, the equation of a trochoid is given implicitly by the pair of equations
$$x = \alpha\theta + \beta \sin \theta, \qquad y = \beta \cos \theta + \gamma.$$

In order that x may have the same period in the trochoid as in the profile of the wave, we must have $\alpha = 1$. We get then by development to the fourth order, choosing γ so as to make the constant term vanish,

$$y = (\beta - \tfrac{3}{8}\beta^3) \cos x - (\tfrac{1}{2} \beta^2 - \tfrac{1}{3}\beta^4) \cos 2x + \tfrac{3}{8}\beta^3 \cos 3x - \tfrac{1}{3}\beta^4 \cos 4x,$$
and putting
$$\beta - \tfrac{3}{8} \beta^3 = a,$$
we get to the fourth order

$$y = a \cos x - (\tfrac{1}{2} a^2 + \tfrac{1}{24} a^4) \cos 2x + \tfrac{3}{8} a^3 \cos 3x - \tfrac{1}{3} a^4 \cos 4x \ldots(22).$$

Hence if y_w, y_t denote the ordinates for the wave and trochoid respectively, we have to the fourth order

$$y_w - y_t = - \tfrac{2}{3} a^4 \cos 2x.$$

Hence the wave lies a little above the trochoid at the trough and crest, and a little below it in the shoulders.

This result agrees well with what might have been expected. It has been shown (p. 227) that the limiting form for a series of uniformly propagated irrotational waves is one presenting edges of 120°, and that the inclination in this limiting form is in all probability restricted to 30°, whereas in the trochoidal waves investigated by Gerstner and Rankine the limiting form is the cycloid, presenting accordingly cusps, and an inclination increasing to 90°. Hence the limiting form must be reached with a much smaller value of the parameter a in the former case than in the latter. Hence when a is just large enough to make the difference of form of the irrotational and trochoidal waves begin to tell, since the limiting form is more nearly approached in the former case than in the latter, we should expect the curvature at the summit to be greater, while at the same time as the general inclination is probably rather less, and the inclination begins by increasing more rapidly as we recede from the summit, the troughs must be shallower and flatter for an equal mean height of wave.

Let us proceed now to the case of a finite depth. As before we may choose the units of space and time so that c and m shall each be 1, and we may choose 0 for the value of the parameter ψ at the surface. Let $-k$ be its value at the bottom. Then since $d\phi/dy = 0$ at the bottom we have from (3) and (4) $dy/d\phi = 0$ when $\psi + k = 0$, and consequently

$$A_i e^{-ik} = B_i e^{ik},$$

whence writing $A_i e^{ik}$ for A_i we have

$$x = -\phi + \Sigma A_i \{e^{i(\psi+k)} + e^{-i(\psi+k)}\} \sin i\phi \ldots\ldots\ldots(23),$$

$$y = -\psi + \Sigma A_i \{e^{i(\psi+k)} - e^{-i(\psi+k)}\} \cos i\phi \ldots\ldots\ldots(24).$$

Putting for shortness

$$e^{ik} + e^{-ik} = S_i, \quad e^{ik} - e^{-ik} = D_i,$$

we have by the condition at the free surface

$$(C + \Sigma A_i D_i \cos i\phi) \{1 - 2\Sigma A_i i S_i \cos i\phi + (\Sigma i A_i S_i \cos i\phi)^2$$

$$+ (\Sigma i A_i D_i \sin i\phi)^2\} - \frac{1}{2g} = 0 \ldots\ldots(25).$$

As the expressions are longer than in the case of an infinite depth, and the problem itself of rather less interest, I shall content myself with proceeding to the third order. We have to this order from (25), on taking account of the relations

$$S_i S_j = S_{i+j} + S_{i-j}, \quad D_i D_j = S_{i+j} - S_{i-j}, \quad D_i S_j = D_{i+j} + D_{i-j},$$

$$(C + A_1 D_1 \cos \phi + A_2 D_2 \cos 2\phi + A_3 D_3 \cos 3\phi)$$

$$\times \left\{ \begin{array}{l} 1 \quad -2A_1 S_1 \cos \phi \quad -4A_2 S_2 \cos 2\phi - 6A_3 S_3 \cos 3\phi \\ + A_1^2 S_2 + 4A_1 A_2 S_3 \cos \phi + 2A_1^2 \cos 2\phi \quad + 4A_1 A_2 S_1 \cos 3\phi \end{array} \right\}$$

$$-\frac{1}{2g} = 0.$$

Multiplying out, retaining terms up to the third order only, arranging the terms according to cosines of multiples of ϕ, and equating to zero the coefficients of the cosines of the same multiple, we get the four equations

$$C(1 + A_1^2 S_2) - A_1^2 S_1 D_1 - \frac{1}{2g} = 0,$$

$$C(-2A_1 S_1 + 4A_1 A_2 S_3) + A_1 D_1 + A_1^3 S_2 D_1 - 2A_1 A_2 S_2 D_1 + A_1^3 D_1$$
$$- A_1 A_2 S_1 D_2 = 0,$$

$$C(-4A_2 S_2 + 2A_1^2) - A_1^2 S_1 D_1 + A_2 D_2 = 0,$$

$$C(-6A_3 S_3 + 4A_1 A_2 S_1) - 2A_1 A_2 S_2 D_1 + A_1^3 D_1 - A_1 A_2 S_1 D_2$$
$$+ A_3 D_3 = 0.$$

A slight examination of the process of approximation will show that whatever be the order to which we proceed, C, and the coefficients A_2, A_4, ... with even suffixes, will contain only even powers, and the coefficients A_3, A_5, ... with odd suffixes only odd powers, of the first coefficient A_1. Writing b for A_1, we may therefore assume, in proceeding to the third order only,

$$C = \alpha + \beta b^2,$$

$$A_2 = \gamma b^2,$$

$$A_3 = \delta b^3.$$

Substituting in the last three equations of the preceding group, which after the substitution may be divided by b, b^2, b^3 respectively, arranging, and equating coefficients of like powers of b, we get

$$-2S_1 \alpha + D_1 = 0,$$

$$(4S_3 \alpha - 2S_2 D_1 - S_1 D_2) \gamma - 2S_1 \beta + S_2 D_1 + D_1 = 0,$$

$$2\alpha - S_1 D_1 + (D_2 - 4S_2 \alpha) \gamma = 0,$$

$$(4S_1 \alpha - 2S_2 D_1 - S_1 D_2) \gamma + D_1 + (D_3 - 6S_3 \alpha) \delta = 0.$$

S.

The substitution for C and the coefficients A_2, A_3, ... of series according to even or odd powers of b with indeterminate coefficients was hardly worth making in proceeding to the third order only, but seems advantageous when we want to proceed to a rather high order. In proceeding to the nth order it is to be noted that the coefficients of C in the group of $n+1$ equations got by equating to zero the coefficients of cosines of multiples of ϕ (including the zero multiple, or constant term), are of the orders 0, 1, 2, ... n in b, so that C being determined only to b^{n-1} in the equations after the first, the terms of the order n in the first equation (which could only occur when n is even) are not required, but this first equation need only be carried as far as to $n-1$. In fact, in proceeding to the orders 1, 2, 3, 4, 5, 6, ..., the velocity of propagation is given to an order not higher than 0, 1, 2, 3, 4, 5, ... in b, and therefore actually to 0, 0, 2, 2, 4, 4, ... since it involves only even powers of b.

The last equations give in succession

$$\alpha = \frac{D_1}{2S_1} \dots\dots\dots\dots\dots\dots\dots\dots\dots(26),$$

$$\gamma = -\frac{1}{D_1^2}(S_2 + 1) \dots\dots\dots\dots\dots(27),$$

$$\delta = \frac{1}{2D_1^4}(3S_4 + 4S_2 + 4) \dots\dots\dots\dots\dots(28),$$

$$\beta = \frac{1}{S_1 D_1}(S_2 + 1)^2 \dots\dots\dots\dots\dots\dots(29),$$

and then by substituting in the first equation of the group on the middle of p. 321, we get

$$\frac{1}{g} = \frac{D_1}{S_1} + \frac{1}{S_1 D_1}(S_4 + 2S_2 + 12)\, b^2 \dots\dots\dots\dots(30).$$

We get now from (23), (24), after rendering the equations homogeneous,

$$x = -\frac{\phi}{c} + b\left(e^{m(\psi + k)/c} + e^{-m(\psi + k)/c}\right)\sin m\phi/c$$

$$-\frac{S_2 + 1}{D_1^2}mb^2\left(e^{2m(\psi + k)/c} + e^{-2m(\psi + k)/c}\right)\sin 2m\phi/c$$

$$+\frac{1}{2D_1^4}(3S_4 + 4S_2 + 4)m^2 b^3\left(e^{3m(\psi + k)/c} + e^{-3m(\psi + k)/c}\right)\sin 3m\phi/c \dots\dots(31),$$

$$y = -\frac{\psi}{c} + b\left(e^{m(\psi + k)/c} - e^{m(\psi + k)/c}\right)\cos m\phi/c + \&c\dots\dots(32),$$

the expression for y after the first term differing from that for x only in having a *minus* sign before the second exponential in each term, and cosines in place of sines. We have also from (30)

$$\frac{mc^2}{g} = \frac{D_1}{S_1} + \frac{1}{S_1 D_1} \, (S_4 + 2S_2 + 12) \, b^2 \dots \dots \dots (33),$$

which gives the velocity of propagation according to one of its possible definitions (see Art. 3, p. 202). In these expressions it is to be observed that

$$S_i = e^{imk/c} + e^{-imk/c}, \quad D_i = e^{imk/c} - e^{-imk/c}.$$

We might of course in the numerators of the coefficients have used expressions proceeding according to powers of S_1 instead of according to the functions $S_1, S_2, S_3 \dots$

Let h be the value of y at the bottom, which is a stream line for which $\psi = -k$, then we have from (24) generalized as to units

$$k = ch \dots \dots \dots \dots \dots (34),$$

so that it remains only to specify the origin of y and the meaning of c. To the first order of small quantities we have

$$x = -\frac{\phi}{c} + b \, (e^{m(\psi+k)/c} + e^{-m(\psi+k)/c}) \sin m\phi/c \dots \dots (35),$$

$$y = -\frac{\psi}{c} + b \, (e^{m(\psi+k)/c} - e^{-m(\psi+k)/c}) \cos m\phi/c \dots \dots (36),$$

and at the surface

$$x = -\frac{\phi}{c} + bS_1 \sin m\phi/c \dots \dots \dots \dots (37),$$

$$y = bD_1 \cos m\phi/c \dots \dots \dots \dots \dots (38).$$

Since y in (38) is a small quantity of the first order, we may replace ϕ/c in its expression by x, in accordance with (37), which gives for the equation of the surface

$$y = bD_1 \cos mx,$$

so that to this order of approximation the origin is in the plane of mean level, and therefore h denotes the mean depth of the fluid. Also since $u = d\phi/dx$ we have to the first order from (3), (4), (6)

$$u = \left(\frac{dx}{d\phi}\right)^{-1} = \left\{ -\frac{1}{c} + \frac{mb}{c} \, (e^{m(\psi+k)/c} + e^{-m(\psi+k)/c}) \cos m\phi/c \right\}^{-1}$$

$$= -c - c \, . \, mb \, (e^{m(h-y)} + e^{-m(h-y)}) \cos mx,$$

and consists therefore of two parts, one representing a uniform flow in the negative direction with a velocity c, and the other a motion of periodic oscillation. To this order therefore there can be no question that c should be the horizontal velocity in a positive direction which we must superpose on the whole mass of fluid in order to pass to pure wave motion without current. In passing to the higher orders it will be convenient still to regard this constant as the velocity of propagation, and accordingly as representing the velocity which we must superpose, in the positive direction, on the steady motion in order to arrive at the wave motion; but what, in accordance with this definition, may be the mean horizontal velocity of the whole mass of fluid in the residual wave motion, or what may be the mean horizontal velocity at the bottom, &c., or again what is the distance of the origin from the plane of mean level, are questions which we could only answer by working out the approximation, and which it would be of very little interest to answer, as we may just as well suppose the constant h defined by (34) given as suppose the mean depth given, and similarly as regards the flow.

Putting $\psi = 0$ in (31) and (32), we have implicitly for the equation of the surface the pair of equations

$$x = -\frac{\phi}{c} + S_1 b \sin m\phi/c - \frac{1}{D_1^2}(S_2 + 1) S_2 m b^2 \sin 2m\phi/c$$

$$+ \frac{1}{2D_1^4}(3S_4 + 4S_2 + 4) S_3 m^2 b^3 \sin 3m\phi/c,$$

$$y = \qquad D_1 b \cos m\phi/c - \frac{1}{D_1^2}(S_2 + 1) D_2 m b^2 \cos 2m\phi/c$$

$$+ \frac{1}{2D_1^4}(3S_4 + 4S_2 + 4) D_3 m^2 b^3 \cos 3m\phi/c.$$

The ratios of the coefficients of the successive cosines in y or sines in x to what they would have been for an infinite depth, supposing that of $\cos m\phi/c$ the same in the two cases, are

$$1, \quad \frac{1}{D_1^2}(S_2 + 1), \quad \frac{1}{D_1^4}(S_4 + \tfrac{4}{3}S_2 + \tfrac{4}{3}),$$

multiplied respectively by

$$1, \quad \frac{D_2}{D_1^2}, \quad \frac{D_3}{D_1^3}$$

for the cosines in y, and by

$$\frac{S_1}{D_1}, \qquad \frac{S_2}{D_1^2}, \qquad \frac{S_3}{D_1^3}$$

for the sines in x. Expressed in terms of D_1, the first three ratios become

$$1, \qquad 1 + 3D_1^{-2}, \qquad 1 + \tfrac{16}{3}D_1^{-2} + 6D_1^{-4},$$

and increase therefore as the depth diminishes, and consequently D_1 diminishes. The same is the case with the multipliers D_2/D_1^2, D_3/D_1^3, S_1/D_1, &c, and on both accounts therefore the series converge more slowly as the depth diminishes. Thus for $D_1^2 = 3$ the first three ratios are 1, 2, $3\tfrac{4}{9}$. $D_1^2 = 3$ corresponds to $h/\lambda = 0\cdot 125$, nearly, so that the average depth is about the one-eighth of the length of a wave.

The disadvantage of the approximation for the case of a finite as compared with that of an infinite depth is not however quite so great as might at first sight appear. There can be little doubt that in both cases alike the series cease to be convergent when the limiting wave, presenting an edge of 120°, is reached. In the case of an infinite depth, the limit is reached for some determinate ratio of the height of a wave to the length, but clearly the same proportion could not be preserved when the depth is much diminished. In fact, high oscillatory waves in shallow water tend to assume the character of a series of disconnected solitary waves, and the greatest possible height depends mainly on the depth of the fluid, being but little influenced by the length of the waves, that is, the distance from crest to crest. To make the comparison fair therefore between the convergency of the series in the cases of a finite and of an infinite depth, we must not suppose the coefficient of $\cos m\phi/c$ the same in the two cases for the same length of wave, but take it decidedly smaller in the case of the finite depth, such for example as to bear the same proportion to the greatest possible value in the two cases.

But with all due allowance to this consideration, it must be confessed that the approximation is slower in the case of a finite depth. That it must be so is seen by considering the character of the developments, in the two cases, of the ordinate of the profile in a harmonic series in terms of the abscissa, or of a quantity having the same period and the same mean value as the abscissa. The flowing outline of the profile in deep water lends itself readily to expansion in such a series. But the approximately isolated and widely separated elevations that represent the profile in very shallow water would require a comparatively large number of terms in their expression in harmonic series in order that the form should be represented with sufficient accuracy. In extreme cases the fact of the waves being in series at all has little to do with the character of the motion in the neighbourhood of the elevations, where alone the motion is considerable, and it is not therefore to be wondered at if an analysis essentially involving the length of a wave should prove inconvenient.

INDEX TO VOL. I.

CAMBRIDGE : PRINTED BY C. J. CLAY, M.A. AT THE UNIVERSITY PRESS.

Printed in the United States
By Bookmasters